T0358181

Discrete Event Simulation for
Health Technology Assessment

Discrete Event Simulation for Health Technology Assessment

J. Jaime Caro
Jörgen Möller
Jonathan Karnon
James Stahl
Jack Ishak

CRC Press
Taylor & Francis Group
Boca Raton London New York

CRC Press is an imprint of the
Taylor & Francis Group, an **informa** business

A CHAPMAN & HALL BOOK

CRC Press
Taylor & Francis Group
6000 Broken Sound Parkway NW, Suite 300
Boca Raton, FL 33487-2742

First issued in paperback 2020

© 2016 by Taylor & Francis Group, LLC
CRC Press is an imprint of Taylor & Francis Group, an Informa business

No claim to original U.S. Government works

ISBN-13: 978-1-4822-1824-4 (hbk)
ISBN-13: 978-0-367-73768-9 (pbk)

Visit the Taylor & Francis Web site at
http://www.taylorandfrancis.com

and the CRC Press Web site at
http://www.crcpress.com

Contents

Abbreviations

Abbreviation	Description	First Mention in Book
AC	anticoagulation	Chapter 2, Figure 2.4, p. 33
AIC	Akaike's information criterion	Chapter 6, p. 207
Algol 60	ALGOrithmic language	Chapter 9, p. 293
AUS	Australian	Chapter 10, Table 10.6, p. 323
BIC	Bayesian information criterion	Chapter 6, p. 207
BMD	bone mineral density	Chapter 1, p. 2
Cath	catheterization	Chapter 5, Figure 5.7, p. 192
CCU	coronary care unit	Chapter 5, Figure 5.7, p. 192
CEAC	cost-effectiveness acceptability curve	Chapter 5, p. 188
CISNET	Cancer Intervention and Surveillance Modeling Network	Chapter 10, p. 312
CRN	common random numbers	Chapter 7, p. 239
DALY	disability-adjusted life year	Chapter 3, p. 113
DES	discrete event simulation	Chapter 1, p. 3
EMR	electronic medical record	Chapter 6, p. 202
EQ-5D	EuroQol 5-dimension	Chapter 3, p. 112
ER	emergency room	Chapter 5, Figure 5.7, p. 192
FPG	fasting plasma glucose	Chapter 2, Figure 2.7, p. 43
GLM	generalized linear models	Chapter 6, p. 208
GPSS	general purpose simulation system	Chapter 9, p. 293
GSP	general simulation program	Chapter 9, p. 292
HbA_{1c}	glycated hemoglobin	Chapter 2, Figure 2.7, p. 43
HIV	human immunodeficiency virus	Chapter 1, p. 17
HR	hazard ratios	Chapter 4, p. 147
HTA	health technology assessment	Chapter 1, p. 1
ICER	incremental cost-effectiveness ratio	Chapter 4, Table 4.25, p. 177
ICH	intracerebral hemorrhage	Chapter 1, Figure 1.4, p. 11
ID	identification	Chapter 4, p. 150
LDH	lactate dehydrogenase	Chapter 4, p. 138
LY	life years	Chapter 4, p. 162
MI	myocardial infarction	Chapter 9, p. 284
NHANES	National Health and Nutrition Examination Survey	Chapter 3, p. 108
NICE	National Institute for Health and Care Excellence	Chapter 8, p. 271
NPI	Nottingham Prognostic Index	Chapter 10, p. 315
NVAF	non-valvular atrial fibrillation	Chapter 2, p. 63

(Continued)

Abbreviation	Description	First Mention in Book
OS	overall survival	Chapter 4, p. 138
PBAC	Pharmaceutical Benefits Advisory Committee	Chapter 8, p. 271
PFS	progression-free survival	Chapter 4, Table 4.1, p. 139
PI	prognostic index	Chapter 4, p. 146
PPS	post-progression survival	Chapter 4, p. 138
PROM	patient-reported outcome measures	Chapter 8, p. 269
QALY	quality-adjusted life year	Chapter 1, p. 6
QoL	quality of life	Chapter 3, Figure 3.11, p. 89
SoC	standard of care	Chapter 4, p. 138
Tmt	treatment	Chapter 2, Figure 2.7, p. 43
TTE	time to event	Chapter 2, Figure 2.9, p. 51
UKPDS	United Kingdom Prospective Diabetes Study	Chapter 2, p. 32

Preface

This book has been a long time in the making. We began experimenting with the application of discrete event simulation to the problems of health technology assessment around the turn of the century, but there has been no comprehensive text that provides a guide to this somewhat unusual variant of a common modeling technique. Until now, we have had to make do with the excellent but slightly inapplicable texts from the engineering and operations research fields. Students and practitioners repeatedly ask us where they can read more about our kind of discrete event simulation, and we invariably tell them that "the book is coming"—well, finally, here it is!

In this book, we have covered all the central concepts of discrete event simulation as they have been molded to make them relevant for health technology assessment. Readers familiar with the technique as it is used in other fields may find this disconcerting. Entities and their attributes are given much more prominence—they are not just objects that trigger events; and the latter are not defined in terms of their system effects. In fact, the quintessential feature of discrete event simulation—queuing for resources—is barely mentioned as it is rarely used in our kind of modeling. Indeed, despite the obvious reality of disease and the health care technologies developed to mitigate its consequences, we simulate a very abstract world without a clear physical counterpart. Nevertheless, discrete event simulation has proved to be a very useful technique for addressing the complex decision problems that arise.

We have written this book with the beginner in mind. All the concepts are described with as little recourse to jargon as possible—yet the requisite simulation terminology is introduced along the way. There are no prerequisites for understanding the material, with the exception of Chapter 6, for which a grasp of basic statistical concepts is necessary. Some familiarity with modeling in this field and with the methods of economic evaluation may be helpful but it is not mandatory. Those accustomed to using other techniques, such as state-transition models, will find an occasional reference to them, but the book is not a discourse on why one methodology should be preferred over another. Nevertheless, it should be fairly obvious that discrete event simulation offers powerful tools and tremendous flexibility for health technology assessment.

By using the chapters in sequence, a reader should be able to achieve sufficient understanding of the technique to be able to construct a realistic model designed to help in the assessment of a new health technology. Chapters 1–3 cover the essential concepts and their implementation; Chapter 4 provides a fully worked out example, using both a widely available spreadsheet program (Microsoft Excel®) and a popular specialized simulation package (Arena®). These examples provide a starting point for investigating software options, but we have not attempted to teach how to use the software. We leave that to dedicated books and manuals. These examples, along with additional implementations in two other software packages (R and TreeAge), can be requested from the authors. In Chapter 5, the approaches to analyzing these simulations, including the treatment of uncertainty, are addressed. Chapter 6 digresses somewhat from simulation per se to tackle the development of the required equations. Although this is more properly the domain of statistics, it is a crucial step in building a discrete event simulation. Some readers will—like some of us—undoubtedly need the help of a good statistician to produce suitable equations, but it helps to understand what is required and the basics of how it is done. Although many of today's laptops offer adequate computing power for most of the simulations in our field, it is always a good idea to ensure models are as efficient as possible; the techniques for doing so are explained in Chapter 7. While efficiency is not a strict requirement for these models, the topic of Chapter 8—validation—ought to be *sine qua non*. Chapter 9 provides a smorgasbord of special topics that are not central to our kind of discrete event simulation but may come in handy. The book concludes with a detailed account of a real case study of a discrete event simulation applied to the problem of screening strategies for breast cancer surveillance.

The use of discrete event simulation to inform decisions about health technologies is still in its infancy. Undoubtedly, the distinctive application of this widespread modeling technique will continue to develop and gain sophistication in our field. Perhaps it will even evolve into a more precise approach expressly designed to address our kind of problems. Until then, however, we hope that this book will provide a solid resource for those seeking to leverage discrete event simulation in their assessments of health technologies.

J. Jaime Caro, MDCM, FRCPC, FACP
McGill University
and
Evidera, Inc.
Boston, MA

Jörgen Möller, MSc (Eng)
Lund University
and
Evidera, Inc.
Lund, Sweden

Jonathan Karnon, BA (Hons), MSc, PhD
University of Adelaide
Adelaide, South Australia, Australia

James Stahl, MDCM, MPH
Harvard Medical School
Boston, MA

Jack Ishak, PhD
Evidera, Inc.
Montreal, Quebec, Canada

Acknowledgments

We very much appreciate the very thoughtful and detailed suggestions provided by Steve Chick, who reviewed a first version, and the extensive copy editing and production help provided by Patti Lozowicki and Nancy Brady.

Authors

J. Jaime Caro, MDCM, FRCPC, FACP, trained at McGill University in internal medicine and epidemiology, where he is now an adjunct professor of medicine as well as of epidemiology and biostatistics. He also teaches DES at Thomas Jefferson University School of Population Health. After founding and leading the Caro Research Institute for more than a decade, Dr. Caro is now chief scientist at Evidera, Lexington, Massachusetts. Dr. Caro chaired the Modeling Task Force, jointly sponsored by the International Society for Pharmacoeconomics and Outcomes Research (ISPOR) and the Society for Medical Decision Making, to produce new guidelines for good modeling practices. Recently, Dr. Caro has been working on developing a new modeling technique, DICE, tailored to problems in HTA. Dr. Caro chaired the expert panel guiding the German government on methods for HTA. He has also helped the World Bank Institute address the growing problem of supreme courts overriding health care system decisions.

Jörgen Möller, MSc Mech Eng, is an associate researcher, Division of Health Economics, Faculty of Medicine, Lund University, and vice president, Modeling Technology at Evidera. He is an expert in health care and logistics management decision modeling using advanced techniques, which he has applied to simulations of various complex Swedish systems. Intrigued by the dearth of simulation in pharmacoeconomics, he joined Caro Research in 2003 as a specialist in DES, creating more than 30 DES for HTA. His focus has been on translating methods from operations research to pharmacoeconomics and on developing guidelines for this type of modeling; he conducts advanced courses in DES and Arena® software.

Jonathan Karnon, PhD, is a professor in health economics, University of Adelaide, Adelaide, South Australia. His PhD focused on a comparison of cohort state transition models and discrete event simulation, resulting in the first of only a few papers that have directly compared these most common modeling techniques for HTA. He has since built a range of DES HTA models, in particular for the evaluation of alternative screening strategies. He has cochaired the ISPOR/SMDM modeling good research practices taskforce working group on the use of DES for HTA.

James Stahl, MDCM, MPH, is an accomplished clinician, researcher, educator, and medical innovator. He is an outcomes researcher with expertise in decision science, health technology assessment, and simulation modeling. His work focuses on health care delivery, process redesign, the development and evaluation of innovation in health care, and improvement in patient experience. He is the director for systems engineering for the Point of Care Testing Research Network and Center for Integration of Medicine and Innovative Technology, as well as the director for the Massachusetts General Physicians Organization/Massachusetts General Hospital (MGH) Outpatient RFID project, senior scientist at the MGH Institute for Technology Assessment, and adjunct professor in mechanical and industrial engineering at Northeastern University.

Jack Ishak, PhD, is executive director of biostatistics and senior research scientist at Evidera. Dr. Ishak specializes in statistical methods for health economics, pharmaco-epidemiology, and observational research, and has been involved in the design and execution of studies in multiple disease areas. His current methodological work focuses on study designs for comparative effectiveness research (e.g., pragmatic and Bayesian adaptive trial designs), methods for adjusting for bias due to crossover in oncology trials, and simulation-based techniques for treatment comparisons (including trial simulation).

1

Introduction

For some time now, it has been recognized that decisions in medicine should be based on the best available evidence, rather than on the opinion of a particular practitioner (Sackett et al. 1996), and many techniques have evolved to gather, analyze, and interpret available data to serve this purpose (Khan et al. 2011). Judgments made by health care decision makers on behalf of populations— whether to cover an intervention, at what price, for whom, and so on—should also be driven by the best available evidence (Eddy 2005). In some respects, this is even more important at the population level because the consequences of a poor assessment can be severe and widespread. Decisions such as those to provide insurance coverage for a new intervention that reduces mortality due to a particular illness, or to limit the availability of a treatment to prevent the severe consequences of major disability, should be fully justified by the available knowledge. Perception by the public that this work has not been done can result in extremely negative press (Hawkes 2008)!

It is nearly impossible for these population-level questions to be answered by a single data-collection study. Such a study would need to provide evidence on the populations of interest (there could be many) under the relevant conditions (which may be highly local and variable) by comparing against appropriate alternative interventions (also subject to local variations in practice) to obtain unbiased information on all health and economic aspects of interest for a long enough period to reasonably cover the relevant outcomes, and do all of these in a timely way to inform the actual decisions. Because of the limitations just listed and others, rather than rely on a single study, health care decisions must be informed by data obtained from a variety of sources. Some of these data sources may be more relevant or valid than others, and combining them in a single analysis forces the decision maker to make various assumptions regarding how the data fit together when estimating the possible impacts of each course of action the policy makers may choose.

The discipline of health technology assessment (HTA) arose toward the end of the twentieth century (Jennett 1992) to meet the demand of bringing together the available data, combining and analyzing it, and interpreting the findings to inform the people who must reach a decision about the technology in question (Emanuel et al. 2007). Although the ethical (Hofmann 2008; Saarni et al. 2008), policy (Schwarzer and Siebert 2009), and stakeholder involvement (Hivon et al. 2005) aspects of HTA have received some attention, much of the developmental work in the quarter century since HTA's beginning has focused on methods to integrate the information into an assessment that is understandable and actionable (Banta 2003).

Putting HTA into a format that is clear and informative requires a framework that can incorporate disparate sources of data and relate them in a mathematical structure that can be used for analysis. This framework must accord with existing knowledge and produce the required estimates of effect, specifically the health and cost implications of each contemplated course of action. This structure, or *model*, consists of *equations* that represent the cause and effect relationships intrinsic to the issue under study. A simple example is: white women over the age of 50 years are at two-and-a-half times the risk of men for experiencing an osteoporotic fracture (Sambrook and Cooper 2006). These equations are fed by a set of *inputs* describing the relevant context of the decision, such as the proportion of women older than 50 years and the distribution of bone mineral density (BMD) in that population; and the *interventions* to be analyzed. For this example, they might be a bisphosphonate, an antiresorptive biological, and an anabolic drug (Rachner et al. 2011). Also required is a means of *computing* and outputting the outcomes of interest (e.g., mean survival, mean quality adjusted survival, and total costs), given the inputs, equations, and other particular demands of the analysis (e.g., its time horizon and discounting rate). These mathematical and computer models are distinct from the physical mockups that many people think of when they hear the term *model*. Physical models are common in many fields, including architecture, aeronautics, engineering, chemistry, physics, and even health care when modeling a specific hospital or other facility, but are rarely practical in HTA.

Given that an HTA model consists of mathematical equations representing the problem, it is tempting to try to solve the equations numerically and obtain an *analytic solution*. For example, if an intervention's only consequence is a reduction in mortality, then the model might involve a single equation that reflects the hazard of death, with a term representing the effect of treatment. In this case, the mean survival can be obtained by integrating the equation with and without the intervention effect and comparing the results (Caro et al. 2004; Ishak et al. 2013; Nelson et al. 2008). When this integration is feasible numerically, a closed-form solution can be obtained and no further action is required. Most HTA problems, however, require much more complex models that cannot easily be solved numerically. In this situation, *simulation* tools are required to implement the calculation of each equation at the appropriate times and in the correct sequences. A well-designed model captures all the pertinent cause and effect relationships of a problem scenario and can calculate what might happen accurately, in the sense that it has external validity because it accords reasonably well with reality (Karnon and Brown 1998). When implementing a model, it helps if there is a standardized technique to follow, one that spells out how to formulate the components, link them together, carry out the computations, and even display the whole structure. Such a technique can help minimize implementation errors and make it easier to communicate with other modelers and with decision makers. The techniques used today to model health conditions and the effects of interventions have all been borrowed from other fields of research. The application of decision trees (Ransohoff and Feinstein 1976; Schwartz et al. 1973), state-transition (*Markov*) models (Beck and Pauker 1983; Siebert et al. 2012), dynamic transmission models (Pitman et al. 2012), and system dynamics (Brailsford and Hilton 2001; Homer and Hirsch 2006) to health care decisions have been addressed in many papers, and there are several excellent books (e.g., see Brandeau et al. 2004; Briggs et al. 2006; Chapman and Sonnenberg 2003; Jacobson et al. 2006; Sox et al. 2013) on these topics.

Perhaps surprisingly given its status as one of the most commonly applied modeling techniques, discrete event simulation (DES) has received relatively little attention in HTA. Outside of HTA, simulation is a well-established tool used by public and private institutions of all sizes across a wide variety of disciplines (Figure 1.1). The original field of military applications remains a very prominent domain (Hill et al. 2001), where it is used for war games, impact estimation, logistic modeling of maintenance, support and service for troops, vehicles, and aircraft. Closely related to military applications are those having to do with logistics and supply chain management (Jain et al. 2013; Tako and Robinson 2012). In finance, it is used from simulations of stock market behavior to simulations of national economies that grapple with whole-country effects of changes in tax rates, employment, and other economic drivers (LeBaron 2006). In civil engineering, it is used to properly dimension the physical components of bridges (Chan and Lu 2012; Wu et al. 2010) and model varying traffic flows and their effect on bridge loads. Other well-established applications include chip design and manufacture, the design of information flow in call centers, and game theory.

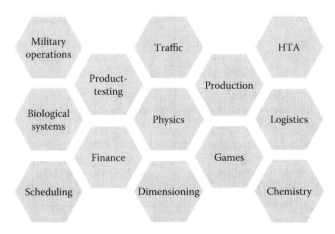

FIGURE 1.1
Some uses of discrete event simulation.

Product testing (Van Volsem et al. 2007), especially of expensive ones, is better done on a simulator. For example, referring to the building of the Airbus 380, a spokesman for the consortium said they were hoping that an evacuation simulator at Cranfield University would reduce the need for live tests. While at present a full evacuation test is required for all new aircrafts, the simulator could be used by Airbus "to do a lot of the research on the safety of an extremely large aircraft." He added: "Obviously if you do full evacuations, you do run the risk of injuring people. In the future there could well be a lot of work done with computer scenarios" (Webster 2013).

In physics, simulations are built to understand things such as the energy level of a bouncing ball, the air flow around various designs of a car, and so on; not to mention, analyses to build particle accelerators, fusion reactors, and such (Cleary and Prakash 2004). The field of chemistry is using various types of simulation to mimic chemical reactions both on the aggregate level and the atomic one (Langston et al. 1994). Although traffic simulation (Lutzenberger and Albayrak 2013; Merrifield et al. 1990) was around very early on, anyone who has been stuck in traffic wishes that simulations would be used more extensively to make sure that roads can handle the traffic and proposed new highways will actually improve and not worsen the situation! In manufacturing, even with a simple production line of one machine feeding another, it is easy to have integration effects, and these become more and more complex with added machines and products. To foresee problems, solve already existing ones, and/or even for planning production, DES is frequently used (Hlupic and Robinson 1998). Indeed, all major pharmaceutical and medical device companies use simulation in their manufacturing for just these purposes (Semini et al. 2006). Even kids nowadays are using DES without knowing it, in games (Alvarez-Napagao et al. 2011) like *Sims, SimCity, Prison Architect, FIFA,* and *Football Manager.* These are just a few of the games using DES concepts as a vital part of their design.

In these areas of application, DES is routinely used to inform real, major decisions with important economic, organizational, and social effects. Simulation can deal with complex, time-dependent problems with many dimensions and can bring together information from many sources, even expert opinion. Those desirable features are what led to its introduction as a modeling framework for HTA.

A few papers discussing and demonstrating the benefits of DES for HTA have appeared in the health care and HTA literature (e.g., see Caro 2005; Caro et al. 2010; Karnon 2003; Karnon et al. 2012; Möller et al. 2011; Stahl 2008), and its use in health care decision modeling is increasing. Existing DES books (e.g., see Banks 2001; Law and Kelton 2000), however, deal only with its engineering uses and other applications, which mostly concern physical systems. And, yet, this approach to modeling offers substantial advantages for addressing health care decisions, allowing for greatly increased accuracy and much better handling of the required complexity.

This book provides a detailed introduction to the concepts and practices of DES as applied to HTA. Those already familiar with DES in other fields may find its specific adaptation to HTA somewhat different than what they are accustomed to. Whereas most other applications involve modeling physical systems (e.g., a facility), the use of DES for HTA rarely does. Instead, the models have to do with the effects of various interventions on the course of a disease or disease-causing condition in human populations. Rather than replicating particular health care systems and populations, the simulations typically address these in more abstract and general terms. This leads to some modification of the terms and concepts from those more commonly seen in operations research and engineering. Throughout the book, the focus is on the HTA-adapted notions, but more traditional definitions are included parenthetically, as appropriate. The material is presented assuming that the reader has some familiarity with modeling to inform health care decisions, but someone completely new to the field should be able to follow along, particularly if they have some mathematical fluency. Throughout this book, DES examples are diagrammed using the flowchart symbols explained in Figure 1.2 (they might seem slightly confusing at first, but the functions they represent will be explained in detail in the next two chapters).

In this chapter, we begin by explaining the HTA context and how it might affect our approach to modeling. We then define DES, specifically as it is used for HTA, and compare it to other popular methods used in our field today. This leads to a discussion of when it should be used and how well accepted DES is today. Finally, some of the pitfalls of addressing HTA questions with DES are presented.

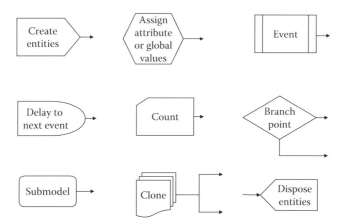

FIGURE 1.2
Flowchart symbols used in this book to diagram a DES.

1.1 HTA Context

There are many books (Kristensen et al. 2001; Sorenson et al. 2008), monographs (Canadian Agency for Drugs and Technologies in Health [CADTH] 2006; European Network for Health Technology Assessment [EUnetHTA] 2011; National Institute for Health and Care Excellence [NICE] 2013), and papers (e.g., see Drummond et al. 2008; Health Technology Assessment International [HTAi] 2014; International Network of Agencies for Health Technology Assessment [INAHTA] 2014; National Center for Biotechnology Information and U.S. National Library of Medicine 2014; O'Reilly et al. 2009) covering a multitude of definitional and methodological aspects of HTA. Here, we briefly mention those of most relevance to our topic. Assessments of new health technologies (or even of existing ones) generally involve estimating their effects (Liberati et al. 1997): the intended ones on the course of the health condition at issue, any unintended—usually adverse—ones, and any other positive or negative secondary ones (e.g., on caregivers). Many of these outcomes are naturally conceived of as events in the sense that something of relevance happens at a moment in time (e.g., a bone breaks, a test is done, a diagnosis is made, and a treatment is started). Others are better thought of as states in the sense that they reflect an ongoing condition (e.g., remaining in remission, experiencing an exacerbation, suffering pain, and being a smoker). The course of the health ailment—whether in terms of the occurrence of events or survival in various states—almost always depends on biological, demographic, socioeconomic, and behavioral characteristics of the persons involved, their experience (i.e., what has happened before), and aspects of the environment they exist in. The representation of such heterogeneity produces a more accurate HTA model, which may, in turn, have important effects on the estimated differences in costs and outcomes among evaluated technologies.

All the consequences must be valued in terms that decision makers find meaningful. For some jurisdictions, it is enough to consider the health effects directly (e.g., fractures avoided), but, for others, this requires applying a weight to the time lived. This weight is supposed to reflect the impact on the person's quality of life of each event or of the time spent in each state. The weights are summed over the time horizon of the model to estimate the number of quality adjusted life years (QALYs) lived by patients receiving the alternative interventions under assessment.

In most cases, the financial consequences of all the effects must also be considered (Russell et al. 1996). This is traditionally done from the perspective of the decision maker or health care system (Siegel et al. 1996) but can consider other stakeholders such as the providers or the patient (Gagnon et al. 2009). To accomplish this requires estimating the resources that are likely to be consumed and how much those cost (ideally, in terms of what is given up if funds are spent for that resource—the *opportunity* cost).

Often, however, the opportunity cost is unavailable and prices or other values that can be more readily obtained are used instead (Luce and Elixhauser 1990). HTA models, then, need to be able to handle multiple valuations and how they change with events and over time. As the valuations may depend on various personal characteristics and factors such as the health care system at issue, the models also need to be able to reflect these dependencies.

Despite the importance of resource use to HTA, most assessments carried out today do not explicitly consider the actual resources available in a particular setting. Instead, based on administrative data, observational studies, guidelines, or expert opinion, the analyst estimates the type and quantity of resources that are required to manage the condition. The assumption is then made that these resources are available whenever they are needed with no capacity constraints or limitations of any kind. In other words, their cost is accrued instantly and no patient ever has to wait for a bed, a test, a doctor, or any other resource that might be needed. The analysis, thus, provides an assessment of what the interventions may yield under ideal health care resource conditions. Although this may be viewed as highly unrealistic, it is the prevailing approach today and the model should be capable of implementing this. More on this topic will be discussed later in this book.

An HTA assessment is necessarily comparative—the interest is in the possible consequences of the technology at issue in relation to those of its alternatives. In order to achieve this comparison, the assessor must conceptualize the course of the health condition at issue over time, and how this course might be affected by the interventions evaluated. Integral to this consideration are the characteristics of the patients and their caregivers, the features of the health care system, and factors such as the environment that may influence the course of the health condition, the effects of the interventions, or how any of these are valued. Thus, a comprehensive HTA model requires handling a complex set of inputs, their interrelationships, and consequences over time to produce clear information for the appraisal.

An HTA is also inherently local. A decision maker is interested in what the consequences of a decision may be in the jurisdiction to which the decision applies. Since populations, practices, costs, and values vary from one place to another, the results of an HTA can be affected by many factors. Even the underlying pathophysiology and therapeutic effects—which we usually assume are common across people with a given condition—can be shaped by local influences. For example, the Honolulu Heart Program found that the impact of various risk factors for heart disease varied in Japanese men living in Japan compared to that in ethnic Japanese men living in the different cultural environments, of Honolulu (Kagan 1996). Thus, to be most useful, the tools implemented in an HTA, particularly the model, should be capable of adaptation to other jurisdictions.

1.2 What Is Discrete Event Simulation?

DES is an approach to modeling that envisages the HTA problem in terms of the course of *individuals* with a health condition, the *events* they can experience, and the effect these have on their health, medical resource use, and other components over time. The adjective *discrete* is somewhat redundant, at least for our field, as we take events to be inherently distinct and occurring at specific points in time. It distinguishes, however, this type of modeling from *continuous* simulation where the modeled processes reflect one or more quantities changing continuously (e.g., glycemia level).

Already in the very definition, the HTA usage departs from the traditional one. The conventional definition of a DES involves the idea of a *system*—typically a physical facility or process—characterized by *variables* that describe the *state* of that system at a point in time and *events* are thought of as occurrences that can alter the *state* of that system. The typical didactic example is a facility like a bank with tellers that are serving customers, characterized by variables like number of customers waiting in line and average wait time, with events such as customer arrival, customer service, and customer departure (Law and Kelton 2000). For most HTAs, however, neither the system nor its states have a clear counterpart. We could think of the health care system characterized by variables like ongoing average quality of life, life expectancy, and mean direct medical costs and take events to be anything that changes those variables, but this is a very awkward conceptualization. In HTA as it is currently conceptualized, we are interested in how a particular intervention modifies the course of illness in individual patients. Although we are less concerned with how interventions change the state of the health care system, clearly the organization of the health care system influences how care is delivered and, thus, the effectiveness of any intervention. For most HTAs, it is much more straightforward to think about states as the conditions patients are in and events as the occurrences that change those states, without being explicit about the real-world system that is being modeled.

In an HTA of osteoporosis treatments (Kanis and Hiligsmann 2014; Stevenson et al. 2005), for example, the interest might be in patients with particular levels of BMD and a given history of fractures. The DES would focus on the next fracture and its site (e.g., hip, vertebral, or wrist), or death in the absence of subsequent fractures. Figure 1.3 displays this simple DES in the form of a flowchart. To start the simulation, we need to start the clock and trigger the logic (e.g., be diagnosed with osteoporosis). The approach to doing this is software-dependent but is symbolized in the diagram by start model denoted in the diagram by the right-pointing pentagon. Each person has characteristics such as the current BMD and the number of prior fractures. Each clinical event happens (if at all) at a specific time for each

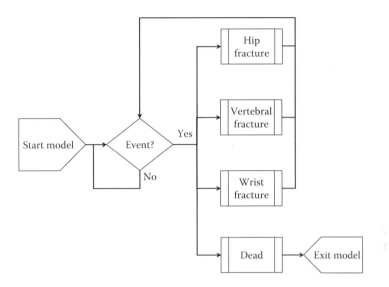

FIGURE 1.3
Simple osteoporosis model structure using DES. After the simulation start, the model checks to see what event occurs (if any) and processes the corresponding one. If it is death, the individual exits the model; otherwise, the model continues checking for the occurrence of fracture events.

individual in the population. That time is determined by the individual's risk for each type of event.

The other components of a DES (Table 1.1), described fully in subsequent chapters, increase the modelers' ability to incorporate features that may be required to properly inform decision makers. Attributes, as opposed to information stored in variables common to the entire model (e.g., the time horizon), allow the simulated individuals to acquire and maintain distinct characteristics (e.g., BMD, history of fractures, age, sex, quality of life, and adherence to treatment) that can affect their experience in the model. The attributes can also be used to store important information (e.g., when did the hip fracture happen? Was there a reaction to the drug?). The clock running continuously in the background enables accurate timing of events and precise consideration of anything that depends on time (e.g., discounting, duration of pain, and time since treatment started). If resources (e.g., hospital bed, nurse, and imaging device) are modeled explicitly—something which, as noted earlier, is unusual in current HTA—then it is possible to consider their optimal deployment and how it affects costs and outcomes. It would also make calculation of costs very transparent. If a resource is given a capacity, then people who need the resource may need to wait for it to become available. The queues that result are explicitly modeled in a DES and could help address the important issue of waiting times in health care delivery, although this is also rarely done in today's HTAs.

TABLE 1.1

Main Components of a Discrete Event Simulation

Component	Description
Events	Things that happen during the simulation
Entities	The objects that are simulated (usually people in HTA models)
Attributes	The relevant characteristics of the entities
Resources	Items that can be consumed or used for some time
Clock	The component that marks the passage of time
Queues	Places where entities wait

1.3 How Does DES Compare to Other Techniques Commonly Used in HTA?

Several taxonomies of model types for HTA have been proposed (Barton et al. 2004; Brennan et al. 2006; Karnon and Brown 1998; Stahl 2008). The current main alternatives to DES for HTA are decision trees, state-transition Markov structures, dynamic-transmission models, and system dynamics.

Decision trees are the earliest type of model used and widely implemented in HTA. Decision trees require that the analyst break all decisions down into a simple, three-part branching structure. One part is the *decision node*—the formal representation of the moment in time when a decision maker makes a choice between competing strategies (typically denoted graphically as a square). For the osteoporosis problem, the decision node might be *cover the new drug* and *don't cover*, or there might be branches for each of the osteoporosis treatments under consideration (i.e., use drug A, use drug B, don't use any drug, etc.). The second part is the series of linked *chance and logical nodes* representing the chain of events resulting from a given choice (typically denoted graphically as circles). This reflects the decision strategy in terms of the specific set or program of actions or events consequent to a decision. For the osteoporosis problem, these nodes would represent the occurrence of the various fracture types and death. The third component is the *value nodes*—the terminal branches of the tree (typically denoted as triangles) that represent the value of the strategy in whatever units are of interest to the analyst (e.g., costs and utilities). The *expected value* of each decision is determined by calculating the weighted average value of the branches emanating from that decision (i.e., *folding back*). Whichever decision offers the greatest weighted average value is recommended. Decision trees can be instantiated as computer models in many widely available software packages (e.g., winDM™ and TreeAge™). A realistic decision tree can get quite complicated (Figure 1.4). At one time, these were a very prevalent

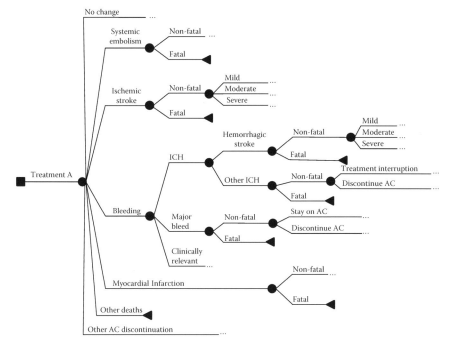

FIGURE 1.4
Initial branches of a decision tree depicting prevention of thromboembolic disease in atrial fibrillation. As denoted by the ellipses, the full tree continues for many more branches, and there would be similar sets of branches for the other interventions considered in the HTA. AC, anticoagulation; ICH, intracerebral hemorrhage.

approach for modeling HTA problems but their popularity has waned with recognition of their limitations, particularly in terms of representing time.

The types of models most frequently used in HTA today are those based on Markov techniques (named after Andrei Markov, a Russian mathematician who lived from 1856 to 1922). Unlike decision trees, state-transition (Markov) models allow the designer to express changes over time in the model. Thus, this technique was proposed for medical decision making as a solution to the time limitation of decision trees (Beck and Pauker 1983; Sonnenberg and Beck 1993). In a state-transition model, the problem is represented by a set of *states* (these typically reflect the illness status in an HTA). The states must be mutually exclusive and comprehensive, in the sense that they cover all possible conditions required by the problem. *Transition probabilities* are specified to describe the possibility of moving from one health state to another. Values, such as costs and quality of life adjustments, are applied to the time spent in each state and summed over the time horizon of the model to estimate outcomes like total costs and QALYs. The time horizon is typically divided into *cycles* of fixed length.

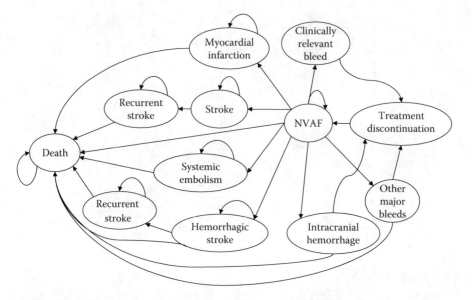

FIGURE 1.5
Diagram of the model of non-valvular atrial fibrillation as a Markov structure. Each oval represents a health state and the arrows show the possible transitions among them. These include recursive arrows pointing to the same state and indicating that some portion of the population remains in that state during a cycle. NVAF, non-valvular atrial fibrillation.

The states in a Markov model may be diagrammed as circles with the transitions from one state to another that happen during each cycle represented by the arcs, or edges, connecting the two states (Figure 1.5). Recursive arcs represent the proportion of the population remaining in a health state (i.e., transitioning back into that state). The directed graph can be restricted in that only certain transitions are allowed. For example, people do not transition to any other state from the *dead* state.

As of this writing, the cohort-based approach remains by far the most commonplace way to implement a Markov model for HTA. In this type of implementation, a group (*cohort*) of people is distributed among a set of health states by assigning a proportion of them to start in each given state (commonly, 100% in whatever is considered the initial state). The model then reflects changes in this distribution over the defined time horizon (Siebert et al. 2012) by using the transition probabilities to move portions of the cohort from one state to another.

The major assumptions of cohort-based Markov models relate to state, time, and memory. People may exist in only one of a finite number of health states. If one were to use this method to examine osteoporotic fractures, for example (Figure 1.6), one might define the following states: *Well with no fracture*, (with) *hip fracture*, (with) *vertebral fracture*, (with) *wrist fracture*, and *dead*. Importantly, in this type of model overlapping health states are not allowed.

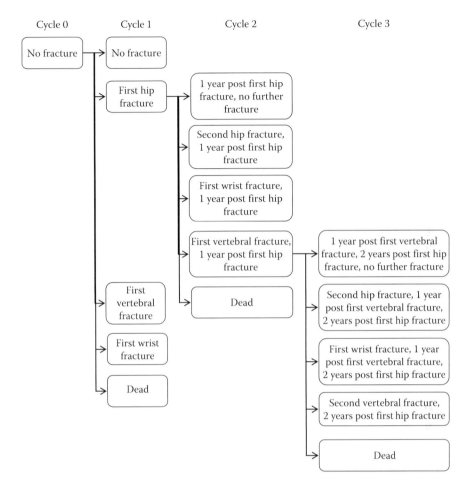

FIGURE 1.6
Cohort-based state transition models: an example of an unmanageable model structure for osteoporosis. The figure shows how, even for a fairly simple model, state explosion rapidly becomes a problem in a state-transition model. Although only one state is shown expanded in each cycle, by the third cycle there are quite a few states already. Expansion of the other states, as is required, considerably increases the number of states.

For instance, a person cannot have both a hip and a vertebral fracture. To represent, this would require a compound state *hip and vertebral fracture*, and the other states would have to be defined as *hip fracture only* and *vertebral fracture only*. This method requires that every possible health state, combinations of health states, and every potential transition be explicitly specified. This unitary state requirement forces the model designer to make simplifying assumptions or have the size of the model structure expand substantially.

These cohort-based models are constructed based on the notion—termed the *Markovian assumption*—that the proportion of the population in a state that transitions to another state does not depend on any other aspects such as the prior states or even the time spent in that state (i.e., persons in a health state at the beginning of a cycle are equally likely to leave that state in that cycle, regardless of any differences in their sequences and durations of preceding states). This Markovian assumption implies that there is no information stored in memory and all patients in a given state are identical. All knowledge of the natural history of disease must be embedded in the structure of the model—subgroups of the population do not carry any history of their disease.

The limitations imposed by the Markovian assumption can be handled to some extent by increasing the specificity of the defined health states (Hiligsmann et al. 2009). As the number of such dependencies increases, however, the volume of health states required to represent the structure becomes unmanageable. Figure 1.6 illustrates the impact of the Markovian assumption with respect to an osteoporosis model structure. The cohort is at risk of bone fracture at three alternative sites (hip, vertebrae, and wrist); multiple fractures are possible at each site; and the type, number, and timing of previous fractures influence the likelihood of subsequent fractures. The number of states required to represent the occurrence of multiple fractures quickly expands. In cycle 2, following a first hip fracture in cycle 1, five states are required to represent the possible pathways (including death). In cycle 3, following a first hip fracture in cycle 1, and a first vertebral fracture in cycle 2, another five states are required. In total, 80 health states ($4 \times 4 \times 5$) would be required for cycle 3, and this number would continue to increase in subsequent model cycles.

Finally, these models advance time using a fixed step (a *cycle*) and constrain the occurrence of state transitions to the end of each cycle. This can introduce errors. For example, individuals may be given credit for living a cycle they didn't live all the way through or be denied some credit for a cycle they did enter. To better approximate the continuous nature of time, many modelers assume that transitions occur in the middle of a cycle. This *half-cycle correction* has been criticized, however, for not properly accounting for time preference and for assuming a monotonic membership function for the states (Naimark et al. 2013). Another option is to decrease the time step until the error introduced no longer matters—even to daily cycling, for example. This avoids the half-cycle correction, but significantly increases the number of cycles in the model.

Implementing the Markov model using individuals instead of as a cohort overcomes some of the problems with the memory-less assumption. Nevertheless, the technique remains focused on states with no direct representation of events; it has difficulties handling competing risks (e.g., if there is a risk of dying, of fracturing a wrist, or breaking a hip, which risk should be applied first) and suffers from the problems of using a fixed time step for advancing time. Moreover, it cannot represent other people of relevance (e.g., a caregiver). A DES only needs to structure the events (in this case, fractures and death) because each individual's health states (e.g., a hip fracture 2 years ago, a vertebral fracture 5 years ago, and a current BMD of 15%) are stored in attributes. The individual's status can be updated as needed, either at regular intervals or upon the occurrence of a new event (see Chapter 2). These stored details are used to inform the next event to be experienced by each individual and the time at which it will be experienced. Competing risks can be handled appropriately, and time runs continuously with no requirement for fixed time steps. Other people, such as family members and caregivers, are readily represented. Indeed, any object that is of interest—living or inanimate—can be incorporated.

1.4 When Is Discrete Event Simulation Useful?

The range of health care problems that can be addressed using a DES is vast. In this book, we address mainly those that are pertinent to HTA: assessing a health technology in order to decide on its use in the health care system. These problems generally seek to estimate the implications of deploying the technology compared to not doing so. The effects considered are commonly limited to impacts on the health of the people concerned (sometimes extending as far as their families and other caregivers) and the costs to those who pay for the technology. This limitation is imposed by convention in our field; DES could easily handle additional effects (e.g., on training of new physicians or on development of new technologies).

For HTA, DES is most useful for representing relatively complex structures that would be difficult to implement using the cohort-based Markov technique. In line with recent guidelines (Caro et al. 2012), DES should be used when the model structure required to capture all important differences in the costs and benefits of alternative technologies requires health states that are so numerous as to be unmanageable to implement as a cohort-based Markov (or state-transition) model. A review of published DES for HTA (Karnon and Haji Ali Afzali 2014) identified features of model structures that imply this condition of too numerous health states.

The most common feature of an HTA that posed problems for the cohort-based Markov approach was the representation of heterogeneity in the characteristics of the population modeled because this requires a large number of states. Cohort-based models require that the population in each state be homogeneous, meaning that all members of the state have identical values for all determinants of the transition probabilities. This is rarely the case in HTA problems, however. With osteoporosis, for example, many factors are known to influence fracture risk, including age, sex, ethnic background, body weight, BMD, family history, smoking status, and alcohol intake (Kanis et al. 2005). Using a cohort model to represent such baseline heterogeneity would require many states—numbering into the thousands—to represent unique combinations of these factors and maintaining that heterogeneity throughout the model. This is tantamount to creating a separate version of the model for each distinct set of characteristics or undertaking separate analyses for each profile. Even so, there remains the challenge of putting all those results back together to represent the costs and effects in the aggregated population, or subgroups. In a DES, individuals are assigned their own unique values for the set of characteristics, including memory of all that has happened. These values—updated as appropriate throughout the simulation—control what happens next to each individual, and the problem is avoided altogether.

A second factor identified in the review concerned the representation of disease progression as a continuous process. Markers of disease progression for many chronic conditions are commonly measured on a continuous scale, for example, glycemia in diabetes and body mass index in obesity. For osteoporosis, BMD is measured on a continuous scale. A particular advantage of modeling disease markers accurately is that such models can be used to extrapolate outcomes from intervention studies that are only powered to detect differences in surrogate disease markers. Cohort-based Markov models can approximate these continuous scales by defining states based on small discrete categories, but such model structures soon become unmanageable, and, thus, cohort-based models tend not to represent disease markers.

A third reason for using DES relates to the representation of time-varying event rates, where the likelihood of subsequent events varies according to the time spent in the current health state. For example, in osteoporosis, the risk of a fracture depends on how long the person has been in the no fracture state. In a Markov model, these time-varying probabilities can be implemented by dividing a health state into a series of states distinguished by duration spent in the health state (e.g., first 6 months after fracture, second 6 months, and so on). The cohorts move through these *tunnel* states until they transition to a different health state. While this can be done for a small number of conditions over relatively brief periods, it rapidly grows in complexity if it is required across many states or for long durations. The implementation of time-varying event rates is straightforward in a DES as time is treated continuously and risks can change accordingly (see Chapter 2).

The fourth feature of problems that commonly led to choosing a DES was cumulative representation of events experienced by individuals over their time in the model. As illustrated in the osteoporosis example (Figure 1.6), this can rapidly increase the number of states while it remains reasonably compact in a DES. Such model structures are most relevant for chronic illnesses that are associated with multiple conditions. It is not surprising then, that in the models reviewed, diabetes was the most common disease addressed using DES.

Little empirical research has been undertaken to compare a Markov model and a DES of the same problem in order to quantify the benefits of implementing more complex model structures using DES. Two studies have shown significant differences. One was a comparison of a single cohort Markov model of Alzheimer's disease with a DES that represented heterogeneity in individual characteristics (Gustavsson et al. 2009). There was more than a 30% difference in the incremental cost per QALY (£46,000 vs. £32,000, respectively). Compared to a cohort-based Markov model for human immunodeficiency virus (HIV) that represented disease progression as a binary measure of viral load (below/ above 400 copies/mL), a DES that represented an additional viral load threshold (<50 copies/mL) and CD4+ T-cell count as a continuous factor better predicted observed results at 5 years (Simpson et al. 2009).

1.5 Acceptance of Discrete Event Simulation

Given the popularity of cohort-based Markov models in HTA, it is instructive to consider the reasons why DES has not been used more frequently.

Although DES has a long history of use in modeling physical health care facilities (Jacobson et al. 2013), such as an emergency room (Konrad et al. 2013), it has not been used nearly as long or as frequently for HTA. A very early simulation in HTA was MISCAN, used for the evaluation of mammography and other screening tests. Although not specifically termed a DES, it employed all the key concepts and simulated the course of individuals down to the tumor level (Habbema et al. 1985). It was followed in 1996 by a smoking cessation model (Warner et al. 1996), and then DES started appearing more frequently from 2000 onward, with a DES for an HTA of migraine treatment written entirely in Fortran published in 2000 (Caro et al. 2000).

Despite the increasing use of DES for HTA, and its clear advantages, more than 30 years after its first implementation, there remain agencies that are resistant to using DES for their evaluations or permitting others to submit models using this technique. Such resistance may be due to limited experience with, and thus understanding of, the method. To draw a parallel, in 1985 one of the more successful Swedish ski jumpers introduced a new way to hold the skis during the jump: rather than keeping them parallel as everyone did at the time, he held them in a V. This led to substantial advantages, making the jumps longer. But, the judges, feeling the pressure from the jumpers accustomed to the traditional ways, deemed the new method less aesthetic and penalized the skier using a V. As time passed, and others recognized the superiority of the technique, the novel approach was used by more and more jumpers and now is the only method seen in championships. In HTA, awareness of DES, as well as the knowledge of where and when it is an appropriate modeling choice (Karnon et al. 2012), is spreading; more HTAs are using DES; and acceptance by agencies worldwide is growing. This extends as well to the understanding of the underlying concepts and how to implement them in an appropriate programming language, of which there are quite a few (see Chapter 9).

The potential going forward is great, especially given the advances in collating and using large datasets, as well as the promising step that various companies have taken to make available the primary, individual-level data collected in all of their clinical trials for further research.

The following sections address some issues regarding the use of DES that are not encountered with a cohort-based Markov modeling approach, which might be holding back the more widespread use of DES in HTA.

1.5.1 Stochastic Behavior

An important aspect of DES—which is viewed as a strength in many fields, but may be regarded suspiciously in HTA—is the involvement of chance in the simulation. *Chance* is used here in the sense that the results of the model can be different for otherwise identical individuals, and they can even be different for identical sets of inputs. This *stochastic* behavior occurs because much of what happens during a simulation is governed by probabilities that specify the frequencies with which alternative events occur. These probabilities are applied in a DES using *random numbers* (a series of numerals where every value is equally likely to be next, comparable to any fair gambling device like a roulette wheel or dice) that allow the model to pick which of the possibilities will actually transpire.

Although ever-changing results can be disturbing, this stochastic nature makes DES useful in representing the real world because the latter is similarly changeable (Table 1.2), with the changes often being unexplainable (another way to define chance). For example, the results of two or more real-world randomized controlled trials governed by identical protocols will always differ because chance enters into every aspect of the studies. Indeed, the calculation of sample size is determined in part by an explicit recognition of this stochastic nature. Thus, DES can reproduce much of what actually happens in reality.

TABLE 1.2

Similarities between Reality and a Discrete Event Simulation

Reality	DES
Over a fixed period of time, a finite number of events occur randomly.	Over a fixed period of time, a finite number of events are generated randomly.
Each real-time period results in a different series of random events.	Each simulated time period results in a different series of random events.
No two time periods will be exactly alike, though overall trends will be similar.	No two simulations (using different random numbers) will be exactly alike, though overall trends will be similar.

1.5.2 Data Needs

Any model requires information to provide specific values for its various inputs. For example, the simple osteoporosis model described earlier needs to know what the rates of the various fracture types are and how they change over time and with antecedent fractures. It also needs estimates of the mortality and how it changes with occurrence of fractures. Since HTA involves comparisons of various intervention alternatives, these rates must be specific to each technology to be assessed. Though not specified in the prior description, there would need to be estimates of all the costs that accrue, and possibly of other types of valuations like quality of life. Almost certainly, the dependence of these inputs on patient characteristics such as age and bone density will need to be considered. This is still only a very partial list for a full model of osteoporosis, but already the reader may be growing concerned that the need for data expands quite rapidly.

Extremely rarely, all the required information may be found in a usable format from one source. Nearly always, however, much evidence will have to be found and processed in order to yield the inputs required for the model. This need for considerable data is not unique to a DES, however. The DES approach does not per se generate additional evidence requirements—it is the problem itself that creates the demand. The reason that a DES may appear to require more data is that the technique renders very transparent what the information needs are by allowing all the requisite components and interrelationships to be represented. Of course, if a valid analysis can be carried out at the cohort level (e.g., there is no heterogeneity to be worried about), then the data demands are significantly reduced because everything that pertains to individualization can be safely ignored. As noted in the review by Karnon and Afzali, however, many disease areas present characteristics that make the use of DES desirable to provide a more realistic representation of the costs and effects of health technologies.

An alternative approach sometimes used by modelers is to consider what data are at hand and then construct a model based on them. If all you have, for example, is a table with the annual probability of each type of fracture, it may be tempting to ignore the influence of prior fractures, the impact on mortality, and the dependence on time and on various patient characteristics. The data needs are certainly much reduced if this tactic is employed, but the credibility of the results tends to be diminished as well, and the question looms large whether this approach yields a model that is fit for the purpose of informing actual health care decisions affecting real populations.

1.5.3 Execution Speed

Another concern sometimes expressed regarding DES is that running the model may take a very long time (Griffin et al. 2006). This fear arises from the number of calculations that a DES must make in order to employ all the knowledge encoded in its equations and chart the course of each individual, record what happens, apply relevant valuations, compute statistics, and so on. The apprehension is heightened by comparison with the near instant calculation of a cohort-Markov model in a spreadsheet or custom software.

If a model takes considerably longer to yield results, this can pose real difficulties for its use in HTA (Caro and Möller 2014). One consequence is that during its construction, the modeler will tend to be more reluctant to execute the model to verify that it is working properly. No one wants to wait for a long time to see results, make changes, and check the results again. This reluctance will increase the likelihood of errors going undetected and may invalidate the eventual HTA results. Another adverse effect of long execution times is that once the model is complete, a reduced set of analyses may be contemplated. In particular, probabilistic analyses that alter many input values simultaneously to see their joint effect on results (Baio and Dawid 2011) may be eschewed altogether as they considerably expand the volume of calculations. This may impair interpretation of the HTA results and lead to incomplete insight or even misleading conclusions.

While there is no doubt that any model executed at the individual level—DES being a case in point—takes much longer to run than a greatly simplified cohort model, these concerns are largely moot with a properly designed DES and availability of modern computing facilities (Caro et al. 2007). In Chapter 7, various techniques for improving the efficiency of calculations in a DES are presented. These should be deployed to the greatest extent possible in every model (not just DES) to maximize the execution efficiency. It is unproductive to complain about long execution times if the model has been structured, programmed, and run in highly inefficient ways. If a maximally efficient model still takes an unacceptably long time to produce results (a very rare occurrence in our experience), then newer computing techniques such as massive parallel processing should be deployed. Given the importance of the HTA decisions, it is difficult to justify unwarranted simplification of a model simply to make it run faster (McEwan et al. 2010).

1.5.4 Law of the Instrument (Wikipedia Contributors 2014)

A danger faced by all modelers is becoming so enamored with the modeling method they have become familiar with that they start to fit all problems to that technique, regardless of its suitability for the questions at hand (i.e., the popular aphorism "if all you have is a hammer, everything looks like a nail"). For example, in the osteoporosis model, a commitment to the Markov technique could lead the modeler to conceptualize clinically aspects that are obviously events—like fractures—somehow as states: pre-fracture state, post-wrist fracture state, and so on. This psychological failing has been recognized as a problem in other fields, such as software development (Ambler 1998), and is quite prevalent in ours.

Although the DES toolbox contains a fairly broad range of tools and it imposes relatively few constraints on the modeler, it is still only one of many methods for modeling a problem, and it is important that it be firmly viewed as such. It would be just another ludicrous case of the law of the instrument if modelers trying to address HTA problems were to begin to see every one of them as requiring a DES. For example, the assessment of a set of antineoplastic interventions where the only outcome of relevance is postponing cancer recurrence and, possibly, prolonging life may be fully addressable using a partitioned survival model (Jackson et al. 2010; Latimer 2011) with parametric failure-time equations that incorporate known predictors. This technique may be suitable and more efficient, without the need to invoke a DES.

As modeling techniques continue to evolve, it will be important that those who (we hope) come to be captivated by DES not get so besotted that they close their minds to alternative techniques. DES is just one tool in a vast, and growing, toolset!

1.5.5 Transparency

In many fields, models are constructed by experts in simulation and in the subject matter under the guiding principle that they should be as detailed and complex as necessary to address the problem posed. The models are used generally to gain understanding not to advocate a point of view. It is presumed that deliberately manipulating the model structure to achieve some desired results—and hiding those manipulations behind dense structures—is not in anyone's interest because such tampering will tend to get discovered when the model fails to match reality sufficiently, potentially with disastrous consequences. Thus, understanding of the model's technical details by lay decision makers who do not possess the required training is not typically a criterion for evaluating the quality of the model. It is understood that forcing the modelers to meet such a standard could lead to such an inaccurate simulation that its results are not usable or may even be misleading.

In HTA, however, the models are often produced for a sponsor with a significant stake in the decision. This may be a manufacturer of one of the products being evaluated or the payer who will have to bear the costs if the results of the model favor covering the intervention. In either case, there can be significant and rational concern that the model may be deceptive, and that these distortions will not be evident to the decision makers, particularly in the typical absence in our field of any serious validation of the simulations against reality. In this context, it is not surprising that transparency gains considerable traction as a criterion for assessing the perceived quality of a model. If those who will use the models to inform their decisions cannot understand what the model is doing in sufficient detail to assess whether these methodological choices are sensible, then the decision maker might dismiss the entire simulation and its results.

To avoid such a draconian conclusion, the sponsors of models for HTA occasionally demand that they be constructed at a sufficiently simple level to enable full review by non-technical people, regardless of what errors this may induce in the structure, analyses, and results. In this case, lack of transparency is fatal, whereas prediction inaccuracies are likely to go undetected. This poses a problem for all modeling techniques but particularly for DES because decision makers are much less familiar with its workings, and some of the other concerns discussed in this section may loom large. Indeed, the perception that maximum transparency is required may bring about a blanket rejection of the technique and a declared preference for cohort Markov models implemented in a spreadsheet.

The perception that DES models are more complex, and therefore less transparent than cohort-based model, is unfortunate because a well-structured and documented DES can be very transparent, even to a non-technical reviewer. An orderly presentation of a DES in a series of flowcharts, using consistent symbols and reflecting clinically sensible concepts, should be readily understood by all concerned. This is further enhanced if the choice of implementation software (covered in Chapter 9) corresponds well to the flowcharts and presents the underlying computations clearly, with a minimum of strange syntax, and the modeler has taken care to use clear labels and avoid obfuscating abbreviations. Of course, the mathematics of any complex equations needed to reflect the problem's interrelationships may require technically adept reviewers, but that is not an issue relating to the choice of modeling technique and should not undermine the overall transparency of the DES. Needless to say, it can be very difficult to follow a DES constructed in a convoluted way, with idiosyncratic meaningless labeling, byzantine shortcuts, poor documentation, and using a multitude of macros to circumvent the limitations of chosen software!

2

Central Concepts

Despite the enormous flexibility of discrete event simulation (DES), a relatively small number of concepts underlie it (Table 2.1). These concepts are: the *events* that define the changes in the entities or in the state of the model; the *entities* that experience those events; the *attributes* that distinguish individual entities from one another; the explicit handling of *resources* and any resulting *queues*; and, perhaps most distinctively, the running of the model in *continuous time*. Other concepts are common across many types of models, not just DES: the use of *information global* to the entire model; *distributions* that represent a possible range of values and the frequency with which they occur; *control logic* governs the overall behavior of the system, including things like entity routing, resource availability, and status tests; *branches* describe alternative routes an individual may take based on either logical or probabilistic conditions; and *sets* are groups of like items, resources, entities, and so on.

In this chapter, each concept is described in detail (implementation is handled in Chapter 3). The chapter begins by addressing the central distinguishing concept of DES: the use of events to articulate the structure of the problem. It continues with the objects that trigger events, the entities, and their attributes. The handling of time in a simulation is considered next. Resources are described in two ways: first in the standard way, which explains the use of capacity constraints, and then in somewhat peculiar manner as used in health technology assessment (HTA) that assumes there are no resource limitations. Although not specific to DES, the use of global information and the concept of statistical distributions are important and are addressed before the chapter concludes with the topic of influence diagrams, a method of graphically representing decision problems.

2.1 Events

A DES conceptualizes reality in terms of the events that can occur. For the purposes of HTA, we define an *event* as something that happens to an entity in an instant of time, after which the entity has changed its own attribute values, its relationship to other entities, or the system as a whole. For example,

TABLE 2.1

Central Concepts of a Discrete Event Simulation

An *event* is something that happens to an entity during a simulation.[a]

An *entity* is a person or thing that experiences events within the model.

Attributes contain information that is specific to an entity.

Time is continuous in a DES.[b]

Resources provide services to an entity. They have a *capacity* to serve a particular number of entities simultaneously. When occupied, they generate *queues*, which are places or lists where entities wait for the resource to become available. *Scheduling* is the logic managing resource capacity at any given time.

---

[a] This departs somewhat from the standard, formal definition.

[b] Continuous in the sense that events can occur at any instant of time during the simulation.

---

a fracture would be considered an event in a simulation of osteoporosis, because its occurrence would cause the entity's quality of life attribute to be adjusted downward, her family member entity to become a caregiver, and an increase in the total cost to the system by the amount required to manage the fracture.

This definition is somewhat different than that typically given in DES books. Entities experiencing events are at the core of all DES, but in HTA the interest is in the experience of the entities themselves, whereas in fields other than HTA the focus is on the *system*, that is, on the collection of entities, limited resources, and other elements that are being modeled. Thus, DES outside of HTA track the *state* of the system as recorded in one or more *state variables* and an event is defined as an instantaneous occurrence that changes that state.

As noted in Chapter 1, however, this definition is correct but awkward for HTA because it requires defining the *system* and its *state*. When the concern is with the course of a clinical condition and how it is modified by one or more interventions, it is straightforward to define an event in terms of what occurs to entities, without specifying the system and its states. For example, a fracture in the osteoporosis model is clearly an event of importance to the HTA. The analyst would want to consider how a fracture affects the sufferer's quality of life, the risk of additional fractures, and, perhaps, even mortality. The cost of treating the fracture would also likely be of interest. Although some of these consequences may alter the *system* (e.g., death removes an entity from the system) and could be captured using state variables (e.g., number of entities alive), this is not necessary—indeed, it is a distraction that takes away from the natural feel with which the DES can replicate the HTA problem.

As the term *event* has a meaning in reality, there may be some confusion between its colloquial use and the technical meaning given to it in DES. Real events can be simple occurrences (e.g., recording of a blood pressure reading) to which a simulation event can correspond directly (e.g., *record blood pressure*).

In real life, however, the term *event* can be applied to what is, in effect, a process (e.g., infusion of a thrombolytic agent to treat a patient with an acute coronary syndrome). These more complex real processes happen over time, whereas a simulated event has no duration—it happens in an instant of time. Thus, real events that are actually processes are represented in a DES via several simulated events which reflect the component happenings.

A formal approach to diagramming the event structure and determining the minimal event set has been described (Buss 1996; Schruben 1983). This *event-graph* approach—also known as *simulation graph*—consists of only two types of elements: *nodes* and *arcs* (Figure 2.1). Each node represents an event, and the arcs (also known as *edges*, or arrows) connect the nodes and reflect the triggering or scheduling of subsequent events. Conditions that must be met for movement along an arc to take place are noted on the edge as is the delay until the subsequent event. Consequences of each event can be listed beside each node. This method is very simple and powerful, but it is not specific to HTA modeling, so details of its implementation are left to the many papers (de Lara 2005; Savage et al. 2005; Xia et al. 2012) and books (Kin and Chan 2010) that address this.

2.1.1 Types of Events

The use of events to structure a simulation is a very flexible concept allowing for natural representation of the problems dealt with by HTA. With the appropriate set of events, a modeler can reflect all the real occurrences of importance, creating a simulation that reproduces reality at a sufficiently accurate level of detail to respond to the decision maker's questions. Different modelers may conceptualize the problem using different sets of events.

There are many types of events included in a typical DES for HTA (Table 2.2). Obvious events in an HTA are the disease-related ones: things

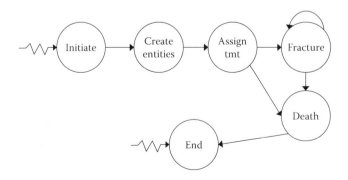

FIGURE 2.1
Example of an event graph depicting a DES. An event graph depicts a DES as a series of circles indicating the events and arcs (edges and arrows) that connect the events in the sequences that may occur. The wavy arrow indicates an event that is scheduled to occur at the start.

TABLE 2.2

Some Types of Events That Are Often Included in Discrete Event Simulations for HTA

Category	Examples
Disease-related	Illness onset, disease progression, new symptom, symptom relief, change in biomarker, sequelae, death
Intervention	Start, change dose, switch, add another, monitoring test, side effect, stop intervention
Clinical activities	Diagnostic testing, physical examination, biopsies and other procedures, surgery, psychosocial interventions
Health care	Admission to emergency department, to hospital, or to intensive care unit; discharge from hospital; transfer to another ward; institutionalization in a nursing home; transfer to another facility
Behavior	Adherence, start or stop smoking, attend visits, diet, exercise, undergo tests, work absenteeism, provide caregiving, wander at night
Specialized modeling tasks	Updating values, duplicating entities, reading from or writing data to a file, recording information, displaying data, time stamping, starting or ending simulation, sorting a list, opening logical gates

such as the onset of illness, progression to a more advanced stage, appearance of a new symptom or sign, relief of a symptom, change in a physiologic factor such as glycemia or blood pressure, occurrence of sequelae, and, of course, death. Just as central are the aspects relating to the interventions: the start of treatment, its continued use, changes in dose or route of administration, addition of other treatments or switches, monitoring tests, and occurrence of side effects. There are clinical activities such as diagnostic testing, physical examination, biopsies and other procedures, surgery, and psychosocial interventions. Closely related are admission to an emergency department or to hospital, discharge from the same, ward transfers, arrival in intensive care unit, admission to a nursing home, transfer to another facility, and so on. As many of these real episodes are processes, their representation in a DES requires several simulation events.

Often, the behavior of individuals and others is important to a problem (Brailsford and Schmidt 2003), and thus, there can be events to represent things like noncompliance; start or stop smoking; diet and exercise, attend monitoring visits, or undergo tests; work absenteeism; practice (or not) safe sex; report contraindications; provide caregiving; wander at night; and many other behaviors. Reasonably detailed representation of many of these behaviors may also require multiple simulation events.

Beyond conceptualizing a version of reality in terms of the events that can occur, a simulation must carry out several additional tasks, including updating values, duplicating entities, reading data from a file, recording of information (and perhaps its output to other software, a display or a printer), time stamping (e.g., when a pain episode starts), starting the simulation by creating the population, ending the simulation because the desired time horizon is reached (or some other ending condition, such as the last person dying, occurs), sorting an event list to figure out which one happens next, and many more that are highly specific to the context of the model. These modeling tasks may be treated as simulation events, despite their lack of a direct counterpart in the reality being modeled, because they are things that need to happen during the simulation. In many software packages, they have to be triggered by one or more entities in order to take place.

One task of this modeling type is entity creation. This can be thought of as an event, named perhaps the *enter* event that creates the *entities*. Although it can correspond to a real occurrence like birth, or the incidence of the illness at issue, or diagnosis via a test, it often just reflects the start of the simulation. In many HTAs, the model begins by creating an arbitrary number of people for whom a decision regarding an intervention will be made. This starting point, although not corresponding to any real occurrence, is modeled as an *enter* event in the simulation. There can be multiple *enter* events in a DES. For example, in a model of osteoporosis, one may have an *enter* event that creates the prevalent population of women with reduced bone mineral density existing at the start of the simulation. If only these women are modeled, then the simulation would be based on a closed population. A more natural approach would incorporate another *enter* event that periodically adds people to the simulation to reflect the incidence of osteoporosis.

Another specialized modeling event that is often necessary when simulating chronic conditions is an *updater* event. In the course of a chronic illness, long periods of time can pass without an event. For example, most women with osteoporosis live much of their lives without the occurrence of a fracture, and most patients with atrial fibrillation suffer neither an embolism nor a hemorrhage. Thus, the interval between the simulated events that correspond to real ones can get quite long and the modeler may wish to see more often the changes in the values of some of the underlying attributes (e.g., age, bone density, and blood pressure) either because they can alter the individual's risk for some events or to be able to record those values more frequently. To force these changes in values to happen regularly, a special *updater* event can be incorporated, whose sole function is to trigger those revisions. This provides a rather efficient way of ensuring that attribute values are updated appropriately.

A particularly interesting specialized event is one that serves as a gate to open or close parts of the model depending on a user setting. As a model involves many assumptions about what may happen, how aspects relate to each other, which equations are used, and so on, it is highly desirable to be able to test the impact of various choices, addressing structural uncertainty (Bilcke et al. 2011). These gates are used to set the model up for subsequent structural sensitivity analyses, and their implementation is covered in Chapter 3.

2.1.2 Event Consequences

All events have consequences of interest (if they don't, then they are not relevant to model and should be eliminated). The consequence may be a single, very simple one (e.g., the total cost is updated to add a new expense) or it can be an extensive, complex list as in the structural gates mentioned above. It is these consequences that result in the outcomes the model is tracking.

Between the start and the end of the simulation, an entity can experience any number of events, each one with its own consequences. These consequences can be effects on other events, or on the values of attributes or global variables. For example, the occurrence of one event may change the likelihood of another event occurring (e.g., a *stroke* event changing the risk of a *death* event); it may also modify the nature of another event (e.g., a *begin anticoagulation* event may change the severity of a subsequent *stroke* event); or it may directly trigger another event (e.g., a *stroke* event leads to a *hospitalization* event). An event can also alter the characteristics of an entity (e.g., the attribute *quality of life* may be changed by the event *side effect*), including those that specify a person's state (e.g., a *stroke* event alters the *disability* attribute). Events can also change the values of global variables (e.g., the value of a variable *QALY* that keeps track of the total quality-adjusted time lived in the model may be modified by a *side effect* event; the cost of a *hospitalization* event will be added to the value of a variable *total cost*). Of course, if resources are explicitly modeled with a limited capacity, then many events will affect aspects of those resources such as the available capacity and will have consequences for any queues that result (e.g., the average waiting time).

Events can also create branch points. For example, after a *hemorrhage* event, a patient may experience a *stop anticoagulation* event or a *decrease dose* event or a *switch treatment* event. Such an event alters the path of the entity through the simulation and can have other consequences. For example, if the entity were to go down the *switch treatment* path and switch to aspirin, then the values of the attributes that relate to the treatment would all be modified (e.g., *treatment* attribute and *dose* attribute). Some variable values would also be modified (e.g., *total treatment cost*). The occurrence of other events could be altered (e.g., *time to stroke* and *time to bleeding*), and even the creation or termination of entities (e.g., *death due to hemorrhage*).

One important consequence of an event may be to remove entities from the simulation. This *end* event could reflect the occurrence of death, cure of the illness, loss of health insurance coverage, or any other happening that signifies that the simulation of that entity is complete. When entities experience this, the model stops considering what happens to them. There can be many *end* events, each one mirroring a different reason for terminating an entity's time in the simulation. Of note, a DES need not have an *end* event because there are other ways for the simulation to stop. For example, the analyst may specify a maximum amount of time for the simulation to run (i.e., the time horizon). When this desired time has elapsed, all entities still active in the model will stop being simulated.

Sometimes an event is invariably a consequence of another (e.g., *hemorrhage* always leads immediately to *stop treatment*), and if the consequence happens immediately, then these events can be considered a single event. Even when real events can be collapsed into a single simulation event, a modeler may choose to make the extra occurrences explicit because it produces a clearer structure and thus facilitates review, validation, and communication. In most cases, transparency should trump compactness of the model structure. That said, the modeler should avoid adding completely unnecessary events—meaning those that do not have any relevant consequences. For example, *becoming dizzy* should only be an event after *hemorrhage* if it is going to alter the attribute *quality of life*, or perhaps change the risk of a *fainting* event—otherwise, it should be left out, even if in reality it is something that happens.

2.2 Event Occurrence

For a simulation to produce results, it must have a mechanism for bringing about the occurrence of the model events when appropriate. In other words, it has to know when to activate the events that define it. The most basic way of activating an event and triggering its consequences is to link it to another event (e.g., *fracture* leads to *admission to hospital*). Of course, when an entity necessarily moves directly from one event to trigger the next, the timing of the subsequent event is obvious, and there is no need to instruct the simulation on how to find it since the event's occurrence will necessarily happen at a given interval after the triggering event.

Many events in a simulation, however, are not set to happen immediately after another event. Their occurrence is subject to chance in the sense that any given entity may or may not experience the event, and, if it does, the precise moment when it will happen can vary from one entity to another, even if they are identical in all relevant respects. For example, two women starting with the same age, bone density, and other risk factor values may suffer a hip fracture at different times, both in reality and in a simulation properly reflecting that reality. We don't know why this differential timing occurs—and it is this that we call *chance*. (If we did know, then we would incorporate this knowledge in determining the times; but we never know precisely when every event is going to occur.) Thus, a simulation has to have a method for establishing when each chance event occurs.

In the standard terminology of DES used in other fields, this method is known as the *time-advance mechanism* because it determines how time passes in the model, but for HTA it is more intuitive to think of this as the *event occurrence mechanism*. There are two main ways of doing this (Figure 2.2).

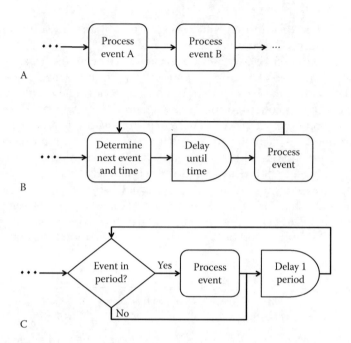

FIGURE 2.2
Three ways of checking for events. In panel A, event A leads immediately to event B. For example, if event A is a fracture, event B might be hospitalization. In panel B, the model determines which event will happen and when that will be. It then waits until that next event time arrives to process that event. In panel C, the model determines whether an event has happened in the immediately preceding period, and, if so, processes it. Either way, it then advances the clock by another period and repeats the check.

The one most commonly used because it is more accurate and doesn't generate other problems is the *time-to-event* approach. The other one is the *periodic* method (or *fixed-step* approach), with the period set by the modeler (for those familiar with Markov models this is equivalent to a cycle, where the model is updated at the beginning or end of a fixed interval of time). The *time-to-event* approach (also known as *next-event-time advance*) is slightly more complicated to implement than periodic checking—mainly due to the calculations involved in determining some of the event times. It avoids, however, the problems associated with periodic checking. The process is also massively more efficient because the simulation only has to check a list of event times. Although periodic checking is inaccurate and can generate substantial problems, it is presented here because many analysts unaccustomed to simulation find it easier to understand.

2.2.1 Time-to-Event Approach

The most accurate way of checking for the occurrence of events is the *time-to-event* approach. This is done by determining in advance when each event

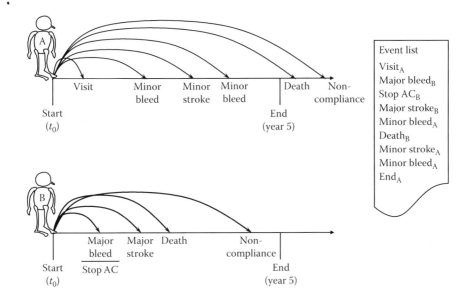

FIGURE 2.3

Illustration of the time-to-event approach in a simulation of anticoagulation (AC) for atrial fibrillation. At the start (t_0), patient A has her times to various events computed. These are indicated on the timeline, beginning with a monitoring visit. When that monitoring visit occurs, the times to other events are recalculated if they are affected by that visit, and they are measured from that point. The model's time horizon of 5 years will be reached before individual A's times of death and of noncompliance, so she will experience neither of these events during the simulation, unless another event changes their times of occurrence. Patient B has a much more severe course, beginning with a major bleed that leads immediately to stopping AC (simultaneous events). This will lead soon after to a major stroke followed by death. Neither noncompliance nor end-of-time-horizon will be reached. The panel on the right indicates the event list for these two patients, as of the initial time. It would change with the occurrence of events during the simulation.

should happen (under the circumstances that are extant at the time of checking, and incorporating the chance element); recording that time; and, then, initiating the event at precisely that time in the simulation (Figure 2.3). To achieve this, an event list is set up at the start of the simulation that contains the times (and types) of all events ordered from soonest to latest. The model then simply looks at the next item on the list and handles that event when the simulation clock reaches the corresponding time. This is also a good opportunity to update any time-based measures (e.g., survival and QALY). It then deletes that event from the list and repeats the process with the next item. The list is modified as appropriate during the simulation. The scheduled time for an event can change (e.g., treatment may delay a disease consequence, such as dialysis in chronic renal failure); a new event can be added to the list (e.g., the start of anticoagulation introduces a time until a bleeding event); or one or more events may be subtracted from the list (e.g., stopping treatment removes all the side-effect events for that individual). These changes to the list are, themselves, consequences of other events (starting or stopping treatment in the examples).

This approach to checking for the occurrence of events is very efficient because the simulation can skip the time between scheduled events in the sense that it does not have to do any calculations during that interval. By definition, no events happen in the interlude. Instructing the simulation to check for event occurrence along the way (as is done in periodic checking) is very wasteful of computation time. In the *time-to-event* approach, the jumps between event times are of whatever duration, typically quite variable, and the only operation that needs to be done is advancing the simulation clock to the next event time.

The time-to-event approach allows a more accurate treatment of any competing risks (Wolkewitz et al. 2014) that operate during the simulation. These risks, by definition, must be able to manifest freely without the expression of one affecting the other. For example, in the osteoporosis model, the individual may face the risk of a fracture and a risk of suffering a treatment side effect. The time of the next *fracture* and the time of *side effect* are determined separately, and thus, the risks are allowed to *compete* properly because the calculation of one does not affect the determination of the other. If periodic checking at fixed time steps is used (see Section 2.2.2), the modeler must decide the order in which to apply the risks. If the events are forced into being mutually exclusive, as is often the case in Markov models where only one transition can be experienced in a given cycle, then the occurrence of one event will shield that individual from the other risks. With time-to-event, there is no need to set a hierarchy of risks, giving some priority over others, and the occurrence of any one event will not introduce error by incorrectly averting the experience of another risk. For an example, see the model constructed using the United Kingdom Prospective Diabetes Study, known as UKPDS (Clarke et al. 2004).

By the same token, since the time of each event is computed separately, there is no reason that two or more unrelated events can't transpire at exactly the same time. This should be quite rare, but if it happens, then the simulation must process the simultaneous events without advancing the clock. If, despite the events not being directly related, one can influence the processing of the other or if the consequences of one affect the other (e.g., one event reduces quality of life by a fixed amount, while the other does so as a percent change), then the simulation must have instructions on the order in which to process the events.

2.2.2 Periodic Checking

An alternative way to determine when an event should happen is *periodic checking*. In this fixed-step approach, the simulation checks whether the conditions required for an event's occurrence have been met over a specific period of time and keeps doing this throughout the simulation, usually with the same time step. Typically, this checking is done at the end of each period,

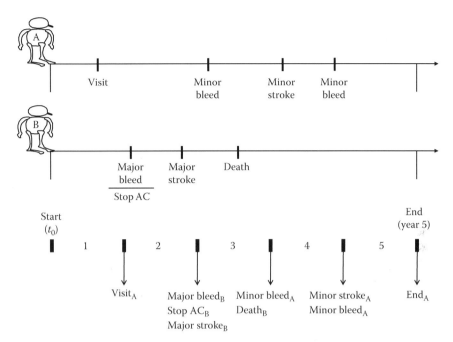

FIGURE 2.4

Illustration of the periodic checking approach in a simulation of anticoagulation (AC) for atrial fibrillation. The time horizon of 5 years has been divided into 5 year-long periods in this model. At the end of period 1, the model determines that patient A has had a visit and processes that event. At the end of period 2, patient B is determined to have suffered a major bleed that leads immediately to stopping AC and also a major stroke, all of which are processed at the end of that period. During period 3, patient A has a minor bleed and patient B dies. In period 4, patient A has a minor stroke and another minor bleed, both processed at the end of the period. Finally, the model's time horizon of 5 years is reached before individual A has died, so she will experience an end event at the conclusion of period 5.

and all the events that were determined to have occurred are assigned the current time (i.e., whatever it is at the end of the interval).

The process of periodic checking is conceptually simple (Figure 2.4). At the end of each period, the model goes through the list of individuals in the simulation during that interval (i.e., all those who started the interval) and determines for each one if any of the events happened during the period. Events that are determined to have occurred are dealt with as appropriate, the clock is moved ahead another period, and the process repeats until the model ends. For example, after a number of women with osteoporosis enter the simulation, the model can check whether any of them suffered a fracture in the preceding month (or other suitable interval). For those experiencing a fracture, the model processes the consequences, including making any changes to their risks of subsequent fractures. Everyone's time-based attributes (e.g., survival and QALY) are updated and then the clock is advanced another month before repeating the process.

Although this approach can seem very intuitive and its apparent simplicity is attractive, it is usually not the best option. First, it can introduce significant error in the timing of events because the event occurrence is specified only at the end of the interval (or other fixed time point, like the middle). Since time is such an important aspect of most HTAs because many of the values (e.g., costs and quality of life) depend on the time over which they operate, errors in timing directly affect the outcomes of interest. This problem can be reduced by minimizing the time step (ideally, no more than a day to avoid introducing additional error produced by longer detection intervals), but this magnifies the second problem. In the majority of HTA models—particularly those dealing with chronic conditions—there are no events occurring in most time periods for any individuals. Most illnesses are uneventful over broad swaths of time. Thus, it is inefficient to force the simulation to go through all the calculations involved in checking for every event in every individual on the list every consecutive period, with a majority of the checks resulting in nothing actually happening. This inefficiency can result in very significant delays in running the model. The delays increase with shorter periods (e.g., days), while the accuracy of the model reduces with longer periods.

Unlike Markov models, where the fixed cycling step can impair the proper consideration of competing risks, this is not the case in a DES because there is no requirement that only one event occur per period per individual. Thus, each of the competing risks can be properly applied and as many events as necessary can be determined to happen to a given individual in a particular period. That said, the imprecise timing of the periodic check can lead to many events being stacked up to occur at the same time. This magnifies the trouble of having to set up rules for sorting out which of the apparently simultaneous events should be processed first. For example, in an atrial fibrillation model that has a stroke event that generates a substantial cost as well as an event for death due to unrelated causes, it is possible for one or more people to meet the conditions for both events in a single period. In this situation, it is unclear whether the death should take precedence or the stroke. If the former is given priority, then the costs due to the stroke (and any other consequences) will be incorrectly ignored. To avoid this, the model would need to have the rule that death is always processed last. With just two events and day-by-day checking, the problem may not be so bad, but with many more events, and especially with longer intervals, this can introduce significant complications.

Fixed-step checking is a reasonable choice when the situation being modeled has naturally occurring periods. For example, if patients are seen at a regular predetermined interval (e.g., at a monthly pre-natal visit) and all consequences are established at those times, then it could be appropriate to use periodic checking. There is, of course, nothing precluding a combination of the two approaches, with periodic checking applied when it fits naturally and time-to-event for the rest of the simulation.

2.2.3 Event Ordering

There is no *a priori* mandatory ordering of events. They can occur in whatever sequence makes sense in reality. Some may have a natural ordering (e.g., a *prescription* event precedes an *ingest treatment* event; *discharge* from hospital happens after *admission*; antibiotics are given upon diagnosing a bacterial infection; and a resource is used before its cost is recorded as an expense). A special case of sequencing is when one event necessarily triggers another. For example, in a model of anticoagulation, a *bleeding* event should prompt a *stop treatment* event. Of course, if the resulting event occurs immediately upon the triggering event, then the two events may be modeled simultaneously as a single event (e.g., although stroke precedes death in the case of a fatal stroke, this can be represented as a *fatal stroke* event). The sequence should be specified by the analyst only if there is a natural order.

Many events, however, occur when their specified time comes up in the simulation and are not subject to any particular sequence. For example, a patient with diabetes may experience the *onset of retinopathy*, a *myocardial infarction*, or *hypoglycemia* at any time. Indeed, another diabetic with the same characteristics may experience these events in a different sequence, or not at all. Since the timing of multiple events varies, it is possible for some unrelated events to occur simultaneously. In this situation, the model must process the events without advancing the clock.

2.3 Entities

For a simulation to produce results, something must trigger the events that define it. At its most general, an *entity* is anything that does this. It commonly corresponds to a physical object that exists in reality (e.g., a woman), and the entity bears the experience specified by the event (e.g., the woman has a fracture). Sometimes, a model requires something to trigger an event (e.g., to initiate reading data from a file), and the entity created for that sole purpose has no counterpart in reality. Although at present most models used in HTA simulate groups of people (i.e., cohorts) without distinguishing their individual members, a DES almost always conceptualizes the *entities* as individuals.

2.3.1 Why Individuals?

In a DES, each entity is an individual in the sense that it can by itself trigger an event and it alone experiences that event. In a model of osteoporosis, for example, each female entity with reduced bone mineral density faces the possibility of a fracture event and, when that happens, only the triggering entity suffers the fracture. Other entities in that model will experience their own fracture events (or not) at their own individual times and will bear their consequences accordingly. Even a group of entities that happen to suffer a fracture simultaneously will each still experience the event separately. This individuality is distinct from the cohort approach where an entire group is modeled at once, with all members of the group in a given state facing the same transition probabilities and portions of the group transitioning to other states through the model.

Individualization is a very important feature of a DES. This simple idea of individually triggering the events has many benefits. It allows for accurate modeling of the times to events; it allows the simulation to properly deal with important differences between people, avoiding the problems that arise when averages are used; it facilitates logging of what events have happened to someone and when they have happened (i.e., retaining *memory*); and it enables assessment of variability in analyses of the outcomes.

As noted in the preceding section on event occurrence, the most accurate way to trigger events is the time-to-event approach: determine when each event should happen and activate it at that time. This is readily accomplished for individuals since each one can experience an event at the appropriate time and any realistic pattern of occurrence can be replicated. Two (or more) entities may share characteristics but they will move through the model separately and may experience different sets of events, at different times, with different consequences, and so on. In reality, for example, one woman with osteoporosis may present with a fracture and suffer another one a short while later, while another woman—despite identical bone mineral density and other characteristics—might be diagnosed and live for some time with no fractures until death occurs for some other reason. A simulation of individuals can reflect these real event patterns in a straightforward way: the first woman would suffer a fracture immediately at the start of the simulation and another sometime later; her duplicate would be simulated without a fracture until her death. Defining the equations that underlie these event patterns may take some effort (see Chapter 6) but implementing them accurately is made possible by simulating individuals.

In the previous example, the women are said to be identical in terms of bone mineral density and other aspects, in order to make the point that they still experience their own individual patterns of events. In real populations, however, people differ in their characteristics. For simulating a real population, it is important to consider the determinants of what will happen, of when it will happen, and of any valuations of the consequences (e.g., their costs). It is very unusual in medicine today to not know something about what drives the occurrence of events in a given illness or condition, and ignoring diversity in these aspects introduces error in the model calculations. By allowing each entity to bear its own set of characteristics and to experience events accordingly, the heterogeneity that exists in a real population is naturally accommodated. The representation of this heterogeneity via individualization facilitates the more accurate portrayal of the eligible population and improves model validity, especially if there are nonlinear relationships between individual characteristics and event occurrence.

For example, age is almost always a factor in the occurrence, progression, and severity of illness—even if it isn't, it is always a determinant of the probability (or time to) death. Say you need to model a population of women with osteoporosis ranging in age from 40 to 80 years, with a mean age of 65. If they are modeled as a group, use of the mean age of 65 years to characterize the group will result in two errors. The mortality rate at age 65 does not necessarily reflect the average mortality of the group because mortality does not change linearly with age. Thus, the death times will be incorrectly estimated. Moreover, as time passes in the model, age changes as well. The mean age of the survivors does not increase linearly with model time, however, because the older individuals are removed sooner (i.e., they die earlier). Thus, the mean age of the surviving cohort may actually decrease over the time horizon of the model (Figure 2.5).

Another aspect of the problem with heterogeneity is that the mean value of any characteristic, even if it correctly reflects the group over time, may not produce an accurate simulation. For example, suppose that in a model individuals are to be treated if their systolic blood pressure exceeds 139. If half of the individuals have a systolic blood pressure of 100 and the other half are at 150, the mean is 125 and the group does not meet the criterion for treatment.

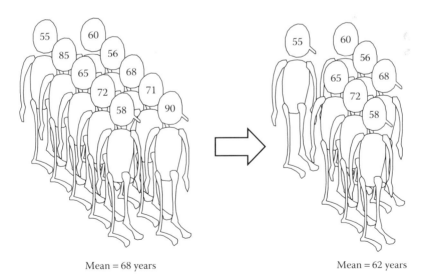

Mean = 68 years   Mean = 62 years

FIGURE 2.5
Illustration of the problem with modeling a cohort known to be heterogeneous with respect to a characteristic that affects the transition probability. In this case, age (written in each figure's head in years) affects the probability of death: the older, the higher it is. Three individuals (aged 71, 85, and 90 years) have died in the 1-year transition depicted. The mean age then goes from 68 years to 62 years, not to 69 years as the cohort technique would assume.

This is clearly wrong as half the individuals should be treated. By simulating the individuals instead of the group, these problems due to heterogeneity are averted.

The number of factors that must be considered because they determine what happens is commonly high in HTA models. As this number rises, the inaccuracies multiply if a cohort is modeled and heterogeneity is ignored.

Keeping track of what has happened to each person during a simulation is important in models of human illness because the past is usually a determinant of the future. For example, the risk of fracture in a person with osteoporosis is different depending on whether a bone has already been broken. This type of memory is difficult, if not impossible, to carry in a group because the history cannot be captured properly at an average level if it is to influence what happens afterward. Thus, knowing the proportion of the population who has had a previous fracture does not necessarily yield the correct risk for the group. This problem gets worse as additional historical features become important—for example, if the location of prior fractures and how long ago they happened matters. Carrying memory properly is best done by modeling individuals, each of whom can *remember* what has happened to them.

Another advantage of modeling individuals is that their outcomes can also be recorded at the personal level. This yields a set of simulated data that reflect the variability of outcomes in a particular analysis. Although unusual in HTA today, this information can be analyzed to assess the spread of results using appropriate statistical techniques. In addition, individualization better provides for analyses to identify subgroups for which an intervention yields better or worse outcomes.

2.3.2 Types of Entities

The concept of entity is very general and flexible. An entity can be anything that is logically able to trigger at least one of the events in the model (Table 2.3). In HTA simulations, there are always people entities. They may be patients if they have an illness already, or they may be people for whom an

TABLE 2.3

Some Types of Entities That Are Often Included in Discrete Event Simulations for HTA

Category	Examples
People	People who may get ill, patients, caregivers, parents, clinicians, therapists, technicians, orderlies
Other life form	Bacteria, virus, seeing-eye dog
Inanimate objects	Scopes of various sorts, ambulance or other transport vehicle, needle, lab report or other test result
Specialized modeling purposes	Object that triggers a recording event, an outbreak, or other special event that no other entity will trigger

intervention is contemplated because they are at higher risk of developing a disease, or they may be people at large for whom a screening or public health intervention is considered. Both types may coexist in a simulation with the type changing, for example, from pre-diabetic to diabetic at some point in the simulation. Various stages of disease could be implemented by creating a type of entity for each one but there is a better way to do this, keeping the entity type as person.

It may be of interest for a particular problem to simulate other types of people entities as well. For example, in a model addressing Alzheimer's disease (Getsios et al. 2012), the caregivers are almost as important as the patients themselves because their health also often suffers and there may be associated economic and quality-of-life consequences. Thus, one might create one or more caregivers for each patient, each of whom could experience many of the model events as well, including some specific to their entity type (e.g., taking a respite holiday). By the same token, in models dealing with pediatric conditions, simulating the parents may be essential to properly reflect the problem. Siblings and other family members may also be of interest. For example, in assessing strategies for preventing infant pertussis, many experts feel that it is necessary to *cocoon* the babies by ensuring that none of their contacts will introduce the bacillus into their environment (Forsyth et al. 2007). Thus, vaccination of the close family members and their behavior need to be simulated (Coudeville et al. 2009). Indeed, in models involving contagion, it is important to include entities reflecting the possible contacts. To select a vaccination strategy for meningitis, for example, it is necessary to simulate roommates in a college dorm, classmates in school, and so on (Caro et al. 2007; Getsios et al. 2004). In an HIV model, entities might include sexual partners, those who might use the same needle for drug injection, and so on (Caro et al. 2007; Rauner et al. 2004).

Another type of people entity is clinical personnel like the doctors, nurses, and therapists (Duguay and Chetouane 2007). If these people are to be simulated explicitly, a decision needs to be made whether to incorporate them as a type of entity or as a resource (see Section 2.6) (Lim et al. 2013). If the reason for modeling them explicitly is that they too will experience events (e.g., a doctor might be attacked by a mentally ill patient), then simulating them as entities makes sense. When they are there just so their cost can be added in or their capacity constraints can be applied, then it is not necessary to simulate them as entities—indeed, it is better to define them as resources. For very complex cases, such as a restricted number of nurses, with different competencies, moving around in a clinic, it may be necessary to model them both as entities and as resources.

In some HTA models, the entities more broadly represent the population, or at least a large segment of it. For example, in models about screening for cancer, the entities are people who are candidates for screening (Ness et al. 2000; Stout et al. 2006). This can be everyone over the age of 50 for colonoscopy or every newborn for some congenital disorders (Ramwadhdoebe et al.

2009). When addressing public health interventions, the entire population may be involved (Fone et al. 2003; Hupert et al. 2002). This would be the case in a model of a dietary salt reduction campaign where everyone may be affected.

2.3.3 Creation and Removal of Entities

By convention in our field, models tend to create all the individuals at the same time at the start of the simulation. Entities enter the model at a special *creation* event, which can generate as many entities as the modeler thinks are needed for the planned analyses. If this *all-at-once* convention is followed, and only one type of entity is created, the model only needs one creation event, and that event can produce all the specified entities at once. This is not a limitation of DES, however: entities of whatever types are required in a model can be created whenever they are supposed to enter the simulation. For example, caregiver entities may enter the simulation when the patients need them. To include more than one type of entity, a separate creation event is required for each type.

In some models, it is necessary to simulate incidence. This can be epidemiologic incidence of a disease; it can reflect individual's newly meeting criteria for entering a population (e.g., reaching a particular age of interest, say 65 years); or movement into the population (e.g., via immigration or joining an insurance plan); and so on. If the model needs to reflect incidence, then additional instances of creation are required. The initial creation events can be *reused* to produce additional entities later on during the simulation, provided that the model structure after the initial creation event works equally well for subsequent entries. Supplementary creation events will be required, however, if the initial pathways followed by these later entrants are different than those specified for the initial creation. For example, whereas at the start of the model the entities are tantamount to a prevalent population—later in the simulation, the incident entities may require confirmatory diagnostic tests or may bear different costs, and so on (Figure 2.6). Either way, a DES can readily model incidence by allowing further creation of entities during the simulation.

In a DES for HTA, there are typically many entities in the model at once. This can be more efficient than modeling one entity at a time because it allows whatever software is being used to intersperse processing of events as they are experienced by whatever entities are due for an event. Moreover, in models with constrained resources, this is required since there has to be more than one entity in the simulation in order to compete for resources.

Entities are removed from the model when it is no longer necessary to simulate them. Just as entities can be created at any time during the simulation, they can be removed from the model whenever appropriate. A common reason for removal is death because, presumably, there is no further reason to track the individual after that. This death event can be a single common one that individuals experience when the model determines their time is up.

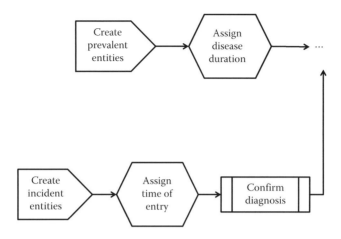

FIGURE 2.6
Two pathways for creating entities. One pathway creates patients who have already had the disease for some time (prevalent entities), while the other creates people who have incident disease but need a confirmatory test before proceeding in the model. They do not need assignment of the disease duration (since diagnosis) because it is zero, but the time of entry into the simulation needs to be recorded in an attribute so that the disease duration can be updated at later time points if needed, by subtracting the time of entry from the simulation time.

Alternatively, if it is necessary to reflect different causes of death, several *death* events can be used. Another common reason for removal is end of the specified time horizon. A single model *end* event will usually suffice in this case. If everyone starts at the same time and there is an overall time horizon that applies generally—the common situation in HTA—then this event will be triggered at that one time. If entities are created at various times, then the time horizon may be reached at different times and the *end* event will need to be triggered accordingly. Other reasons to remove an entity arise from changes in their disease (e.g., progression to a more advanced stage that is no longer of interest to the simulation), or in their characteristics (e.g., at the sixty-fifth birthday triggering a change in insurance). Another possibility is that the person leaves the population (e.g., via emigration or loss of health insurance). End events can be implemented to reflect whatever reasons there may be to remove an entity.

2.4 Attributes

One of the important features of a DES is the one that helps make each entity an individual. This element, known as an *attribute*, is very simple and yet very powerful. At its most basic, an attribute is a characteristic of the entity that helps identify it. This attribute belongs to each entity in the sense that that information is associated with the entity throughout the simulation. Each entity has its own value for each of the attributes that characterize it. An obvious example for people-type entities is age. Almost always in a model, it is important to know each entity's age because it is a determinant of the occurrence of many of the events, may affect the effectiveness or safety of interventions, and may alter the valuations of outcomes. Thus, the simulation must be able to retrieve the value of each entity's age, and this remains one of the entity's characteristics from start to end of the simulation. Every one of the entities will have an *age* attribute but the value stored in that attribute will differ from entity to entity, and over time.

2.4.1 Attribute Values

An attribute can contain any value that is appropriate to the characteristic it represents. An attribute like *age* will store the positive number corresponding to the entity's age at the time the value was recorded. Even if a real number provides the full representation of an attribute (e.g., *t*-score for bone mineral density ranging across the entire scale of real numbers), it may be sufficient to store it in categories (e.g., *normal*: *t*-score of −1 and above, *osteopenia*: *t*-score between −1 and −2.5, and *osteoporosis*: *t*-score −2.5 or below). Some attributes may be inherently categorical (e.g., sex). Occasionally, several attributes may be required to fully represent a characteristic (e.g., *smoking* represented by *current status*: never smoker, ex-smoker, smoker; plus *amount smoked*, *type of smoking*, *time since last smoked*). Whatever type and number of values may be needed, a DES can represent these features using attributes. In many DES software, the values are encoded using numbers but there is nothing in principle that prevents the use of characters or other symbols. Numbers, however, are easier to use in formulas that refer to the value of an attribute and as indices to variables or other grouped items.

Just because an attribute has been defined does not mean that it must contain a value. For example, in an osteoporosis model, there may be an attribute *time since last fracture*. For an entity who has not yet had a fracture, this attribute would have no value (zero would be incorrect as that would imply that a fracture just occurred), and it would remain empty so long as that entity has not yet experienced a fracture—the attribute would exist for those entities but its value would remain undefined. Indeed, it may stay that way until the end of the simulation if a fracture never occurs.

As implied by the age example, the value stored in the attribute can change at any time. Indeed, *age* would be updated periodically as time passes in the simulation. This would be the case for many attributes, the values of which would have to reflect not only the passage of time, but also changes in the condition being modeled, and the occurrence of events during the simulation. For example, in an osteoporosis model, the value of *bone mineral density* would change as the disease worsened; the *time since last fracture* would start at zero upon a fracture occurring and then would increase until the next fracture occurs; and the start of treatment would be recorded in its own *on treatment* attribute (along with details of the type, of treatment, its dose, etc.) and would modify other attributes such as the *decrease in bone mineral density*. Regardless of the frequency and pattern of changes required to maintain the values of attributes, a DES can manage this.

Sometimes a specific value needs to be stored unchanged because that information is important either for subsequent calculations (e.g., the value of glycosylated hemoglobin at the time of calculating cardiovascular risk), for reporting as part of the results (e.g., the age at the start), or for other model purposes (e.g., for making a treatment decision based on the age at the time of first relapse). In this situation, the attribute containing the changing information won't suffice. One or more additional attributes need to be specified to keep the information. For example, the entities would have *starting age, current age,* and *age at first relapse* as attributes.

Values for attributes are set and updated as a consequence of an event (Figure 2.7). Attribute value-setting can be instigated in many ways. Often, this happens at a specific related event (e.g., a *fracture counter* is updated by the occurrence of a fracture; systolic blood pressure is recorded at a *doctor's visit* event). Attribute values can be set conditional on the results of a test (e.g., *diabetes* attribute is turned on if *fasting blood glucose* exceeds a certain value). They can be modified by a decision (e.g., *treated* is turned off if patient decides to discontinue use of the drug). Branch points can lead into attribute-setting events (e.g., if patient is admitted to hospital, then change *location of care* to hospital). Attribute values may also be updated at specific

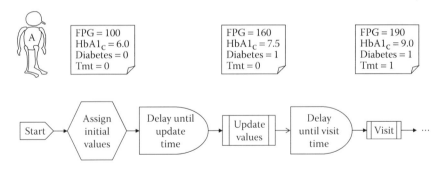

FIGURE 2.7
Illustration of the assignment of attribute values and their updating at events. Diagram illustrates the model logic and the values of four attributes in the square with the upturned corner. In this case, individual A begins the simulation without diabetes (=0) and thus, with a normal fasting plasma glucose (FPG), normal glycated hemoglobin (HbA1c), and no treatment (Tmt). After some time, an update event is triggered as nothing else has happened. Individual A's FPG is now somewhat elevated as is his HbA1c, so he is deemed diabetic, but as this is as yet undiagnosed, he is not being treated. After some further time passes, a visit to the doctor ensues, at which point his values have worsened with no treatment, but the doctor now diagnoses the diabetes and starts treatment.

events designed for that purpose and set to occur at particular points in time. A modeler may want, for example, to ensure that no more than three months go by without updating certain attributes (e.g., the value of glycated hemoglobin, HbA_{1c}, in a diabetic). An *update* event would be set to occur after three months if nothing else has triggered an updating of attribute values.

2.4.2 Types of Attributes

Any characteristics of the entities that are pertinent to the model should be represented as an attribute (Table 2.4). A characteristic is pertinent if it affects some aspect of the simulation. A common example is a determinant of the occurrence of an event (often termed a *risk factor*). Another example is things that affect quality of life or costs, which are important aspects in HTAs. Many items stored as attributes have to do with outcomes (e.g., the number of fractures a woman with osteoporosis has suffered). Thus, attributes can reflect demographics (e.g., age, sex, and socioeconomic status),

TABLE 2.4

Some Types of Attributes That Are Often Included in Discrete Event Simulations for Health Technology Assessments

Category	Examples
Demographic	Age, sex, socioeconomic status, postal code, race, ethnicity, height
Physiologic	Blood pressure, glucose level, urinary output, weight
Personal history	Educational level, age at menarche, family size
Medical history	Prior illnesses, allergies, prior surgeries, age at diagnosis
Family history	Early heart disease, breast cancer in maternal relatives
Other personal	Environmental exposures, living situation (e.g., in college dormitory)
Behavior	Smoking status, duration, intensity, type of smoking, adherence, drug use, sexual orientation and activity, number of partners
Disease	Diagnosis, subtype, extent, severity, symptoms, signs, stage, score, affected biomarkers, comorbidities, genetic profile
Intervention	Type, intensity, features, monitoring, duration, side effects
Caregivers	Number, type, availability, payment
Health care	Location, duration, discharge destination, payer, copay
Health outcomes	Years lived, disability, quality of life, quality-adjusted life years, discounted values
Cost outcomes	By resource use type, economic category, payer, discounted values
Times	Age at events, episode durations, delays in queues
Counts	Of events by type, of doctor visits, other health care encounters
Specialized modeling purposes	Identification number, age at start, last event, next event, time until next event, time since last event, identities of clones, random numbers, structural flags, entity type, animation picture

physiologic quantities (e.g., blood pressure, glycemia, urinary output), personal history (e.g., age at menarche, educational level, number of children), medical history (e.g., prior illnesses, allergies), family history (e.g., of heart disease, of breast cancer in maternal relatives), and any other features that might be important (e.g., body mass index, air pollution in the environment).

One aspect that can play a key role is the entity's behavior. For example, many attributes might be used to characterize smoking: *duration*, total *pack-years* of smoking, *type of tobacco* products used, *time since last smoked*, and so on. Another important behavior often incorporated in models has to do with whether the patient uses a treatment as prescribed. Attributes to represent compliance might include proportion of *days covered*, *amount* of treatment *taken*, *persistence* with treatment, and even more precisely, the *time at which treatment is taken*. In a model of HIV infection, the *number of partners*, *drug use*, *sexual behavior*, and many other such aspects could be included among the attributes. Thus, any behavior of the entities that is pertinent to the model should be described by one or more attributes whose values are set and updated as necessary during the simulation.

Another aspect of people-type entities that is often relevant in models concerns their caregivers. Not only can caregivers generate costs, but depending on who furnishes the care, the progression of illness can be altered as can the effects of treatment and other aspects of the patient's course. A patient suffering dementia, for example, may be more or less likely to be institutionalized depending on who provides care and for how long. Thus, attributes might reflect the *number* and *type of caregivers*, *when care* is provided, *whether* they are *paid*, and so on. Of note, if the caregivers are themselves modeled as separate entities, then some of these features would be incorporated as attributes of the caregiver entities.

Models constructed for HTA typically consider the costs accrued, and one of the aspects that tends to affect those costs is where the care is provided. For example, the cost of infusing a drug is quite different if this happens in hospital, in an outpatient department, in a freestanding facility, or at home. Thus, attributes pertaining to the location of care are often carried in these models. These attributes might identify the specific *location*, the *time spent* in that location, *from where* the patient came, *where to* he or she goes next, and so on. For example, a model of pneumonia might track the patient's *location* of care from nursing home to emergency department, to a general ward, then to the intensive care unit, back to a ward, perhaps to radiology, and then back to the nursing home, but perhaps at a different level of intensity of care.

Detailing the disease characteristics, not only at the start of the model, but updating them throughout the simulation as events occur, is usually an important component. Many such characteristics may be required depending on the disease at issue and the purpose of the model. These can cover details of the diagnosis, the specific form of the disease, extent and severity, manifestations, sequelae, and any other aspects that may be relevant. For example, in a model comparing treatments for diabetes, there could be attributes covering: *type* of diabetes, *duration* of disease at baseline and throughout, extent of *microvascular involvement* in each organ system, occurrence of *macrovascular disease*, degree of *hyperglycemia* (fasting, post-prandial, and glycosylated hemoglobin), *hypoglycemia* (frequency and severity), and *body mass index*. There might even be attributes addressing *insulin resistance, beta cell function,* and hepatic production of *glucagon*. In many diseases, scoring systems are used to assess the extent of disease. For example, there is the tumor-node-metastasis system in cancer and the expanded disability status scale in multiple sclerosis. The elements adding up to the score might be represented as separate attributes or the aggregate score might be carried instead. Many additional attributes may be used to represent aspects of any disease sequelae (e.g., *proteinuria* and *creatinine clearance* to reflect renal failure in diabetes). Other attributes may carry information about relevant comorbidities.

Apart from the features of the disease, HTA models must represent the characteristics and effects of the technology that is under assessment. These properties are carried in attributes if they vary from individual to individual (otherwise, they can be stored as global variables). The nature of these attributes will differ substantially depending on whether the technology is a drug, a device, a test, a procedure, a process, a program, and so on. For example, for a drug, there might be attributes representing dose and frequency of use, route of administration, response to treatment, occurrence of side effects, degree of adherence to the regimen, and so on. There may also be attributes addressing drug–drug interactions, and increasingly those having to do with genetic make-up that affects metabolism of the drugs (e.g., CYP2C19 for clopidogrel or CYP2C9 and VKORC1 for warfarin). For a device such as a pacemaker, the attributes would have to do with the type, date and complications of implantation, battery life, functions, breakdowns, effect on arrhythmias, and so on. For a screening test, it might be important to carry information about its sensitivity and specificity given the person's characteristics, data on prior results, time since last screen, and so on. Specific lists of attributes would be compiled for each type of intervention depending on the context.

HTAs are interested in the total benefits and costs accumulated over a period of time. Thus, it is necessary to keep track of the quantities involved. This can be done at the individual level by using attributes to store the accruals. A very simple example is the total survival time for each person, which can be recorded in a *survival* attribute that is updated periodically and processed in analyses at the end of the simulation. Many analyses will adjust this time according to some estimate of the quality of life to produce QALYs. As the quality of life changes over time according to what is happening, the adjustments are best made along the way and the results accumulated in a *QALY* attribute. For both of these, as well as for other items whose valuation is affected by the time of occurrence because of time preference, the discounted quantities can be recorded in additional attributes, with the discounting taking place at the time of the accrual (see Chapter 3). The same applies to costs, which can be accrued according to their category (e.g., direct and indirect), their type (e.g., drug, laboratory tests, hospital, and doctor visits), who pays for them (e.g., insurer and patient), and so on. Whether any of these results should be accumulated in attributes is addressed in the next section.

As nearly all HTA models involve the passage of time, there are many timing aspects that may be recorded in attributes. Age itself is one of these, as noted already. Others might include durations of various episodes, when events occur during a simulation, delays due to waiting for a resource to become available, and so on. These are detailed in Section 2.5. Apart from timing, the simulation will need to record what has happened, what the last event was, how many of a given type have occurred, and anything else about the experience of the individual that needs to be remembered. This includes the state the entity is currently in (e.g., second remission from cancer) and any states that have been experienced already (e.g., first remission).

Various items may need to be stored during a simulation purely for modeling purposes, and if these are information that differs among individuals, then they should be recorded in attributes. For example, when implementing a time-to-event approach, it is important to keep track of what the next event is for each individual and when it is supposed to occur. This can be done in two attributes. An even more detailed event sequence can also be stored if this sequence is determined all at one time. For example, it may be the case that upon suffering a myocardial infarction, a patient is to be transported to the emergency room; from there the patient will be admitted to the cardiac catheterization laboratory; then transferred to the coronary care unit; and, if still alive, eventually to a ward. This entire sequence could be entered in an individual's attributes upon the triggering event (i.e., the heart attack in this case).

Finally, it may be useful to store a unique label for each entity so that it can be identified as needed, particularly during debugging of the model.

2.4.3 What Information Should Be Stored as an Attribute?

In many situations, the model designer faces the question of how to store a particular type of information, especially regarding outcomes. For example, should an attribute *QALY* be created to accrue each individual's quality-adjusted survival or should this quantity be accumulated globally for everyone (perhaps by intervention and other stratifying features)? If the data are stored individually, then the simulation can produce a dataset that contains the relevant information for each entity, and this dataset can be analyzed like any other to estimate the mean, median or other percentile, confidence intervals, range, and any other distributional measures that are desired. It may also enable analyses of the results for subgroups of the population, which can be established at any time without having to redo all the work.

Storing all the outcomes individually may be interesting, but if it is not directly relevant to the problem at hand, then the additional computer memory required and increased runtime are wasteful. For a single attribute—even for a few—this is not going to be a noticeable problem, but it may become one if the modeler is not careful to avoid specification of unnecessary attributes. Moreover, most analysts do not enjoy wading through extensive amounts of data to find the items of interest.

The alternative to recording results in attributes is to store the information globally (Figure 2.8), with each entity contributing to the corresponding accumulator whenever appropriate. This produces the totals required with much less use of memory and without burdening the modeler with reams of output. Of course, the model loses track of who contributed what when, and it is more difficult, even impossible, to derive distributional measures. Another consequence of this is that analyses by subgroups that were not conceptualized *a priori* force the modeler to rerun the simulation.

A closely related design decision is whether a particular type of information should be included in a simulation or left out altogether. If it has no role in determining any of the model endpoints, then it should not be incorporated. For example, in most models eye color is not a factor in the occurrence of any of the events; it does not affect the valuations of the events in terms of quality of life or costs, nor does it make any difference to any of the decisions or structural branch points. Although it is perfectly possible to have an attribute *eye color*, it would be a waste of memory and computing time (and of researcher time to obtain the input information and process any results). Sometimes, it is difficult to be sure whether an attribute might be necessary. For example, features like age and sex are so commonly stored as attributes in DES for HTA that they are almost there by default. It is best to exercise restraint, however, and only incorporate information when it is reasonably certain that it will have an important impact on model outputs. It is also worth noting that, depending on the software used, it can be relatively easy to add an extra attribute later if the need arises.

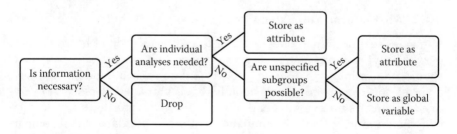

FIGURE 2.8
Algorithm for deciding whether to store outcome information as an attribute or as global information.

2.4.4 Memory

In the vast majority of problems addressed by HTA, the history of what has happened is an important aspect of the model because it can affect the occurrence of future events, their consequences and valuations, the decisions that are made, and many other aspects of the simulation. An obvious example is the response to prior treatments, including adverse reactions. Lack of response could prompt a decision to use an alternative treatment, and this would definitely be the case if there was an anaphylactic reaction. The number, type, and severity of prior events often help determine what happens next. Compliance, risk-seeking behaviors, propensity to seek medical care, and many other experiences may be of importance to the simulation.

This type of information is inherently individual as each entity will have a personal experience in the model. Memory for this historical material can be created by specifying attributes where the information can be stored. These attributes remain empty until the corresponding event has occurred and then are filled with the appropriate data. For example, in a model of a disease where relapses are an important event, many things may depend on the number of relapses that have already occurred. An example is multiple myeloma, where the response to treatment declines as the relapses increase. The disease also becomes more aggressive and the symptoms and sequelae may not return to baseline, even if remission is attained again. Thus, a model in this disease area would likely have an attribute *relapse number*, and possibly *severity of last relapse*, *response to last treatment*, and *time since last relapse*.

There may also be a need to know about things that have happened prior to the entity entering the model. For example, in a model of a recurrent event, like asthma, the number of exacerbations already having been experienced at the start of the model would be recorded in an attribute. That attribute might be incremented as new exacerbations occur during the simulation, or might stay unchanged to keep that information, with a separate counter set to log new events.

2.5 Time

The passage of time is a central component of a DES: attribute values change, events occur over time, results accrue accordingly, and so on. Simulation time reflects time as it elapses in the real world. Just as time in real life flows uninterrupted, simulation time runs continuously, in the sense that an event can happen at any moment during the simulation. Simulation time begins running at the start of the model (there is an exception to this discussed in

Chapter 3) and continues to go by until the run ends, unaltered by any of the events, entities, or other components of the model. The degree of precision with which simulation time is specified is at the discretion of the modeler, but for most HTA purposes, it is not necessary for the units to be finer than days—smaller fractions implying unrealistic precision for many of the components.

To track the passage of simulation time, every DES has an explicit clock. This clock shows current time in the simulation (of note, it does not correspond to the computer's clock). Timings of importance to the simulation or to the results are derived from the simulation clock. For example, to calculate the length of stay in hospital, the *admission* event would record the time (and date) in the corresponding entity's attribute, and, upon *discharge*, that event would look up the *discharge time* on the clock and subtract from it the one stored in the *admission* attribute. In this case, the precision would probably be in days. Any time-related input or output parameter can be derived by consulting the simulation clock as necessary.

Since the modeler is only interested in the happenings that are relevant to the HTA problem, it is not necessary to mark the passage of every simulated instant. Instead, the clock can be managed efficiently by advancing it from one relevant moment to another. How this is done depends on the *event occurrence* mechanism used.

As noted already (see Section 2.2.1), most simulations for HTA are constructed using time-to-event approach. With this approach, the interest is only in the times at which events happen and not in between (during the in-between time, the entities are alive in the simulation, but nothing is happening to them). Thus, the simulation clock can be allowed to advance from the time of one event to that of the next one (Figure 2.9). This interval when nothing is happening (e.g., while waiting for the next outbreak of an infectious disease) may be quite long (e.g., months or even years) but may reduce to days, or even minutes, during a rapidly evolving event (e.g., treatment of a cardiac arrest).

If the model is constructed using periodic checking, then the simulation clock is advanced in uniform discrete time steps corresponding to the desired periods (Figure 2.9). This is similar to the minute hand in a wall clock that doesn't display seconds: it jumps from marking one minute to the next, although the seconds are passing in the background. The modeler is free to set the time step and may specify that it changes depending on the circumstances. As with the time-to-event approach, time is running continuously in the background, regardless of what time steps are set.

Many things may need to occur during a simulation at the same instant in model time. For example, values for attributes may need to be derived and stored in the appropriate places; times until next events may need to be calculated; logical checks may have to be carried out; information may need to be read from files output to displays, storage devices, or printers; and so on. For all of this activity, the software running the simulation needs computer

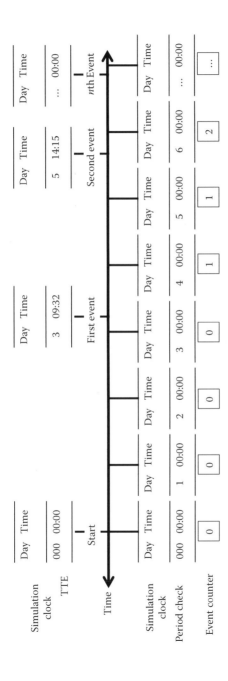

FIGURE 2.9

What happens with the simulation clock depending on the type of event checking that is implemented. Continuous real time is represented by the double-headed arrow in the middle. Above the arrow is the time-to-event (TTE) approach, showing snapshots of the simulation clock at the start and when each event occurs. Below the arrow is the periodic checking approach with snapshots of the simulation clock at the end of each period (one day) when event occurrence is checked. The counter reflects the cumulative occurrence of events in the periodic checking approach.

time to process the instructions. Thus, the computer's clock may be advancing, but the simulation clock should not move forward—it is *paused* to allow for these things to take place. The length of the *pause* depends on the capacity of the processors. This is an aspect of the computing device, but not of the simulation.

2.6 Resources and Queues

In the real world, resources are nearly always subject to a capacity constraint. An emergency department, for example, has a limited number of beds (Raunak et al. 2009); a laboratory can only process a particular number of tests per hour; a given physician can see a maximum number of individuals during her office hours (Swisher et al. 2001); there are a fixed number of operating rooms (Cardoen et al. 2010); and so on. Indeed, it is this limitation of resources that creates the economic problem that HTA tries to address: if one spends resources in one area, then there is an opportunity cost in the sense that one has to forego spending them in other areas. The constrained availability of resources may result in individuals having to wait for a resource to be available if it is fully occupied—a situation all too familiar in many health care systems (Barua et al. 2010; Comas et al. 2008)! Unlike state-transition models, DES is able to represent such resource constraints and facilitates the management of queues for the use of such resources. In fact, many modelers would think of this as the main use of DES.

A resource in a DES is a special kind of event where a service is provided to an entity. In health care models, resources commonly reflect provision of health care services. They can be beds in an emergency room or hospital ward, examining rooms in a clinic, diagnostic or imaging machines, surgical suites, devices of various sorts, monitoring machines, and so on. Clinical personnel can also be thought of in this way if there is no reason for them to have their own attributes (in which case they would need to be entities). Special resources may incorporate the ability to transport entities (e.g., an ambulance or a gurney). While some resources can be used many times (e.g., a mammography machine), others are meant to be used only once (e.g., a urinary catheter). These disposable resources can also be modeled in a DES.

The number of entities that a resource can serve simultaneously is the *resource capacity* (Figure 2.10). For example, an operating room has a capacity of only one. If it is in use, then it is not available to other entities. In contrast, a nurse in the recovery room may be able to take care of several patients at once. The capacity of a resource may change over time. The capacity of a disposable resource diminishes as it is consumed unless the stocks are replenished; the capacity of an imaging device may be more constrained

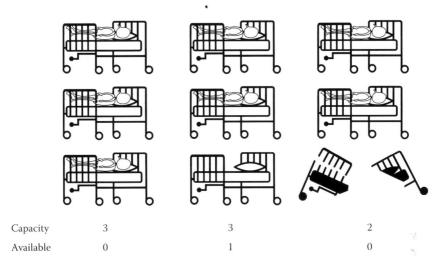

| Capacity | 3 | 3 | 2 |
| Available | 0 | 1 | 0 |

FIGURE 2.10
Illustration of changing capacity and availability of a resource. This facility has three beds. In the first column, all beds are occupied so nothing is available were a new patient to arrive. In the second column, one patient has been discharged, making a bed available. Unfortunately, that bed breaks, reducing the capacity to two in the third column, with no availability.

on weekends; the capacity of a doctor's office may be zero after business hours, and so on. Capacity may also change due to other circumstances. For example, the recovery room nurse who ordinarily takes care of five patients may have a reduced capacity of one, if that patient is in severe distress or on a medication that requires constant monitoring.

Resources typically have a *cost*. This can be a simple amount every time the resource is used, or it can be a value that is determined according to various rules. These can range from time-based rules (e.g., cost per hour of use), to more complex ones that take into account when the service is provided (e.g., during or out of office hours), to very complicated ones that consider any number of other aspects, such as how many times the service has been provided, how constrained the capacity is, who pays, where it is provided, and what discounts or copays apply.

Providing a service usually takes some time. If a resource is occupied when another entity needs it and no other similar resources are available, then the requesting entity must wait until the resource is free (if it is reusable). Thus, a line will form (Figure 2.11). This *queue* is any place in the simulation where an entity waits for a resource to become available. These places can be physical or virtual. For example, when individuals arrive to see a doctor, they typically have to remain in a physical waiting room until the doctor is free. Patients who need a transplant, however, wait on a list for an organ to become available (Stahl et al. 2007). Such queues are virtual rather than physical.

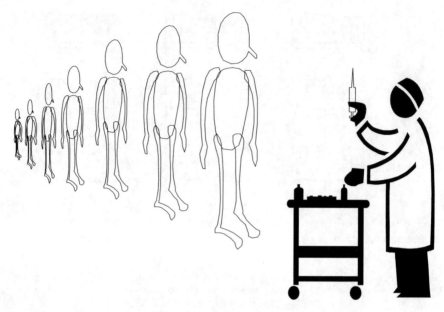

FIGURE 2.11
Queue forming in front of a clinician administering a vaccine.

Whether physical or virtual, queues have logic. In the waiting room, the receptionist may follow a *first in first out* (FIFO) logic—everyone is given the same appointment time and whoever gets there first sees the doctor first. Another common type of logic is *highest value first* (HVF) where the next person served is the one with the highest priority. This is often the case in the waiting room of an emergency department (Konrad et al. 2013) or on the waiting list for transplant (Shechter et al. 2005). There are other possibilities as well (e.g., last in first out [LIFO] logic). Simulations can track and report many aspects of a queue. These can range from its length at various times, the average waiting time, the longest time, who is in the queue, why it is lengthening, and so on.

Queues may have a capacity as well. For example, an emergency room may close when it cannot handle any more patients in need of life support. Capacity may be specified in terms of duration in the queue. Waiting lists in some countries, for example, have a mandated maximum time, after which the individual is allowed to take some other action (e.g., go to another country for care). Thus, when the queue reaches capacity or the waiting time maxes out, the decision rules controlling the queue must specify what happens next.

Most resource-constrained health-related DESs have addressed questions relating to the organization and delivery of health services (Jun et al. 1999) and have represented outcomes as measures of throughput, rather than clinical or patient-reported outcomes. For example, a DES-based evaluation of a proposed reorganization of surgical and anesthesia care found that the new system improved patient throughput at reduced cost (Stahl et al. 2004). Another DES study evaluated four alternative locations for a trauma system helicopter base, finding that one location resulted in more completed missions, and lower mean times to the scene and to the medical center (Clark et al. 1994).

Despite this resource-constrained context being the norm for modeling in other fields, HTA models—curiously given their purpose—generally do not reflect capacity constraints. As noted in Chapter 1, most HTA models today ignore the reality of capacity constraints. Instead, they assume that whenever a resource should be used, it is immediately available to whoever needs it, regardless of how many require it at that time—they take the capacity of all resources to be infinite. In these unconstrained models, it is not necessary to represent the resources explicitly—they can be implemented in a DES by simply accumulating the corresponding costs as the resources are used (see Chapter 3). Whether these models without resource constraints ignore resources altogether or set the capacity of all resources to infinite, no simulated person ever waits for a resource to become available and all related aspects (e.g., queues and waiting times) are moot. As unconstrained models are a simplification of constrained ones, there is no extra difficulty in constructing them as a DES.

In these unconstrained HTA models, the effects of real resource constraints may not be entirely ignored. The extent to which they are incorporated is very indirect, however. It has to do with the fact that the values used as inputs for certain parameters have been estimated by studying a real health care system. Thus, they reflect the resource capacity constraints in that system. For example, the parameter estimates taken from an observational study used to predict disease progression following the experience of a multiple sclerosis relapse reflect the timeliness of intervention for that relapse in the health care system where those data were collected. The relevance of any differences in resource capacity between the time and location of the observational study and the time and jurisdiction to which the model is applied is often not considered, however. Moreover, it is difficult though not impossible to test for these potential effects. The appropriate modeling of resource constraints could get quite complex. For example, the model would have to represent the resource constraints in the time period and jurisdiction from which the data used to estimate the model's parameters were collected. Additional parameters might also be required to represent the effects of earlier, or later, access to relevant resources. A sophisticated approach to sensitivity and constraint analysis is required but can provide quite profound insights as to the success of any given policy in the real world.

By not representing resource capacity explicitly within a model, it is impossible to model any impact an intervention may have on outcomes. For example, a treatment that reduces asthma exacerbations may free up resources such as emergency room beds. As these can then be used for other patients, waiting times in the emergency room may be reduced and the queues may become shorter. The possible benefits in this regard are, thus, completely ignored in most HTA evaluations, leaving a potentially substantial category of effects un-investigated.

Although adjustment for capacity constraints may not be considered, differences in availability of key resources may well affect the estimated cost-effectiveness of the health technologies. For example, in evaluating cardiovascular preventive strategies, a DES represented the availability of angiograms, angioplasties, and bypass grafts, all high-cost interventions requiring significant capital investment, and for which there are usually waiting times (Cooper et al. 2008). Another DES (Jahn et al. 2010) assessed the impact of alternative numbers of daily stent placements, and associated wait times for surgery, on the cost-effectiveness of drug-eluting versus bare metal stents in patients with coronary artery disease. Repeat surgery, and hence extra demand and longer queues, is more likely with bare metal stents. A third DES (Crane et al. 2013) represented the parallel processes of disease progression and pathways of care at a hospital-based glaucoma service, noting current appointment delays due to constrained service capacity. Alternative scenarios regarding the frequency of follow-up, the length of the booking cycle, and the earlier use of laser treatment in the clinical pathway were tested. With respect to the technologies assessed, an alternative management algorithm of using laser therapy prior to pharmaceutical intervention resulted in the gain of additional QALYs at lower costs (i.e., it was a dominant strategy).

Even if there are no capacity constraints for any of the resources (and therefore, no queues will form), the modeler may still designate places (often called *holds*) where entities wait for a required event to happen before they can proceed. For example, a patient suffering an accident or stroke (Stahl et al. 2003) may need to wait for an ambulance to arrive before proceeding to the emergency room. Initiation of treatment may require waiting for a test result, such as that of a magnetic resonance imaging, to be available. When modeling an episodic infectious disease, the model may have entities hold until an outbreak occurs before continuing the active simulation of their course. While these holds are not queues per se, they do have many of the same properties, and simulations can track and report many of the same aspects.

2.7 Global Information

Although not specific to the DES technique, global information is an important concept for modeling. In contrast to attributes, which are meant to store details that are pertinent and possibly unique to each entity, other information in the model is global in the sense that it is not tied to any one entity but rather is carried for the model as a whole (Table 2.5). While attributes

TABLE 2.5

Some Types of Information That Are Often Stored in Global Variables in Discrete Event Simulations for Health Technology Assessments

Category	Examples
Distributions	Of age, sex, and other initial values for attributes
Resources	Capacity, schedule, uptake
Medical	Proportion of clinicians following guidelines
Economic	Inflation rates, discounting rates, unit costs, payer table
Health outcomes	Survival, quality-adjusted life years, event counts by intervention, and other subgroups
Cost outcomes	Total, by type, payer, by intervention, and other subgroups
Counts	Of events by type, of doctor visits, and other health care encounters
Specialized modeling purposes	Time horizon, current time, number of entities at start, current number of entities in the simulation, structural gate flags, number of replications to run, and current replication

generate as many copies of the data as there are entities in a particular run, there is only one copy of each item of global information, regardless of how many entities are simulated. If this lack of specificity suffices, then storing information globally is more efficient than using attributes.

Global information can be accessed or modified by any event in the simulation. For example, at a *mammography* event in a model of breast cancer screening, a woman entity may be assigned her risk of a positive finding from a value stored globally, and her probability of a false positive mammography would be computed using global information combined, perhaps, with some of her own data stored in her attributes. This facility to use global data is not restricted to entities. Resources (e.g., their capacity), decisions (e.g., proportion of physicians who follow guidelines), calculations (e.g., inflation rates), and other components, all have access to the global information. By the same token, many of the components can modify the values stored in the global variables. A very simple example is a counter of the number of entities in a simulation. This counter is increased by one when an entity enters the model and is decreased whenever one leaves.

There are many items for which a global value is sufficient. The most basic are those which specify various details that control the simulation in some way. An important case of global information is the clock that keeps time for the simulation. A related example is the *time horizon*. Although some models are run until all the entities die or otherwise exit the simulation, many specify a maximum simulation time or *time horizon* (e.g., simulate for 5 years). This value, stored globally in *time horizon*, will tell the simulation when to stop. Another example is the number of entities to create, which determines the size of the population that is simulated. The discount rate, which is used in calculations that incorporate time preference, is global information, as is the inflation rate, which is used to adjust values to other years, and so on.

Another basic set of components in economic models like those used for HTA are the unit costs (i.e., *prices*, although the term *costs* has a very precise technical meaning) that are to be applied to the consumption of resources. These unit costs are typically general to the entire model as the price does not vary for each entity. In some cases, the cost accrued by the entity may be individual (e.g., if the dose given is by weight) but the unit price (e.g., the cost per ml of drug) is global. In some jurisdictions, the cost may vary according to the payer—the private insurer may pay a different amount than the government or the individual covering the cost privately—but all of these data can be stored in a global table. This also applies to any discounts, copays, and other intricacies of coverage and reimbursement.

A somewhat related set of data that can be stored globally has to do with who pays for what. In economic analysis, the perspective taken (i.e., from whose point of view the analysis is done) is important. For example, the government insurer may pay for care in the hospital but not at home. Instead of rerunning the simulation and all the analyses for each perspective of interest, a DES can do it all at once if the payer information is provided in a global table. Whenever a resource is consumed, the model can attribute its cost to the appropriate account(s), which are kept in several global variables, and the results can be produced simultaneously for all the specified perspectives at the end of a run.

In budget impact models, an important aspect is typically the market share of each product before the introduction of a new intervention (Sullivan et al. 2014) and how these are modified by uptake of the latter. To run this simulation, the initial distribution would be specified globally. The uptake may require additional details, such as how it changes over time, or in various patient subgroups, but all of this would be stored as global information.

Many of the attributes in a simulation may have their values assigned by selecting from a range indicated by the modeler (see Section 2.8). For example, the age at the start in an osteoporosis model may be chosen from a range of 50 to 80 years, with a mean of 65 years. These distributions of values are specified by one or more parameters (e.g., the normal distribution is given by its mean and variance). It is natural to store such distribution parameters as global information since the distributions are applicable to every entity or other component that needs to select a value. Even the selected distributional form to use in a particular run may be specified in global variables. By storing this information, the impact of these structural choices can be analyzed.

As noted in the preceding section, many of the quantities that must be accrued to produce the results can be accumulated into global variables. This is the case for survival, QALYs, total costs, and cost breakdowns, and the discounted versions of all of these. Many other things can be accrued, such as the number of uses of a specific resource, the proportion of patients discontinuing treatment, the frequency of cases in an epidemic, the ongoing case fatality rate, and prevalence of various states.

2.8 Distributions

A distribution depicts the spread of values for a particular quantity (e.g., an input parameter for a model) in terms of the relative frequency with which each value occurs in the population of interest. Familiar examples are the proportion of people in a population who are male and the range of ages, as summarized by a mean and standard deviation; a pie chart depicting causes of death as a visual representation of the distribution; and a life table giving the numbers of people alive at different ages. Although this concept is not specific to DES, it is central to simulation because it underpins the individualization of entities and their experience in the model. Distributions are used to represent the range of values for various components of the model, such as sociodemographic characteristics and disease status, time-to-event calculations, and event types. Distributions are also key to the fitting of the controlling equations (see Chapter 6) and are used to represent uncertainty around the values of inputs (e.g., clinical effects of the interventions, costs, and utilities) to inform the analysis of parameter uncertainty (see Chapter 5). Selecting values from distributions is covered in Chapter 3 and illustrated with an example in Chapter 4. The ability of DES to work in continuous time allows more direct use of empirically derived distribution data, such as the times to an event like a stroke or fracture recorded in a registry. This, in turn, can allow the model to be truer to reality, instead of being forced to make simplifying assumptions about the distribution that represents the data.

2.8.1 Specifying a Distribution

There are several ways to specify a distribution (Law 2011). The simplest is to itemize the observed values and the frequency with which they occur, yielding an *empirical* distribution. If the values are distinct and disconnected (i.e., categories), then the distribution is labeled *discrete*. For a discrete distribution, a value can only be chosen from the specific categories allowed. For example, when describing the distribution of sex in a population, there are only two options: male or female. There is nothing in between (not addressing sexual orientation or genetic disorders that blur that distinction).

Discrete distributions are typically specified by the proportions that fall into each of the allowed categories. This can be done separately for each category (e.g., 45% male, 55% female), or it can be cumulative (e.g., 45% male, the rest up to 100% female). Graphically (Figure 2.12), the former is like side-by-side bars, one 45 units high and the other 55 units, while the latter is like stacked bars, the bottom portion being 45 units high and the top portion adding another 55 units to make a total height of 100 units. These specifications are entirely equivalent but the latter turns out to be more useful for simulation because it facilitates selecting a specific value by using a random number (e.g., assigning sex to an entity), especially when there are more than two categories.

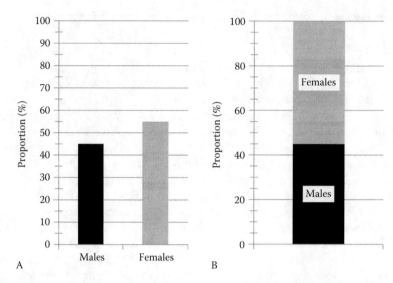

FIGURE 2.12
Example of a discrete distribution. Panel A displays the discrete distribution with separate bars for each sex, while panel B displays the same information in terms of a cumulative distribution.

If the permissible values are not disconnected, in the sense that any value in between any other two is allowed, then the distribution is known as *continuous*. For example, when describing the age distribution in a population, every value within the observed or defined range is allowed—the level of precision is dictated by the needs of the simulation. A continuous distribution can also be specified empirically by describing the observed values for a parameter and their frequency (Figure 2.13).

A continuous distribution is sometimes specified by categorizing the values and expressing the proportion that lies in each category. A familiar example is the table in reports of clinical studies that lists the proportion in each age group, defined by 5- or 10-year intervals (Table 2.6). Needless to say, this way of specifying the distribution is an approximation of its underlying continuous nature, and the broader the categories, the less accurate it is. This approximation can also be described cumulatively, with the proportion in each category adding to the total from preceding categories (Table 2.6).

Most continuous distributions used in DES for HTA are bounded, at least at one end, in the sense that no values are permissible outside a particular limit. For example, age is naturally bounded at zero (and probably at some upper end also, though the precise limit for humans is not known). Truly unbounded distributions are mostly for expressing uncertainty around statistical parameters (e.g., around the risk ratio of a new intervention compared to an existing one, as derived using data from a clinical trial).

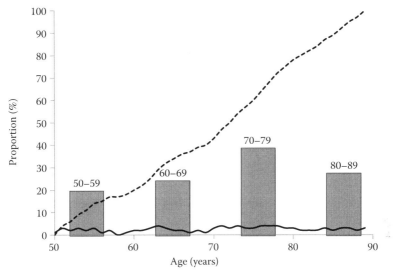

FIGURE 2.13
Graphical display of a continuous distribution. The solid wavy line reflects the distribution of age values without accumulation, while the dashed line shows the cumulative distribution of those age values. The bars display the same information but with the values aggregated into categories.

TABLE 2.6

Distribution of Continuous Age Expressed
in Categories

Age Category (Years)	Proportion (%)	Cumulative Proportion (%)
50–59	18	18
60–69	22	40
70–79	35	75
80–89	25	100

Whether discrete or continuous, the empirical distribution may be a useful way to characterize what is known about the particular parameter—in some situations, it may be the only way to depict the available information. In most cases, however, it is possible to impose a shape to it (e.g., the familiar *bell* curve of the normal distribution). This formulation is intended to describe the underlying distribution from which it is assumed that the observed values came. The shape should be selected based on knowledge of the nature of the factor or similar factors. This is beneficial because it allows interpolation and extrapolation of values to areas where observations are lacking. It also smoothes out irregularities in the observed values that are thought to be due to chance, and it provides a compact form for entering the distribution in the model (instead of listing many values describing the empirical distribution).

For example, the age distribution in a clinical study that enrolled adult individuals defined as 18 years or older may not have observed anyone aged 22 years and the oldest person may have been 77 years old. These observations don't necessarily imply that neither 22-year-olds nor 78-year-olds are part of the population—just that none of those ages was observed in that particular study. Similarly, there may have been quite a few 30-year-olds, again without implying that that observed concentration is a good description of the real population of interest. Of course, if it has been established clinically that people with that illness are invariably younger than 78 years, then that bound would become an appropriate limit, but in most situations, this is not the case and the gaps in data are just that.

Imposing a particular shape to a distribution is advantageous because it yields a specific standard form that can be characterized mathematically. This shape is given by a mathematical function that describes the frequency with which values are observed. For discrete values, this is known as the *probability mass function* and for continuous ones, as the *probability density function*. These functions relate particular values to their frequency using one or more constants (known as *parameters of the distribution*). Unfortunately, these parameters are given various, and somewhat confusing, names (e.g., shape, scale, and location). Also, they are represented inconsistently, with various Greek letters depending on the writer or software package. It is important when employing a formal distribution to carefully understand how the terms and symbols are being used.

Once a standard function is selected, it can be used to sample specific values in a simulation (see Chapter 3). As with the basic approach described above, this is conveniently done by expressing the function cumulatively (Figure 2.13) so that zero corresponds to the lower bound and 100% to the upper bound (the cumulative density function). Although enabling the sampling of values is the main use of distributions in a DES, other handy measures can be drawn from the function. These include the familiar mean (or *expectation*), the median, variance, and various ranges such as the interquartile range and extreme value ranges, such as the 95th percentile. All of these are descriptors of the distribution.

2.8.2 Commonly Used Distributions

A large number of standard distributions have been formally specified (Johnson 2013) and many are included in software packages specifically meant for simulation, and even in spreadsheet applications. The standard distributions can be classified in various ways, such as discrete or continuous, parametric or nonparametric, and bounded or unbounded. Table 2.7 lists some of the distributions commonly used in DES for HTA.

TABLE 2.7

Common Standard Distributions

Continuous			Discrete	
Parametric		Nonparametric	Parametric	Nonparametric
Unbounded	Bounded	Bounded	Bounded	Bounded
Normal	Log	Uniform	Poisson	Empirical
Logistic	normal	Triangle	Binomial	
	Gamma		Negative	
	Weibull		binomial	
	Beta			
	Dirichlet			

2.9 Using Influence Diagrams

Influence diagrams are a very useful way of representing decision problems (Parnell et al. 2013). In influence diagrams, the required problem concepts are represented as shapes (typically ovals) with arrows connecting them. For example, the design of an atrial fibrillation model might start with ovals representing the concepts: *non-valvular atrial fibrillation* (NVAF), *thromboembolism*, and *death* (Figure 2.14). Arrows are then drawn connecting the concept ovals to indicate the *influence* of one concept on another, with the concept at the beginning of the arrow (the *parent* node) influencing the one at the head (the *child* node). This influence may be causal (e.g., *NVAF* causes *thromboembolism*) or it may simply reflect whether or not a factor is known to be relevant to another, without implying causality. For example, the direction of the arrow may simply indicate a temporal relationship, with the *parent* concept preceding in time the occurrence of the *child*.

In an influence diagram, the presence of an arrow connecting nodes carries meaning and so too does its absence. Therefore, unless specifically indicated, relationships are assumed *not* to exist. This makes influence diagrams useful for explicitly depicting the relationships between concepts (e.g., thromboembolism) and a given outcome (e.g., disability and quality of life). Some implementations of influence diagrams allow arrows to point to other arrows, indicating modification by one factor of the relationship between two others. For example, the arrow from anticoagulation to the one between thromboembolism and disability indicates that disability due to a stroke depends on anticoagulation, because it is known that disability varies when a thromboembolic event occurs despite anticoagulation (Schwammenthal et al. 2010).

One advantage of representing problems with influence diagrams is that these are usually easily understood by stakeholders. Thus, as a means of

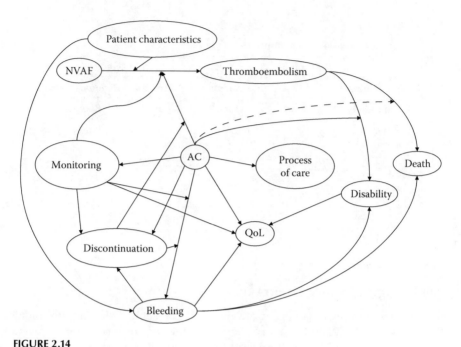

FIGURE 2.14
Influence diagram for a discrete event simulation of anticoagulation to prevent thromboembolism in non-valvular atrial fibrillation. The circles represent the concepts that the modeler has established as necessary to incorporate in the simulation. The arrows reflect how one concept affects (or *influences*) another but without necessarily implying causality (dashed arrow implies that the relation is uncertain). AC, anticoagulation; NVAF, non-valvular atrial fibrillation; QoL, quality of life.

conveying a design and providing a basis for discussion among the stakeholders, they can be very useful. Often, these discussions unearth disagreements that would have been difficult to reveal otherwise. If this happens at the design stage, any resulting changes are much more palatable than when the differences of opinion are uncovered in light of the results of a completed model. Although influence diagrams can become the analytic structure in and of themselves (Kjærulff and Madsen 2012; Srinivas and Breese 2013), this is not recommended as they can be difficult to evaluate.

3

Implementation

In order to produce results that inform a health technology assessment (HTA), a model design must be converted into executable computer code that performs the required calculations. Implementation refers to the actual processes involved in taking the model from design to working software. A discrete event simulation (DES) can, in principle, be constructed using paper and pencil. But apart from its didactic benefit, this is extremely cumbersome even for a very simple model and not feasible for a realistic HTA. Fortunately, the advent of computers made it possible to carry out complex DES, and, thus, since the 1950s there has been very active development of specialized software for DES (see Chapter 9), with many options available today (Greasley 2008; Schriber et al. 2013; Tewoldeberhan et al. 2002). In addition to dedicated simulation software, it is also possible to use general programming languages (Blunk and Fischer 2014; Garrido 2013; Nutaro 2011). This requires, of course, manually coding many of the procedures that are provided as ready-built tools in DES software, but some feel that the gain in flexibility is worthwhile. A third option is to try to use spreadsheet software (Greasley 1998; Klein and Reinhardt 2012; Seppanen 1998). While this appeals to many in the HTA field because of the ubiquitous availability of the programs and the feeling of familiarity with their workings, the spreadsheet approach to representation of a problem and to calculations does not readily accommodate the longitudinal and complex nature of realistic DES for HTA. Nevertheless, many are willing to try and pointers are given throughout this chapter on how to implement features in spreadsheets.

In this chapter, the components and steps required to implement a DES are described. This is done first on a general level and independent, as much as possible, of any particular software. Nevertheless, suggestions are provided for each category of software—be it dedicated DES software (Schriber et al. 2013), a general programming language (Eldredge et al. 1990), or a spreadsheet (Van Gestel et al. 2012). The details of actual code will depend, of course, on the specific software package selected for implementation, but the descriptions given here should allow anyone familiar with the syntax of their chosen software to translate their design into a full working model.

The chapter starts with description of the logic that controls a DES implementation. It then addresses the use of distributions—a topic that is central to many of the implementation tasks—before delving into the handling and specification of events. Creation of the population to be simulated and assignment of the entities' attribute values are addressed next, followed by the approaches to handling time, an important aspect of DES. Since HTA models are about the effects of health care interventions, their implementation is covered at some length. Recording of the results is very software dependent but a general description is given. Although most HTA models do not consider resource constraints, the chapter concludes with a description of the explicit implementation of resources.

3.1 Control Logic

The components of a DES are put together into a working model by connecting them into a logic that governs what happens. The precise way to specify this logic is very dependent on the software chosen for the implementation, with many variations in the way to name and connect the elements, the available prebuilt modules or tools, and so on. Nevertheless, the basic components are fairly common and, in some software packages, such as ARENA® and Simul8®, correspond directly to the concepts described in this chapter.

A flowchart provides a convenient way to illustrate the basic elements of the control logic for a DES in HTA (Figure 3.1). The flowchart consists of module symbols using the graphical convention laid out in Chapter 1 and arrows connecting the modules. These arrows have no implication other than that an entity follows that direction in moving from one module to the next. Each of the modules corresponds to a component of the DES—how it is implemented depends on the specifics of the software used.

The top line of Figure 3.1 describes the setup of the DES. First, the people who will be simulated because they represent those who receive the comparator technology (e.g., current standard of care) are created. Each entity is then assigned initial values for the attributes that characterize him or her. Part of this process involves assigning each person the event times given the comparator intervention. Each entity is then copied to generate identical individuals, one for each technology to be evaluated. Some of the attributes of each duplicate person will need updating to reflect the effects of the technology received. At a minimum, an attribute will identify which technology the cloned individual will receive. Almost certainly, some of the event times will be modified as well. Once the model is set up, the simulation searches the event list for each entity to identify the initial event for each one. These events are then placed on the event calendar. The clock then moves forward to the time of the earliest event in the calendar. For the entity experiencing that event, the attribute values are updated to reflect whatever has happened in the intervening time (e.g., costs that have accumulated). Depending on the type of event that is occurring at that point, the entity is transferred to the appropriate event, where the consequences of the event—including effects on costs, quality of life, and the likelihood of subsequent events—are processed.

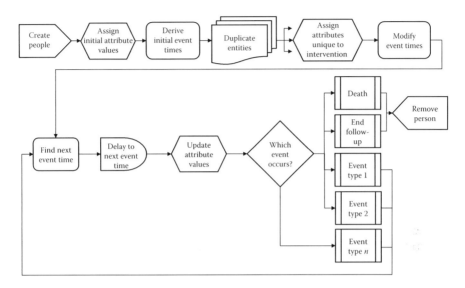

FIGURE 3.1

Example of simulation control logic. The rectangular shapes with vertical lines represent events of various types, and rectangles with rounded corners represent submodels involving several calculations or logical steps. The pentagrams are specialized events that create entities or remove them from the simulation. Hexagons are where values are assigned or modified. The diamond is a branch point where an entity may take one of several paths, according to which event it is experiencing. The rectangle with an oval end indicates a forced delay where an entity is held until the time of the next event. The shape looking like stacked pages represents a special event where entities are duplicated.

Any event times that change as a consequence of that event occurring (e.g., death may now occur sooner) are updated for the entity experiencing the event. If the event is a death, or an alternative event that signifies the end of the simulation for that entity (e.g., end of model time horizon or an entity's 100th birthday), then the entity will leave the model. Following other types of events, the model cycles back to identify the time of the next event for that entity and place it on the event calendar. This loop continues until all entities have left the model or some other condition stops the simulation. This process is illustrated with a detailed example in Chapter 4.

3.1.1 Branching

A particularly important component of the logic in a DES is the branching function, represented in Figure 3.1 by the diamond labeled *Which event occurs?*. Generally, branch points evaluate a condition and, depending on the result, take entities down one path or another: IF *condition X* IS *true* THEN *do something/go somewhere*, OTHERWISE *do something else/go somewhere else*. The *condition* is a logical (i.e., Boolean) check that yields a true or false response. For example, if attribute *age* is greater than or equal to 65 years, trigger the event *retire*, if not, then follow the logic for *continue working*. This logical check can incorporate a probabilistic aspect by assessing whether a random number drawn for the entity is below a threshold level. If true, the logic or event will be triggered (see example in Section 3.2).

Another option is to have more than two branches emanate from a branching point, as happens in Figure 3.1. This is really just a compact way of implementing a series of dichotomous branch points placed in series. This is shown in Figure 3.2, where the top panel diagrams the compact form, while

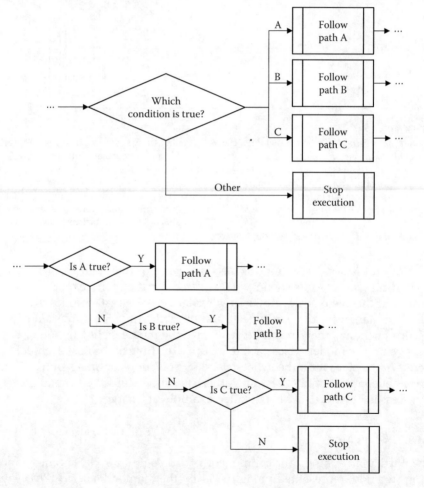

FIGURE 3.2
Example of a Boolean branch point that tests a condition. In the top panel, the branch point first tests condition A, and if it is true, the entity takes path A; if not, it tests condition B, and if it is true, the entity takes path B; finally, a third condition is tested, and if true, the entity takes path C. In the bottom panel, this same process is illustrated as a series of dichotomous branch points, each testing one condition.

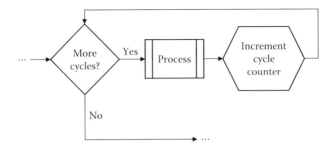

FIGURE 3.3

Branch point used to track number of cycles. This branch point forces entities to go a certain number of times through a particular process. If this kind of logic is implemented, care has to be taken to avoid trapping the entity in a cycle where no other events can be experienced.

the bottom panel shows the series. The figure also illustrates the use of a catchall *Other* branch condition, which provides for the possibility that none of the conditions A, B, or C is true. If this occurs and no other possibility was expected, then something went wrong somewhere. This is a good way to build in error checks at branch points, where an entity selecting that *Other* branch triggers a message, a stop, or writing to an error log.

An interesting use of a branch point is as a gate to open or close certain pathways through the model. Whether the gate is open or closed depends on either a condition that is tested during the model run (e.g., if *day* is *Sunday*, then *outpatient clinic is closed*) or a flag set at the start of execution (e.g., in this run, use a constant mortality). The latter is particularly useful for structural uncertainty analysis (see Chapter 5).

A special case of a branch point is to force entities to go through a specific logic sequence: a set number of times based on a counter (Figure 3.3). This counter should be an attribute in nearly all cases so that each entity stays in the loop the specified number of times. If a global variable were to be used as the counter, then the execution of the loop would depend on how many entities are in the loop at the same time and if there have been entities previously in that loop. For example, if a patient has to go through three cycles of chemotherapy before continuing in the simulation, then the number of cycles already experienced would be stored in an entity attribute and used to determine if additional cycles are required.

3.2 Using Distributions

In Chapter 2, the concept of a distribution was explained. A distribution describes a range of values in terms of how frequently each one occurs. This is very useful for a DES because it enables the modeler to select and assign a specific value for any model component that requires one (e.g., an attribute, an event time, and a particular item of global information), in a systematic way that is in accordance with available knowledge. Broadly, distributions are used in a DES to sample values at an individual level or for the population as a whole. At an individual level, values are obtained from specified distributions to assign attribute levels or to control an entity's course through the logic. For example, a starting age may be assigned to each individual by sampling a value from a distribution representing the spread and frequencies of ages in the eligible population. This use of a distribution typically recurs many times during a simulation.

At a population level, distributions are used to represent the uncertainty around the true value of input parameters (e.g., the probability of experiencing a particular treatment-related adverse event). For each execution of the model, a value can be selected from the distribution for this parameter. This informs uncertainty analyses. For example, values at the lower and upper 95% confidence intervals may be selected for a deterministic analysis looking at extreme possibilities, or values may be sampled from the entire distribution to inform a probabilistic sensitivity analysis. When distributions are used in this way, the sampling typically takes place once at the beginning of a replication.

The distributions selected for relevant population parameters or for individual values are incorporated in the model logic, including any dependencies (e.g., the distribution of body weight may vary according to age and sex of the person), and a method must be specified for sampling values when they are needed during the simulation. These steps are covered in the following sections.

3.2.1 Selecting a Distribution

In Chapter 2, a summary of commonly used probability distributions was tabulated, and Chapter 6 details the specifics of determining and applying distributions to the various types of data, including time-to-event information involved in the equations for use in the simulation. Here, the steps that should be taken for selecting a distribution are described briefly. Full details regarding the selection of distributions are given in many papers (Law 2011), statistical texts (Johnson 2013; Pratt et al. 1995), and simulation books (Law and Kelton 2000).

The first step in selecting an appropriate distribution to represent a range of values is to decide whether the values are best characterized by a continuous or a discrete distribution. Knowledge about the factor at issue may indicate that a discrete distribution should be used. For example, the distribution of sex in the population is clearly discrete, while age is evidently continuous. In many cases, however, the choice is less obvious. For example, disease severity may be thought of discretely (e.g., mild, moderate, and severe), or it might be rated on a continuous scale (e.g., using a visual analog scale ranging from 0 to 100). Even when the parameter is known to be continuous, the modeler may choose to categorize the values (e.g., into intervals of age) instead of representing it fully continuously. Visual inspection of frequency plots of the data may help with the decision. This will require consideration of the type of parameter and how it relates to others in the model. For example, the severity of a disease may be described better in terms of specific levels (e.g., mild, moderate, severe, and very severe) than a continuous score, because its effect on outcomes is not constant along a point scale of severity.

The next step is to decide whether to define an empirical distribution (i.e., one that simply describes the observed data without making any assumptions about how they disperse) or fit a standard parametric distribution. Generally, it is preferable to find the best fitting distribution from among well-characterized standard distributions because this allows for both interpolation and extrapolation, it evens out irregularities, and it describes the data in a compact form. It must be borne in mind, however, that the chosen distribution may misrepresent the values involved, the underlying biology or other ramifications, and, therefore, validation of the choice is important (see Chapter 8). Widely available software (e.g., Kenett et al. 2013) can test the fit of a broad range of alternative standard distributions to observed data, essentially automating the choice of distribution.

In many cases, especially in the HTA field where data over time may be quite limited, many distributions will appear to fit equally well and the choice must be based either on parsimony (i.e., use the simplest distribution that fits) or, even better, on knowledge about the factor and what makes the most sense. For example, the exponential distribution for mortality may appear to fit well when survival has been observed for only a relatively short time (e.g., only until 15% of the study population has died). Although the exponential distribution is very simple, it is usually inappropriate for human mortality because it never reaches zero percent surviving. This implies very long survivals, beyond known human life span, for some proportion of the population. Given this understanding, the prudent modeler would choose a different distribution for mortality—perhaps the Gompertz—that despite being more complicated provides for projections of survival that are much more in accord with what is known about human mortality.

3.2.2 Incorporating the Distribution

Incorporating a distribution in a simulation requires specifying what type it is and the corresponding parameters that define the distribution. How this is done depends on whether the distribution is empirical or one of the standard parametric ones and on the specific software used. If the software includes the selected distribution as one of its standard functions, then it may be sufficient specification to select the software's function that implements that distribution (e.g., NORM for the normal distribution in popular spreadsheet software) and give it values for the parameters that define it (e.g., the mean and standard deviation for a normal distribution). In doing this, it is very important to ascertain that the parameters supplied are in the form that the software expects. For example, when specifying the exponential distribution, some software require the constant that goes in the exponent, whereas other software want the reciprocal of that value! Greek letters are often used as symbols for the parameters but they are not given consistent meanings so it is not enough to provide, say, *beta*. One must look into what the software interprets as *beta*.

If the software has a built-in function for an empirical distribution, incorporating it in a simulation is easy, though if there are many segments, it might be tedious. Most simulation software allow the modeler to specify an empirical distribution in terms of its cumulative distribution function. This involves denoting the boundaries of each interval across the entire range of the distribution and the cumulative frequency of values up to each boundary. These cumulative frequencies correspond to the heights attained by the segments in the stacked bar analogy provided in Chapter 2. For example, the distribution of sex with 45% males and 55% females would be specified by stating that this is discrete, and from 0% to 45% of the distribution the value *male* should be assigned, whereas from 45% to 100%, the value *female* should be designated. The specific syntax for doing this is dependent, of course, on the software. In many specialized software, it is something like: *DISCRETE (0.45, male; 1.0, female)*. The first term identifies the type of empirical distribution (i.e., discrete in this case), and the items in the parentheses give the cumulative borders and the name or value of the segment.

Similarly, the cumulative distribution function of a continuous distribution of age specified using three intervals (Figure 3.4), say, 20% in age 55 years up to 65 years, 50% in age 65 years up to 74 years, and the remaining 30% in 75 to 85 years, would be specified as *CONTINUOUS (0.0, 55, 0.2, 65, 0.7, 75, 1.0, 85)*. By convention, each upper border belongs to the next interval, but it is wise to verify how the borders are handled in the specific software being used. In the example, all ages below 65 years fit within the 20%, while 65 years, itself, is part of the next interval containing 50% of the values. Within a segment of the empirical continuous distribution, the values are considered to spread in a linear fashion. (Note that if this was specified as a discrete distribution instead, then only the interval of age would be known but not its precise value.) This is tantamount to interpolating linearly between the ends of the interval where the selected random number falls.

For example, for an entity that was given the random number of 0.45 to select its age, the software would determine that the value falls in the second interval (i.e., 65 to 75 years because 0.45 is between 0.2 and 0.7), and linear interpolation would generate a value of 70 years as follows:

$$\text{Entity age} = \left[\frac{75 - 65}{0.7 - 0.2} \times (0.45 - 0.2) \right] + 65 \tag{3.1}$$

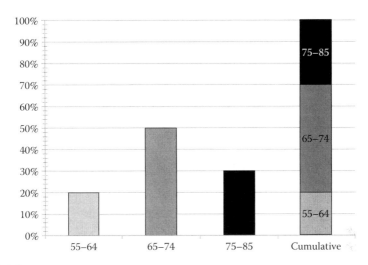

FIGURE 3.4
Example of a continuous distribution of age specified using three intervals. Stippled bars reflect the distribution without accumulation, whereas the stacked bar provides the cumulative distribution.

The fraction in Equation 3.1 gives the slope of the line segment between the borders of the interval (20); this is multiplied by the probability distance from the lower border to the random number to give the amount traveled within the interval (5); and this is added to the lower border of the interval (65) to yield the required value.

If the chosen software does not have built-in functions for empirical distributions, then the specification involves providing the table, or array, that contains the boundaries of the intervals and the proportion of the distribution that lies within each segment. It will also be necessary to note whether the table should be interpreted as discrete or continuous, although this can be incorporated directly into the sampling method. Finally, equations like the one above need to be written to carry out the actual sampling.

If one of the standard parametric distributions is chosen but the software does not include a built-in function for the selected distribution, then the equation defining it must be provided along with the values of the parameters. Typically, these equations are given in textbooks in terms of either the probability density or mass function. For DES, it is preferable to specify the cumulative distribution function, which gives the cumulative frequency of the distribution up to a given value (e.g., for the value age 75 years, what proportion of the population are at or below that age; 70% in the example above). Given that the use for most distributions in a DES is selecting a specific value, it is best to specify the inverse of its cumulative distribution function (also known as the *quantile function*) so that the specific value is a function of the cumulative proportion to that point. In other words, in the age example, the function would read the value at which 70% of the population is at or below that age (i.e., 75 years). This inversion of the cumulative distribution function is algebraically tractable for some standard theoretical distributions, but not all.

In implementing the distributions, it is important to bear in mind that the values of one factor may relate to, or depend on, those of another one. For example, body weight is known to be related to sex. Thus, the distribution of body weight used in a simulation should take into account the entity's sex. Indeed, one distribution might be specified for males and a different one for females. There are many statistical approaches to assess potential bivariate correlations among factors, and their full description is beyond the scope of this book. These are extensively dealt with in texts (Diggle and Diggle 2002). If primary data are available (i.e., on individual subjects), scatter diagrams plotting one set of values against another can be used to visually inspect for the possibility of a relationship, and formal statistical tests such as the rank von Neumann ratio (Bartels 1982) can be run. Alternatively, relevant literature can be reviewed to identify potentially important correlations and possibly to obtain the corresponding quantitative estimates of the relationship.

Correlations may extend across several factors. For example, weight may depend on not only sex, but also age, ethnicity, and socioeconomic status, and some of these factors may depend on each other as well. Multivariate or joint distributions can be used to represent these more complex relationships. Such distributions are generally fitted to individual-level data describing all of the relevant factors. Observed relationships may or may not be causal, but lack of causality or ignorance about its existence does not excuse the modeler from implementing the relationships. In order to accurately reflect the real world, the simulation should represent as many correlations as are observed.

3.2.3 Selecting Values from a Distribution

The main reason to implement distributions in a DES is to enable the selection of values for various purposes, for example, assigning baseline characteristics to entities or defining the time to an event. This selection is conceptually quite straightforward. Since a distribution specifies the frequency with which alternative values occur, the process involves selecting a particular frequency and *reading* the corresponding value.

Figure 3.5 illustrates the process of sampling values from a discrete probability distribution using the quantile (or inverted cumulative distribution) function. This function provides the value from a given distribution corresponding to each cumulative frequency from 0% to 100%. In the sex distribution represented in Figure 3.5, *male* is the value corresponding to any frequency number from 0 to less than 45, whereas any value above 45 (up to 100) indicates a *female*. To determine whether an entity is male or female, each entity samples a value between 0 and 100 from a uniform distribution (i.e., all values in the range from 0 to 100 are equally likely to be sampled). Random

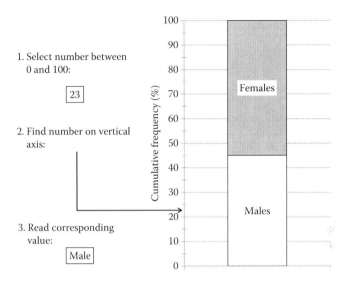

FIGURE 3.5
Graphical depiction of the selection of a value using the quantile function of a discrete distribution. Drawing random number 23 and finding its place along the *y*-axis indicates that this entity will be of male sex.

sampling assures no systematic distortion, but the degree of accuracy in representing the full distribution also depends on the number of selections made from the distribution. As the number of selections increases, the accuracy improves. For example, with only 10 draws, the result may be 6 males and 4 females, quite different from the underlying distribution, whereas with 1,000 draws it will come much closer to 450 males and 550 females.

This process can be seen graphically in Figure 3.5, where random selection of a number from the range 0 to 100 gave the number 23. This number, 23, is then found on the vertical axis of the chart and the corresponding value—*male* in this case—is read. If an entity had triggered this process, then its attribute *sex* would now contain the value *male*.

Another way to think about this selection process is to consider a pack of 100 cards, where 45 of them contain the term *male* and 55 contain the term *female*. For each entity requiring a value for its *sex* attribute, a card is selected from the pack and whatever the card reads will be the value assigned to *sex*. The card is replaced each time so that the next entity making the selection is choosing from the entire pack of 100 cards. Note that if only a few entities make this selection, it is quite likely that the proportion who selects a male versus female card is quite different from 45%.

In specialized DES software, syntax for the quantile function for discrete distributions is usually included. For example, if the distribution of severity is specified as *DISCRETE* (0.3, mild; 0.9, moderate; and 1.0, severe), then the software will select a random number between 0 and 1 (note that it doesn't matter if the range is given as 0 to 1 or as 0 to 100, or some other such range as long as this is specified consistently), find its place in the cumulative distribution of severity, and return the corresponding value. Thus, a random number of 0.56 would yield a *moderate* severity. Commonly used spreadsheet software does not include a discrete quantile function; thus, the modeler must specify how to make the selection. This can be done by tabulating the distribution (Table 3.1) and using a lookup function to find the desired value.

Although it is somewhat awkward, for many spreadsheet applications, the tabulation needs to have as the first column or row the cumulative frequency and the values as the second one, because the lookup function checks only the first column or row of the table. In addition, the values need to be shifted so they start at 0. Thus, the severity example would be entered as shown in Table 3.2.

The lookup function then uses a random number from 0 to 1 and finds the place along the first column or row that is less than or equal to the random number and returns the corresponding severity value. Thus, a random number of 0.2 would yield a severity of *mild* in Table 3.2 because 0 is the highest entry in the first column that is less than or equal to 0.2. If the random number were 0.3, then the lookup function would yield *moderate*, and any number of 0.9 or higher would yield *severe*. If Table 3.2 were entered into a Microsoft Excel® worksheet, then the required equation would be specified as =VLOOKUP(RAND(),A1:B3,2,TRUE), assuming the table starts in cell A1.

If the distribution is continuous, then the process is very similar except that each frequency number corresponds to a different value in the distribution. In Figure 3.6, for example, the graph depicts the cumulative distribution of expected age at death estimated for females at the time of birth in the United States. If the frequency number 82 were chosen at random, then

TABLE 3.1

Discrete Cumulative Distribution of Severity

Severity	Cumulative Frequency (%)
Mild	30
Moderate	90
Severe	100

TABLE 3.2

Spreadsheet Tabulation of a Discrete Cumulative Distribution

Cumulative Frequency	Severity
0.0	Mild
0.3	Moderate
0.9	Severe

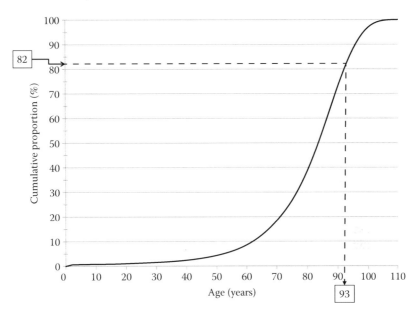

FIGURE 3.6
Graphical depiction of the selection of a value using the quantile function of a cumulative distribution. The random number 0.82 (82%) was drawn. Its position is found on the vertical axis, and by reading across to the curve representing the quantile function, the age of 93 years is obtained.

that newborn female entity would be assigned a time until death of 93 years. For any frequency number that might be chosen, the quantile function can be used to select the corresponding value from the distribution.

If the continuous distribution has been specified using a standard theoretical distribution, then there may exist a closed-form expression for the quantile function. In this situation, there is an equation that directly provides the value for any given frequency number that might be chosen. In other words, the process depicted graphically in Figure 3.6 can be accomplished mathematically. This works for some distributions commonly used in DES for HTA, such as the *exponential*, the *Weibull*, the *Gompertz*, *logistic*, and *log-logistic*. For example, for the *exponential* distribution, the quantile function is given by Equation 3.2.

$$\text{Value} = -\frac{\ln(\text{random number})}{\lambda} \tag{3.2}$$

where λ stands for the parameter that describes this distribution.

The modeler just needs to provide the value of λ, and either the name of the distribution or the quantile function and the sampling proceeds via straightforward calculation.

For other commonly used distributions, such as the *normal*, *beta*, and *gamma*, there is no closed-form solution, but the solution can be obtained using a set of differential equations that have already been worked out. For yet other distributions, it may be necessary to use an algorithm to find the solution. Fortunately, many of the software available for DES, including spreadsheets and general programming languages, offer built-in quantile functions for most of the desired distributions.

Sometimes it may be necessary to translate a distribution that has been specified empirically as a series of intervals and corresponding proportions of the values (such as age in Figure 3.4) to a continuous distribution. In such cases, it is necessary to interpolate within a segment in order to obtain the desired value. Figure 3.7 represents this process for the quantile function for the age distribution specified previously in Figure 3.4. Each of the three line segments represents the portion of the distribution in each interval. Since we are not told how the values are distributed within each interval, the function is drawn as a straight line connecting the two borders of the interval. The first segment goes from a cumulative frequency of 0 at age 55 years to 20% at age 65 years, the second segment extends linearly from there to a cumulative frequency of 70% at age 75 years, and the final segment takes it to 100% at age 85 years. This linear interpolation may be used as the basis for sampling an age for entities: if the frequency number 0.44 was drawn, then 69.8 years would be assigned as the age.

This is tantamount to using the quantile function to represent a uniform distribution between the interval ages.

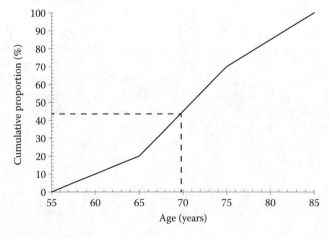

FIGURE 3.7
Graphical depiction of the selection of a value from a continuous distribution of age that has been categorized into three strata (solid line). The random value 44 was drawn and this is indicated by the horizontal dashed line. Where it hits the solid line indicates the value of age for this individual: 69.8 years.

For the distribution of age between 65 and 75 years, this can be implemented as

$$\widehat{\text{Age}} = 65 + 20 \times (\text{random number} - 0.2), \text{ for random number } (20,70) \quad (3.3)$$

or, more generally,

$$\widehat{\text{Age}} = \text{interval lower border} + \text{slope} \times (\text{random number} - \text{cum freq}) \quad (3.4)$$

This can be implemented in a spreadsheet by tabulating the intervals (Table 3.3) with the cumulative frequency in the first column and computing the slope of the interpolating line in each interval (difference between upper and lower border divided by difference in cumulative frequency). The lookup function then finds the interval in which the random number falls, reads the interval's lower border and slope, and implements Equation 3.3.

3.2.3.1 Assigning Sociodemographic Characteristics and Disease Status

As noted in Chapter 1, a common feature of DES in HTA is the representation of heterogeneity in the sociodemographic characteristics of the target population. Individuals vary in attributes such as age, sex, and smoking status, and this variation should be reflected if these characteristics influence the likelihood and timing of events to be experienced during the simulation, or they help determine other values such as utilities and costs. Similarly, DES may need to represent relevant stages of disease (e.g., progression in cancer) or risk factors (e.g., bone mineral density or cholesterol levels). Thus, entities have attributes to represent the values of these characteristics at entry into the model, and these values can change over the time horizon of the simulation. Values for these attributes should be sampled from appropriately selected distributions.

The best approach to modeling the sociodemographic and disease characteristics is to consider the distribution of profiles, rather than each factor separately. By doing this, any correlations among factors that exist in the population are correctly implemented, regardless of how complex they may be. For each model run, at each point at which values are to be assigned, a single distribution reflecting all the relevant profiles is required. For example, the age and sex distributions shown graphically earlier might be presented in a

TABLE 3.3

Spreadsheet Tabulation of a Continuous Cumulative Distribution

Cumulative Frequency	Lower Border	Upper Border	Slope
0	55	65	50
0.2	65	75	20
0.7	75	85	33.3
1.0	85	–	–

TABLE 3.4

Proportion of the Population According to Age and Sex

Age	Females (55%)	Males (45%)	Overall (%)
55–65	14.5	26.7	20
65–75	56.4	42.2	50
75–85	29.1	31.1	30

single table instead of specifying them separately, with the age distribution given for each sex (Table 3.4).

This information can then be combined into a single distribution of the profiles, creating six age-sex intervals (Table 3.5). Sampling from this joint distribution using the equations given earlier would assign an entity a particular age and sex.

This idea can be extended to any number of characteristics. For example, the three categories of disease severity could be added to Table 3.5 to describe 18 profiles of age, sex, and severity. Clearly, as more characteristics are added to form the profiles, the number of possible combinations increases rapidly. Eventually, there are almost as many different profiles as there are individuals in the population.

Often, primary data describing a population, such as information obtained from the subjects in a clinical trial, are available. In cases where these empirical data adequately cover the relevant range (i.e., there are few or no gaps across the factor space), the empirical profiles can be treated as a joint, discrete, non-parametric distribution and sampled directly for use in the simulation, with each sampled profile describing one simulated entity. Alternatively, if a full covariance matrix can be defined, a parametric multivariate probability distribution may be used (e.g., the multivariate normal distribution), or individual correlation parameters describing the relationships between pairs of parameters may be defined, and individual probability distributions for each parameter can be sampled sequentially. Either way, this presumes that

TABLE 3.5

Joint Distribution of Age and Sex

Sex	Age Interval	Cumulative Frequency
Female	55–65	0.08
	65–70	0.39
	70–85	0.55
Male	55–65	0.67
	65–70	0.86
	70–85	1.00

the clinical trial population is applicable to the HTA problem at hand (see Section 3.5).

In the absence of a single, suitable, relevant empirical data source, separate distributions for each factor must be defined. Values are sampled from discrete and continuous distributions as described earlier in this section, with additional parameters specified to represent observed or assumed correlations between individual parameters. Techniques such as the Iman-Conover method (Iman and Conover 1982) can be used to generate correlated sampled values across any set of probability distributions based on specified rank-order correlations between parameters.

3.2.3.2 Sampling Time to an Event

The distribution of times to an event can be conveniently represented as the cumulative proportion of people not yet having experienced that event (equivalent to 1 minus the cumulative distribution function or *complementary cumulative distribution function*). This is often called a *survival* curve because death is one of the most commonly represented events, and the complementary cumulative frequency represents the proportion of persons still surviving over time. In other fields, the name is generalized to a *reliability function*, or its complement, the *failure-time* curve. Since many events in a simulation are neither deaths nor failures, it is better to use the more general term *time-to-event curve*, with the understanding that this actually represents the cumulative frequency of entities not yet having experienced the event.

Values from time-to-event curves are sampled in the same way as from other distributions. The sampled value between 0 and 100 represents the percentage of the population who remain event-free at the time at which the entity experiences the event of interest. This is illustrated graphically in Figure 3.8, where the randomly sampled value of 16% leads to the corresponding entity being assigned a time of 37 months until that particular event.

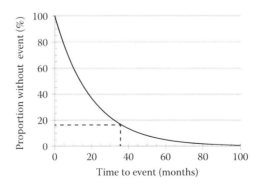

FIGURE 3.8
Sampling a time from a cumulative time-to-event curve. The random number 0.16 (16%) was sampled. Finding its place on the vertical axis and reading across to the quantile function gives an event time of 37 months.

If a standard theoretical distribution is specified and it has a closed-form quantile function, then the time-to-event is computed by entering the random number into the appropriate formula. For example, the exponential curve shown in Figure 3.8 is given by

$$\hat{t} = -\frac{\ln(\text{random number})}{0.05} \qquad (3.5)$$

Thus, for a random number of 0.16, the formula yields 36.65 as the time.

3.3 Event Handling

As noted earlier, events are the central concept of a DES. An event is, in general terms, a point in time when something happens that affects another component of the model. Usually, events in a DES for HTA reflect something happening to an entity (e.g., a person becoming ill), and this can change the value of one or more attributes (e.g., a *disease exacerbation* event decreasing the value of attribute *quality of life*) or of a global variable (e.g., a *hospitalization* event changing the value of *total* cost). Sometimes, an event is specified purely to change attribute or variable values, even though, in reality, nothing has happened to any entity. In order to activate an event and process its consequences, the model needs to know when the event happens.

The easiest way to implement events is to create a list that describes the events that will happen over the course of the simulation. This *event list* is ordered from soonest to latest event. For each event, this list contains the type of event and the model time at which it is supposed to occur. For example, a very simple list might be the following: entity 1 *becomes ill* at day 90; entity 1 *dies* at day 361. When simulation day 90 arrives, the model processes the consequences of that entity becoming ill (these might involve changing attribute or global variable values, or might trigger several additional events such as a *doctor's visit* where treatment is started). On day 361, that entity suffers a death event and is, presumably, removed from the simulation. It should be noted, however, that the occurrence of the first event (*become ill*) could change the time of the second event. If it is a very serious illness, for example, the time of the *death* event might now change to happen, say, on day 91. Part of the consequences of the first event, then, would be updating the *event list*.

Each entity in the simulation has its own *entity-level event list* (left-hand side of Figure 3.9), consisting of the anticipated events currently scheduled to happen and when those are expected to occur. In a DES, there are usually many entities in the simulation at once (an essential feature if there is competition for resources), and so the model must integrate all the entity-level

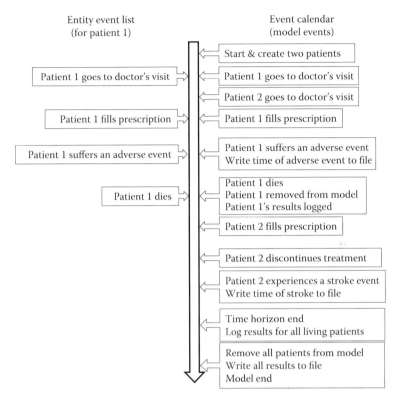

FIGURE 3.9
Depiction of entity event list and model event calendar. The vertical arrow indicates time, flowing from the earliest moment at the top. On the left-hand side are the events on the list for entity 1. The right-hand side shows the event calendar for the two entities currently in this model.

event lists into an overall model-level list (right-hand side of Figure 3.9). This master list of all anticipated events and their scheduled times is called the *event calendar*.

Although some events are very simple (e.g., *death*), many real happenings can be, depending on the purpose of the model, quite complex. These may require multiple chained simulated events to represent them adequately. For example, the real occurrence of developing appendicitis might involve a visit to the emergency department, an exam by the doctor, a diagnostic test, admission to hospital, undergoing surgery, and discharge from hospital (likely a much more detailed sequence in reality). Whereas the simple event is clearly singular, the complex one is really a series of events triggered by one initiating event. When this is the case, it is not necessary to put the entire sequence into the *entity-level event list* because the logic of the model may specify what will happen after the initiating event (e.g., *get appendicitis*) occurs (e.g., *enter emergency department, doctor's visit, diagnostic test, admission to hospital, surgery,* and *discharge from hospital*). These further events are placed on the *event calendar* as a consequence of the instigating event and based on its time. Apart from triggered event sequences, any events defined purely for modeling purposes (e.g., *create entities* and *write results to file*) are also included in the *event calendar*.

3.3.1 Event Calendar

Time advances in most DES from one event time to the next one, rather than in fixed steps, with the time between events ignored since, by definition, nothing of interest is happening there. Thus, the event calendar controls the simulation—it is constantly being scanned and updated to determine what the next event to process is. If the clock is advanced in fixed steps, the event calendar still applies but it contains only a *next check* event with its timing determined by the time step. All the things that happened to each entity during the interval between *next check* events are assumed to happen at the time of the *next check*, so they are added to the *event calendar* simultaneously, at that time and processed immediately.

When using DES-specific software, the modeler institutes and manages the entity-level lists, while the software handles the *event calendar* (Figure 3.10) and the simulation clock. Events that are next on the entity lists are placed

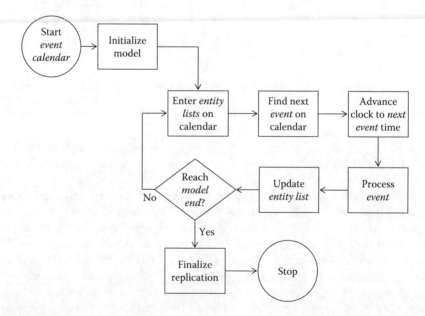

FIGURE 3.10
Processing of event calendar. At the start of the simulation, various items have to be initialized, including the event calendar and the simulation clock. As the entities acquire their event lists, these are entered on the event calendar. The model then finds the earliest event on the calendar and advances the clock to its time. The event is then processed, the entity list is updated, and the process repeats unless the condition to terminate the simulation has been met.

by the software on the *event calendar*. It then advances the simulation clock to the time of the next event and initiates its processing. That event may modify the times—and thus the ordering—of events already on the list or may add new ones to the *event calendar*. The software takes care of these changes and removes the just-processed event from the list.

If an event's time is beyond the end of the *time horizon* or after an event that removes the entity from the simulation (e.g., *death*), then that event will never be processed. This can be a handy way to keep an event on the entity-level list in case it is needed, but making sure it won't process until it makes sense. For example, if a patient *stops treatment*, then the time for *side effect* event is no longer relevant but it may become so after a *treatment re-started* event. When the entity experiences the *stop* treatment event, the model sets the event time for *side effect* to the very distant future so that it will never be placed as the next event on the *event calendar* because that entity's simulation time will be sure to end first. For example, if the model *time horizon* is known to be no longer than say 100 years, then setting the time of *side effect* event to 10,000 years will ensure it cannot occur. Then, when the entity experiences a *treatment re-start*, making the deleted *side effect* event possible again, its time is altered back to an appropriate value and it will reappear on the *event calendar*.

If a general programming language is used to implement the simulation, then the modeler will usually need to write code to create and handle the *event calendar*. Creating and updating the simulation clock will also be necessary. The specifics of this code vary with the choice of software but all of the components listed in Figure 3.10 need to be implemented. It turns out that handling the *event calendar* can be a very laborious task, with considerable computation time required if there are many events and changes to the list. The reason is that processing this list requires repeatedly finding the correct place for events. Either the list is maintained in a sorted state with the next event enumerated first or it is unsorted with events listed in some order like priority of execution upon simultaneous occurrence. If the former approach is used, then to insert a new event on the calendar or relocate an existing one, the list needs to be searched until the correct spot is found or resorted after the time is listed. If the second approach is implemented, then inserting a new event is easy but the list needs to be searched constantly to find the next event. Dedicated DES software takes care of this process automatically, a major advantage of using one of those packages. The handling of event lists is an area of active research, with many algorithms having been developed to try to improve the efficiency of the processes (Wang et al. 2010).

In a spreadsheet, the event calendar may not be needed because each entity's set of events is handled separately. This is the case so long as entities do not need to interact (e.g., because they need the same constrained resource at the same time), and the event structure is simple enough that use of supplementary macros can be avoided. In this situation, the simulation clock becomes implicit in the times recorded for each entity's event list. If a general macro programming language is necessary to implement parts of the structure, then the issues with general programming languages will apply to the spreadsheet implementation as well.

3.3.2 Entity-Level Event Lists

The focus in the remainder of this section is on handling of the *entity-level event lists*. The easiest way to implement the *entity-level event list* is to use attributes to store the required information. This means that each entity will carry around its own list and have complete access to the information in it, which is appropriate since these events pertain only to that entity. Depending on the software used, there are two main ways to set up this *entity-level event list* using attributes. Some software support indexed attributes, or *arrays*, that can be given a specific name. This name represents a set of memory slots that hold the times for the set of events, with each type of event assigned a specific spot in the array (e.g., given by the index numbers in Table 3.6). If this utility is available, then it is possible to implement an attribute *event list* that contains the times of all the events for that entity. The index for *event list* identifies which time corresponds to which specific event. In Table 3.6, for example, the time of 175 days corresponds to *event type 1*, 920 days to *event type 2*, and so on. For the model to determine which event happens next for this entity, it needs to look for the minimum value in the indexed attribute. In Table 3.6, the third *event* occurs before any of the other four and would be placed on the *event calendar* for day 90.

A key that relates the index to the event type can be specified as a global variable since it is common to all entities. In a model of osteoporosis, for example, index 3 might refer to hip fracture (Table 3.7).

An analogous approach to indexed attributes is to group the attribute values in a set. These sets have an implicit index that enumerates the members of the set in order. This makes it possible to find a specific attribute value and determine its place in the set. In Table 3.8, for example, a search of the *grouped event time set* for the lowest value would indicate that the fifth member of the set is the next event for this entity as it is scheduled to occur on day 12.

TABLE 3.6

Indexed Attribute *Event List* Containing Event Times in Days for One Entity

Index	1	2	3	4	5
Time (days)	175	920	90	128	372

TABLE 3.7

Key for *Event List* Stored in a Global Variable

Index	1	2	3	4	5
Event	Stop treatment	Death	Hip fracture	Start treatment	Spinal fracture

TABLE 3.8

Grouped Attributes Containing Event Times
in Days for One Entity

Grouped Event Time Set				
75	740	45	28	12

Again, the key to these events would be needed. If it contained *doctor's visit, death, stroke, bleeding,* and *start treatment,* then the model knows that the next event set for day 12 is *start treatment.*

To use *entity-level event lists,* the model needs to check each *entity-level event list* for the next event time and type and then put each entity's next event on the event calendar at the appropriate place. This process is repeated for each entity after an event is experienced. Regardless of whether the *entity-level event list* is stored in an indexed attribute or in a set, it has to be searched for the lowest time, and its position in the ordered structure has to be recorded. For an indexed attribute, the steps to accomplish this are given in Table 3.9.

After checking every index value, the attribute *Next Time* will contain the lowest time value (or if there are several identical minimum values, the first one found) and the attribute *Next Event* contains the index value for that event. This tells the model at what time this entity will have its next event and also, by referring to the key, what kind of event must be processed at that time. Search of a set will follow the same process.

The actual syntax of the commands to accomplish the search of the indexed attribute or set varies with the specific software. In specialized DES software, there are often special functions such as SEARCH that help with these tasks. Using general programming software requires implementing in code the steps in Table 3.9, although again there may be built-in functions that help. In a spreadsheet, the event times for each entity are stored in a sequential series of cells (typically in a row, one for each entity). This series of cells is an array, and there is usually a SORT function that can be used to ensure the array is

TABLE 3.9

Steps to Finding the Next Event for Each Entity Using an
Indexed Attribute *EventTime*

Step	Instruction
1	Set *Index* to 1 and *Next Event* to 1
2	Let *Next Time* = $EventTime_{Index}$
3	If *Next Time* > $EventTime_{Index + 1}$ then *Next Time* = $EventTime_{Index + 1}$ and *Next Event* = *Index* + 1 Else make no change
4	*Index* = Index + 1
5	If *Index* > max index stop Else return to Step 2

TABLE 3.10

Spreadsheet Containing Event Times in Days for One Entity

	A	B	C	D	E	F
1	Event type	Doctor's visit	Death	Stroke	Bleeding	Start treatment
2	Event time	75	740	45	28	12

ordered from lowest to highest value. Sorting, however, requires that the correspondence between the event type and its time be maintained somehow. It may be easier, therefore, to keep the array unsorted so that the position of each cell serves as the index. Then, a MINIMUM function can be used to find the lowest value and a MATCH function can be used to detect at which position in the array that value was found. If Table 3.10 were entered in a spreadsheet, for example, and the cells storing the values were B2 through F2, then MIN(B2:F2) will yield the value 12 and MATCH(12,B2:F2,0) yields 5. The next event is identified as *Start treatment* using INDEX(B1:F1, 1, 5), and it can be scheduled to happen at day 12. After processing of that event (e.g., by changing *treatment* attribute to *On* and recalculating whatever times are affected by treatment onset), the value in F2 would be changed to a very large number, say 1 trillion, to ensure that treatment is not started again.

3.4 Specifying Events

A DES for HTA contains many types of events. A typical HTA model will have events of the types listed in Table 3.11 (see Chapter 2 for full discussion of event types). Most obvious are the clinical events the person can experience, but there are others that can also affect the person or the outcomes, and

TABLE 3.11

Event Types in a Typical HTA Model

Entity Events
　Create entities
　Duplicate entities, one for each intervention
　Clinical events
　Monitoring events
　Death
　Assign and update attribute values

Model Control Events
　Access inputs
　Output results
　Model end

there are events that have to do with control of the model itself. Given this range of event types, the specifications of what is to happen with each one will vary considerably. Indeed, only broad guidance can be given about how to specify events, with the details of what should be done as a result of triggering a particular event having to be worked out depending on not only the type of event but also the context and purpose of the simulation.

Specifying an event in a DES means indicating what its consequences are and what should happen when an event of that particular type is triggered. For example, for the event *Death*, the simulation needs to know what steps to take when the time for such an event is reached. This might be as simple as removing the entity from the simulation: If *Death* occurs, then *delete* entity. But, it can also involve recording that the event happened: If *Death* occurs, then increment *Total Deaths* by 1, or accruing the cost of that death, and so on. One possible algorithm for processing *Death* is given in Figure 3.11.

In specialized DES software, the implementation of a *Death* event as specified in Figure 3.11 depends on the built-in functions offered. Sometimes, a single module may serve the purpose. For example, most DES software

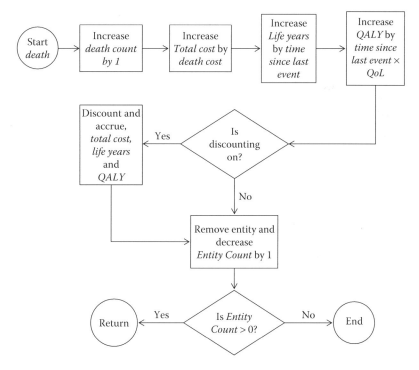

FIGURE 3.11

Algorithm for processing a *death* event. At the start of the death event, the death is counted (possibly by treatment group) and its costs are processed. The survival time and quality-adjusted life years (QALYs) are accrued, together with their discounted values, if appropriate. Since this is a terminal event, the entity is removed from the simulation. QoL, quality of life.

offers an *End* module that removes entities from the simulation. If that is all that the *Death* event requires, then a single module would do the job, but to fully implement the specifications in Figure 3.11 would typically require several modules, each one carrying out some aspects of the process.

When using a general programming language, event processing is written as a subroutine that is called from the main program at the appropriate time. The steps diagrammed in Figure 3.11 are written in the applicable syntax. Instructions that are repeatedly executed throughout a simulation (e.g., the discounting of values in Figure 3.11) can be subroutines of their own. In a spreadsheet, implementation of the event specifications can take advantage of the way spreadsheets process information. For example, if there is a column that records a 1 in an entity's row when *Death* happens, then increasing the death count is done by summing that column, which gives the number of deaths. As the consequences of the events in a DES get more complex, it becomes increasingly difficult to contain the entire DES within the spreadsheets themselves. At some point, the modeler is forced to implement at least some of the event specifications using a macro language, often available in the spreadsheet software, such as VBA® in Microsoft Excel®. At this juncture, the same considerations as for a general programming language apply, except that the spreadsheet's calculation, storage, graphing, and other features can still be leveraged.

3.4.1 Use of Submodels

Specifying the simulated event or events that reflect real clinical events can be fairly simple, consisting of only a few steps, or it can get very complex and detailed. This depends on the type of event and the purpose of the model. For example, in a model that only needs to report the number of fractures that occur during the simulation, it is sufficient to count them as they occur, and if recurrence is considered, update the time to the next fracture according to the appropriate equation for subsequent fractures. Another modeler might be interested in what happens when a fracture occurs and, thus, would incorporate in the consequences of the fracture event details about care, whether medical attention was rendered, where and whether surgery took place, and so on.

Most DES software includes the possibility of hierarchically structuring the model so that complex clinical episodes can appear at the top level of a model as a single module. This module groups the episode's events in a submodel, making them visible at a deeper level in the model. This way of arranging the logic of the model allows the modeler to organize each clinical episode in a separate submodel, thereby making the overall model logic easier to follow and review. It is similar to the use of subroutines in general programming software. Submodels can also be nested inside other submodels. There is a risk, however, of losing sight of how the submodels relate—a good rule of thumb is not to exceed five levels of nesting.

3.4.2 Monitoring and Treatment Change Events

The specification of a monitoring event—often a visit with a health care provider—has the same features as a clinical event. It can be very simple (e.g., a point-of-care test of glycemia) or quite complex (e.g., for a model of HIV infection with treatment switching, the logic for the decision regarding which drug is most appropriate in the next line of treatment after a failure is rather substantial and relies on various attribute values, including those that record the experience of the entity to that point). If the DES includes resource constraints (see Chapter 2), then the monitoring event involves the appropriate resources, with capacity constraints, and queues and their behavior defined.

Another clinical event pertains to changes in treatment. For example, a patient may discontinue treatment for various reasons. This is modeled by placing a *discontinuation* event on the entity event list, with a time drawn from the corresponding time of discontinuation distribution. When this time is reached, the event is triggered and the entity stops using the treatment. Since this almost certainly changes the risks of various clinical events, those event times will need to be re-calculated. Other attributes such as treatment cost will also need to be reset. A discontinuation event can also be triggered as a direct consequence of an adverse event.

3.4.3 Specialized Modeling Events

Beside the clinical events, various specialized modeling events may be needed in a DES. These events have to do with things such as assigning and updating values, controlling how inputs are read, how results are outputted, and how the simulation is terminated. There are also various specialized events that improve handling of the entity events.

3.4.3.1 Events to Assign Values

In a DES for HTA, the values of attributes are of considerable importance as they drive many of the event occurrences and often record important outcomes. Thus, assigning those values and keeping them up to date is essential. To ensure this, a frequently used type of event in a DES for HTA is one that is triggered solely to a*ssign values*. This kind of event provides for entering or updating attribute values (or global variables, as appropriate). There is typically an event of this kind placed early in the logic of the DES to trigger the initialization of attribute values. The initial assignment of values to attributes is covered more fully in Section 3.5. There are often additional *update* events set to occur at various points to provide for updating the attribute values, often assigning values to many attributes and variables at the same time.

The approach to specifying this kind of event is simple. For example, to assign age 25 years to an entity's *age* attribute, the *assign* event would have a statement of the form *age attribute* = *25*; and to assign the price of a chest X-ray as $75, it would be *price chest X-ray* = *75*. Of course, what is on the right-hand side of the equal sign can be a very complicated equation and often involves the use of random numbers to make selections from distributions (see Section 3.2.3). Many specialized DES software include modules designed for this purpose, which offer various options to facilitate choosing the attribute to be updated, how the new value should be chosen, and so on. When using general programming software, one or more subroutines can be created to assign and update values. These are just a collection of assign statements that are called when the *assign* event is triggered. In a spreadsheet, the assignment of values takes place either by the analyst entering a value directly (or reading values from some other file) or as a result of invoking a function that yields a value.

A rather efficient way to implement the updating of attribute values is to incorporate logic that forces all entities to pass through an *update* event immediately before triggering any scheduled clinical or other event on the *entity event list* (Figure 3.12). By doing it this way, all entities experiencing a scheduled event will have their attribute values updated as appropriate. These updated values are then available as inputs to equations that determine the time of subsequent events. Another advantage of forcing passage through an *update* event before every scheduled event on the *entity event list* is that the time since the last update is easily computed (i.e., the difference between current time and time of previous update, which has been stored in an attribute). This interval of time is important in HTA models because it has value (e.g., survival time, quality-adjusted life years [QALYs], and cost of being in that state) and is used in other calculations (e.g., discounting).

In some HTA models, the intervals between scheduled events may be quite long for some entities. If it is important to consider the impact on event times of any underlying changes in attributes (e.g., ongoing increase in body weight affecting time to diabetes onset), then it may be useful to force the regular occurrence of an *updater* event. This event is inserted into the entity event lists whenever the gap between other scheduled events exceeds a desired limit, with the sole purpose of getting the entities to update values and recalculate event times as appropriate, despite no other event happening. This is also useful to keep current any information that is displayed on an ongoing basis (e.g., current mean body weight in the population) during the simulation. The steps in an *updater* event are given in Table 3.12.

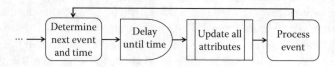

FIGURE 3.12
Introduction of an *update* event. The update event is inserted before processing any scheduled event to ensure that all attribute values are current at the time the next event is processed.

TABLE 3.12

Steps in an Updater Event

Step	Instruction
1	Compute interval since last update
2	Add interval to last *age* value
3	Accrue increase in survival (add computed interval)
4	Apply any values to the interval (e.g., quality of life adjustment) and accrue (e.g., in *QALYs lived*)
5	Apply any economic valuations to the interval (i.e., costs) and accrue (e.g., in drug costs)
6	Compute discounted interval and repeat steps 3–5 for discounted versions of those quantities
7	Update other attribute values affected by the interval
8	Update any event times affected by the interval
9	Store time of update

A variation of the *updater* event may be needed to ensure that time-lagged attribute changes take place at the appropriate point. For example, the quality-of-life changes occurring after a clinical event may have been defined as happening at fixed times after the event. This has been an assumption in schizophrenia models where immediately after a relapse there is a large drop in the quality of life, but after 6 months it improves substantially, and does so again at one year. The initial drop in quality of life would be implemented in the *relapse* event, but a specialized *updater* event is required to change the value after 6 and 12 months (Figure 3.13). This *update quality of life* event is added to the entity event list, and its time is set to be 6 months afterward.

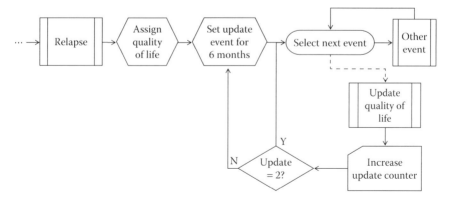

FIGURE 3.13

Use of a specialized updater event to ensure that quality of life is updated at 6 and 12 months after a relapse. After a relapse event, the quality of life is modified accordingly and an update event is placed on the event calendar 6 months later. When that time arrives, the quality of life is updated (other events may have happened in between, hence the dashed arrow), a counter is increased, and the process repeats until the required number of updates is complete.

When the first *update quality of life* event occurs, the attribute is modified and another *update quality of life* event is set for another 6 months into the future. At the second *update quality of life* event, the additional change to the attribute is made and no further updating event is set. The same concept applies to fixed titration of dose and can be used whenever there is a known lag to a change in attribute values.

3.4.3.2 Events to Prompt Input and Output

The methods for getting data into a model and for outputting results are very software dependent and may even vary depending on the purpose and implementation of the model. Most specialized DES software allow the user to enter information directly in the model itself, but it may be more efficient to read data from any number of external sources such as spreadsheets, text files, or databases. The approach to linking to those external sources and reading in the information will vary from package to package but will often require the specification of an event to prompt reading of the inputs. That event is scheduled at the start of the simulation. If the DES uses a general programming language, then the method for reading and storing information is also provided by the specific software. If the model is programmed in a spreadsheet, then the data can be included in the same file on other worksheets.

The outputting of results typically follows the same approach as the accessing of input values, except, of course, in reverse. In some software, results may be displayed on the computer screen as they accrue during the simulation, or they can be automatically written out at the end of the simulation using built-in reports, or they may be sent directly to a printer. The methods for doing this are also software dependent. Additional processing of the output (e.g., to derive a cost-effectiveness acceptability curve) may then take place in whatever software the analyst deems appropriate.

3.4.3.3 Events to End the Simulation

There are several ways to stop a simulation. For an individual entity, the end occurs if an event that removes the entity from the DES is experienced. This event may reflect a clinical one (e.g., death) or may be specified as a specialized modeling event to remove the entity from the model (e.g., because he or she has reached a maximum eligible age). When all the entities have been removed from the model—either due to death, loss of eligibility, or for other reasons—and no new ones will be created, the entire simulation ends. It is not mandatory, however, to have an end event in a DES. If there is no *end event*, then there has to be another way to stop the simulation.

Often, the analyst wants to terminate the simulation after a specified time is reached even if there are still entities in the model. For HTA, it is common to set such a *time horizon*. Implementing this requires a way to halt the execution of events on the *event calendar* and freeze the *simulation clock* (if all entities entered the model at the same time, as is customary for HTA).

Some specialized DES software provide a built-in way to do this. A terminate time is specified as a special global variable, and when the *simulation clock* reaches this value, the software stops the execution. If this function is not provided, or if a general programming language is used, then the comparison of the specified termination time with the time on the simulation clock must be coded as an instruction. A convenient way to do this is to place a *terminate* event on the *event calendar* at the desired time. This *terminate* event is not on the *entity-specific event lists* but will be detected as the next event at the appropriate time and the execution will be stopped.

If the DES ends because it reaches a *terminate* event at the end of the desired time horizon, or because another terminating condition is met (e.g., a particular number of events have occurred), there may still be entities in the model. As long as all the required outputs have been recorded, the model can proceed to write them out as the final task of the simulation. More commonly, however, there is output information that has not yet been recorded—costs may have accumulated since the last recording, additional survival time has accrued, and so on. In this situation, an explicit *model end* event can be incorporated to force all remaining entities to record any pending results. The modeler needs to make sure that at the end of the time horizon, but before the simulation terminates, all entities have a *model end* as the next event in their *entity-level event lists*. This *model end* event differs from the *terminate* event. The latter only stops execution, while the *model end* event processes entities before removing them from the model to collect relevant outputs before the simulation ends because all entities have left.

3.4.4 Event Sequences

In a DES constructed using the time-to-event approach, event sequencing is created by the repeated ordering of the event times in the entity-level event list. The list can contain as many events as necessary—30 to 40 events is not uncommon in an HTA DES. Some of these may be set to happen at a fixed time (e.g., *model end*), some at fixed periodic intervals (e.g., monitoring events), while the majority are ordered based on the event times drawn from their time-to-event distributions. This corresponds to the apparent random ordering of events across people in the real world.

Sometimes an event sequence is mandatory in the sense that it is replicating a series of occurrences that follow one another in real life. It is unnecessary to enter the entire series in the entity-level event list to set the sequence. For example, if an entity has a *fall* event that leads to a *fracture* event, followed by an *X-ray* event, and then a consultation with an orthopedist to get treatment based on the results of the X-ray, a triggered event sequence (Section 3.3) will allow the modeler to split the episode into a chain of events. These are assumed to occur at the same time, but in a particular order. The advantage of decomposing the sequence is that it increases transparency and facilitates modifying attributes such as *quality of life* and global variables such as *total cost*.

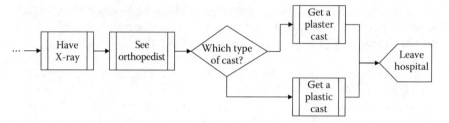

FIGURE 3.14
Simple mandatory event sequence with a branch point. The processing of a fracture event is represented by a sequence of component events from the initial X-ray through the visit with the orthopedist, the choice and placement of a cast, and departure from hospital. One event triggers the next in this sequence and no simulation time passes.

Moreover, such sequences may have branch points that send the entity down one of several paths (Figure 3.14).

A common modeling challenge arises when a clinical occurrence may lead to death (i.e., the illness is acutely fatal). One approach to this problem might be to specify different time-to-event distributions for fatal and non-fatal clinical events (panel A in Figure 3.15). These are not independent distributions, however, and thus, this approach artificially inflates parameter uncertainty if correlations between the times to events are not represented. Besides, it may be difficult to obtain data to estimate separately these distributions.

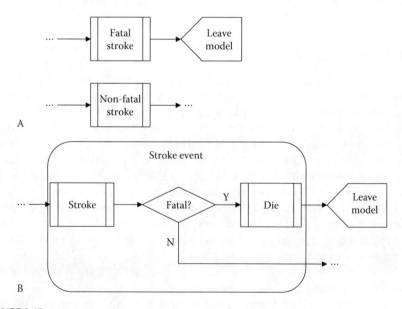

FIGURE 3.15
Two approaches to modeling a possibly fatal event. In panel A, the fatal stroke event is treated as a different event from the non-fatal one and their times would be selected from different distributions. In panel B, the stroke event is singular with death as one possible outcome.

An easier approach is to use a triggered event sequence to represent the clinical episode. One simulated event reflecting the start of the clinical episode is defined, regardless of whether the episode will end up being fatal or not. Then, a mandatory sequence with a branch point that leads to either a *death* event or continuation in the simulation is specified (panel B in Figure 3.15). A binary discrete distribution is required to represent the probability that the clinical occurrence is fatal or non-fatal, and this kind of information is much more likely to be available as it corresponds to the case fatality rate. Of note, the distribution of fatal and non-fatal events may change over time. For example, it may be more likely that early clinical events are fatal, with case fatality rate declining over time. In that case, it will be necessary to specify how the case fatality rate changes with time.

3.4.5 Composite Events

In clinical trials, it has become commonplace to designate the outcome in terms of the occurrence of any of a set of related clinical episodes. For example, the composite endpoint of myocardial infarction, sudden death, and hospitalization for other cardiac reason is often used in cardiovascular studies. In such a study, the information about time-to-event may be reported for the composite endpoint but not for the separate components (e.g., in a single Kaplan–Meier curve), and only the overall proportions each component represents are given. If appropriate, this can be replicated in a DES by aggregating several related event types and specifying a single time-to-event distribution for the composite event. For example, a standard function such as the Weibull might be derived from the clinical trial data and used to sample a time to measure acute cardiovascular event, without considering which specific cardiac crisis occurs. Then, an additional distribution is defined (e.g., a discrete distribution with three options corresponding to the three components), and it is sampled to inform the type of cardiac event experienced at that time. This can be problematic, however, if the distribution of event types changes over time. For example, it may be more likely that early events are *sudden death*, whereas later ones are *hospitalizations for other cardiac reason*. Maintaining a single event type distribution that is invariant over time will introduce inaccuracy but may be unavoidable if data are lacking with which to estimate a more complex function that can handle time-varying distributions for the type of event experienced.

3.4.6 Duration of Events

In standard definitions of DES, all events are instantaneous in the sense that they happen at a particular (discrete) point in time during the simulation. In an HTA, many real occurrences can also be handled as instantaneous events. A change in state, for example, happens with no time passing. Before the event, the individual is in one state and afterward in another state—the only change is in whatever attribute(s) indicate the state. For example, the patient with cancer in remission suffers a *progression* event that indicates that that patient once again has active disease. This event has no duration—it only reflects the transition. Most of the events serving modeling purposes are also instantaneous. The *updater* event, for example, changes one or more values but does not take up any of the simulation time in doing so.

In real life, however, it usually takes some time for something to happen. This can be very short (e.g., occlusion of a coronary artery) or extend over many days (e.g., an asthma exacerbation), and it depends to some extent on how the clinical episode is defined. For example, the clinical occurrence may be defined in terms of an episode of care (e.g., a hospitalization for asthma exacerbation) because the concern is largely with the resulting cost that is accrued. In other cases, the clinical occurrence might be defined more by the onset and resolution of symptoms (e.g., a migraine). Regardless of how the clinical episode is defined, its real-life duration must be handled by the simulation, despite the fact that simulated events are always instantaneous (i.e., have no duration).

These real episodes can be simulated by implementing two model events: one that signals the start of the episode and another its end. For example, when a migraine start is scheduled, the entity would trigger the *start migraine* event, where *pain, quality of life,* and other relevant attributes would be modified as appropriate, and an *end migraine* event would be scheduled. When the time comes for the migraine to end, the corresponding attribute values would be modified again and the duration of the migraine episode would be recorded, if needed. The steps in this clinical episode are given in Table 3.13.

TABLE 3.13

Steps in Simulating a Clinical Episode That has a Duration of Interest

In the Event **Start Clinical Episode**	
Step 1	Increase corresponding event counter by 1
Step 2	Store time of the event in corresponding attribute
Step 3	Set *location of care* attribute to appropriate value (if needed)
Step 4	Modify other attribute values affected by the start of the event (e.g., quality of life)
Step 5	Select end time for the event and place *end clinical event* on the entity-specific event list
In the Event **End Clinical Episode**	
Step 6	Accrue costs for the event (possibly depending on time since start event)
Step 7	Set time of next event of this type to appropriate value (well beyond *time horizon* if recurrence is not allowed)
Step 8	Reset *location of care* attribute to appropriate value
Step 9	Modify attribute values affected by the end of the event (e.g., quality of life and time of other events)

Often, the timing of the *end clinical episode* event is set by the starting event (e.g., discharge from hospital will occur a number of days after admission, corresponding to the length of stay). Between those two events, other things of relevance to the HTA can happen (e.g., a hospital-acquired infection), and these can be incorporated as simulation events as well (Figure 3.16, panel A). Indeed, some of those intra-episode events may change the timing of the episode's end (e.g., initiation of treatment for the migraine or getting an infection in the hospital) or might completely alter the individual's course (e.g., death in hospital).

If the clinical episode involves a change in the person's physical whereabouts, this should be handled without confining the entity to a location in such a way that it cannot experience other events. For example, for a hospitalization, the *admission* event would set a *location of care* attribute to *hospital* and would calculate the time of the event *hospital discharge* and place it on the entity-level event list. When the time for discharge comes up, the value of the *location of care* attribute is reset to *home* or wherever the patient is discharged to. During this interval (i.e., while that patient is *in hospital*), the entity continues

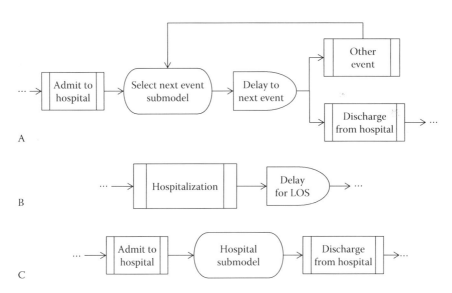

FIGURE 3.16
Three ways of handling an event that has a duration. In panel A, the *Admit to hospital* event is processed and the entity returns to the *Next event* logic so it can be exposed to the other event risks during the stay at the hospital. The discharge is handled like any other event. In panel B, the interest is only in the duration of hospitalization itself (and perhaps in counting the event and its cost, which are assigned when the event is processed), but a single event plus a delay corresponding to length of stay will shield the entity inappropriately from what may happen during the hospitalization. In panel C, the onset of hospitalization is signaled by one event and its termination by another. In between, there is a submodel that contains the logic for what happens in hospital, but if it allows the corresponding length of stay time to pass, it will also inappropriately shield the entity.

through the logic without any hold, and any other scheduled events on the entity event list can happen as appropriate. The length of stay is easily computed (i.e., it is the difference between the time of the event ending the episode and that of its starting event), and any costs or other values can be applied as appropriate. In addition, any things that should happen during the hospitalization (e.g., angioplasty) can be added as events on the entity event list.

A tempting alternative might be to represent the episode as a single simulation event and reflect the passing of time by assigning the corresponding time interval as the event's *duration* (Figure 3.16, panel B). This is not a good idea, however, not only because it departs from the definition of a simulation event as instantaneous but also because it can generate serious problems. An obvious one is that this precludes modeling anything of relevance that happens during the episode (e.g., complications that arise during surgery). Even if no details of the episode need to be simulated (e.g., the only thing of interest for the HTA is the duration of the pain episode), this approach does not work well because during that time that is not explicitly modeled, the individual is shielded from experiencing other events, unrelated to the episode at hand, that may logically occur during that time. For example, during a *hospitalization* event, a patient may still suffer a side effect from medications that were being used before admission, may develop a new complication of their underlying illness, or may even die due to unrelated causes.

Another alternative is to create a submodel to simulate the clinical episode in the DES and have the entity enter the submodel at the start of the clinical episode (e.g., have the entity go into a *migraine* submodel); stay there during the clinical event, allowing the time to pass for the duration of the episode; and then exit the submodel at the end of the clinical incident (Figure 3.16, panel C). For HTA models, this can pose the same kind of problems as using a single simulation event with a duration. If the submodel contains a delay, then anything that happens during simulation time needs to be reflected inside the submodel, and this is a very inefficient way to implement the clinical episode. If that is not done, then holding the entity in the submodel prevents the occurrence of other events scheduled in the entity event list, even if they were to happen during that delay. For example, a stay in a rehabilitation facility after a fracture would protect the entity from another fracture, death, and all other scheduled events until discharge. This will distort the results, particularly if the delays are long ones. In setting up the submodel, if one is used, the modeler must take care that the start of the episode does not inadvertently isolate the person from the rest of the simulation.

Thus, the DES should keep to the stricture that simulation events do not have duration. It is better to trigger the event that signals the start of the episode and then allow the entity to continue on its way until the next scheduled event, which may or may not be the end of the episode. A submodel can still be used to organize details of the event start but no time should pass in that submodel. This allows other unrelated events to occur and enables any required parts of the incident to be modeled as their own events.

3.5 Creating the Population

Every DES needs at least one entity to trigger the events that constitute the model. For HTA, the main entity type is an individual who may receive the intervention being assessed (i.e., someone who is part of the *target* or eligible population). Thus, it is necessary for the DES to recreate a simulated population that reflects the characteristics of the target population as closely as required to address the questions at hand, and in sufficient numbers to deal with uncertainty, and to represent any subgroups of interest. Depending on the purpose of the model, this population may be created at one time or people may enter the simulation at various time points. The current practice for HTA models is to start all of the entities constituting the target population at the beginning of the simulation (i.e., at model time zero). This means creating the people and giving them initial attribute values when the model starts running. Alternatively, there can be periodic entries of a number of people representing staggered cohorts (e.g., the people needing care at various successive times in a budget impact analysis) or of individuals starting at varying times when they meet the conditions for the simulation (e.g., to represent incidence).

Whether there is a single entry at the start, or entries over time, the modeler needs to specify either how many entities enter at each time point, how many entries there will be, or how often an entry will occur. The type of entity entering (e.g., patient and caregiver) also needs to be specified. In specialized DES software, there is usually a module that implements the creation of entities. This module usually assigns two attributes automatically: the *entity identification* and *entity type*. The entity identifier is a unique number that allows specific entities to be identified during the simulation. This can be useful for model debugging purposes. It can also be used to link entities. For example, entities may be cloned so that comparisons of technologies are not confounded by variations in patient characteristics. Each clone can store the entity identifiers of his or her fellow clones so that their paths can be synchronized if appropriate (see Chapter 7). The type of entity is a label that categorizes the entity and may be used at various points in the simulation to determine what should happen. For example, if both pediatric patients and their parents are in the model, then a check of which type is experiencing a clinical event could alter the consequences: a parent entity might miss work, while a child entity might miss school. In this case, the simulation may also store the child entity identifier in a parent attribute so that they can be recognized as a family unit when needed. If general software is used, then the entities are created using a loop that steps through the create instructions, commonly placing the *entity identification* and *type* in an array that contains all the entities in the simulation. Alternatively, separate arrays may be used to hold the different types of entities. In a spreadsheet, creation of the population can be conveniently done by using the rows on one page, with each row representing one entity and the columns reserved for the entity's attributes, including its *entity identification* and *type*. The identifier is not really necessary, however, as it can be proxied by the row number in the spreadsheet. Type can also be left out if the tables for different types are placed on separate worksheets.

If all the entities will be created at the start of the simulation, the modeler may be tempted to generate one entity at a time instead of initiating all of them at once. The experience of this entity is simulated until its death, it experiences another type of removal event, or it reaches the end of the time horizon. Then, another entity is created and the process is repeated (Figure 3.17). Although this will yield the population, one member at a time, it can be a very inefficient way to run the model (see Chapter 7). Moreover, if resource constraints are implemented in the model, creating and simulating one entity at a time won't work correctly as there will be no possibility of competition for the resources.

3.5.1 Dynamic Creation

Whether one at a time or all at once, starting all of the entities at model time zero is very artificial and does not correspond to the real world where people enter (and leave) the target population over time. The dynamic approach, where there is an ongoing flow of people into the model, is more realistic and enables proper simulation of what happens at physical locations of care, like a hospital or a clinic; or even of contexts like transplantation where constrained resources and demand are spread over time. It also allows proper modeling of financial flows, as when trying to analyze the budget impact of different interventions,

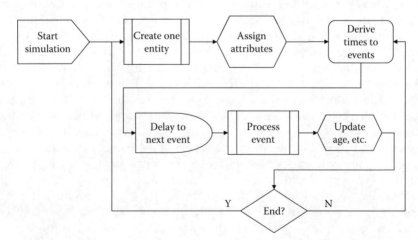

FIGURE 3.17
Creating and simulating one entity at a time. In this DES, a single entity is created and allowed to experience all of its events until its time in the model ends. Then, a new entity is created and the process repeats. At any one time, there is only one entity in the model.

where usage can vary from year to year. For customary HTA models concerned solely with the estimation of cost-effectiveness in the absence of resource constraints, analysis of a dynamic population may be unnecessary.

The dynamic nature of a population can be implemented by creating entities at defined constant intervals (e.g., every month) or at varying fixed points in time (e.g., at 3, 6, and 12 months, and annually thereafter). This amounts to putting a *create* event into the event calendar at the desired times. In specialized DES software, the module that implements the *create* event often includes a schedule function that places additional create events at whatever time periods are specified. Alternatively, the modeler can add *create events* for each of the times when new entities are to enter the model. This offers more flexibility as the additional *create* events may be followed by their own logic, while use of a single module with a scheduling function feeds the newly created entities into the same logic. Either way, the modeler needs to specify the number of entities entering at each time point and their type. In general software, the create loop is a subroutine that can be called at the appropriate times in the simulation, specifying the number of entities to create at each time point and type. In a spreadsheet, the periodic creation can be mimicked by adding an attribute *start time* as the third column after *entity ID* and *type*. For entities starting at the beginning of the simulation, the *start time* is set to 0, and for those reflecting the periodic entries, it is set to the appropriate times.

Fixed periodic creation is appropriate when the process that is being simulated has such entry patterns but often this is not the case. The best approach for handling a dynamic population is to define one or more distributions that represent the influx of entities. These distributions specify the start times or the time between starts. The model then uses the distribution of such *inter-arrival* times to set a delay between the creation of one or more entities and the next group. Epidemiologic incidence functions are an example of this approach. Arrivals at a health care facility are another. Of note, the create function need not operate throughout the time horizon. For example, a clinic may allow new arrivals only until an hour before closing. In specialized DES software, the create modules typically provide for specifying these distributions and for how long they will operate. In general software, the distribution must be specified and times must be drawn at which the creation loop will be called. In a spreadsheet, the value in the *start time* column is drawn from a distribution rather than set by the modeler.

3.5.2 Using Super-Entities

Occasionally in an HTA, a large number of people are to be modeled, but for some portion of the simulation, there is little interest in their individual experience. For example, when generating a population that meets certain clinical conditions, it may be useful to conceptualize and implement the real source population to keep the numbers in the target population accurate. The sole function of that source population, however, is to reflect over time the number of people in it, classified into a few categories (e.g., 10 groups defined by age interval and sex). What happens to individuals in the source population is not of interest except when they meet the eligibility criteria for the target population. In such a situation, where very many entities need to be created to represent a source population, and changes over time are monitored (e.g., aging), it may be helpful to group entities by risk level or other overarching characteristics into *super-entities* that represent an entire category of people (see Chapter 7). Each *super-entity* has an attribute that reflects the number of people in the category. It experiences events that release the individual entities and possibly to transfer some of its members to another *super-entity*.

3.5.3 Creating Other Entity Types

Although people are the ubiquitous type of entity in HTA models, a DES can include many other types of entities. These can be other living things. For example, a bacterium like *Neisseria meningitidis* might be simulated as an entity because it can undergo an event that changes its serotype. Inanimate objects (e.g., a colonoscope, a lab test, an ambulance, and a needle) may also be simulated. The decision whether to model these inanimate objects as entities rather than as resources depends on whether it is necessary for them to be able to experience events. For example, it may be better to simulate a colonoscope entity if these tools can undergo different types of sterilization.

Just as with events, there can also be a need for an entity type that is created to serve specific modeling purposes. These special entities are implemented when it is necessary to trigger a particular event that is required for the running of the model but none of the *real* entities will experience that event. For example, if it is necessary to read information (e.g., from an external file) and then write the results at the end, a single entity of type *read-write* can be created, and it will experience two events (apart from creation): one to read the data and another to write out the results, such as the total cost or accumulated survival, to an external file (Figure 3.18). While doing this it is important to make sure that the structure of the model does not inadvertently allow the special entity into the broader simulation—it should only be allowed to experience the special events. In specialized DES software, this can be done by creating a separate self-contained structure that implements this special logic. That structure should not connect to the main structure of the model.

Another use for a special model entity is to force a change in particular information used by the model. For example, the rate of inflation may need to be adjusted periodically. A special entity could trigger an adjusting event annually.

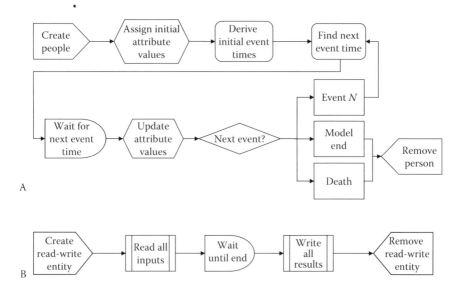

FIGURE 3.18
Flow diagram showing use of a special *read-write* entity. The main logic for this DES is depicted in the modules in panel A. In panel B, a special entity is created whose sole purpose is to trigger the reading of inputs at the beginning of the simulation and the writing of all results at the end.

An interesting example of a special entity was implemented in a model assessing various vaccination strategies to prevent bacterial meningitis epidemics (Caro et al. 2007). The validation work for that model showed that the outbreaks of disease occurring in the simulation were less frequent than those observed in available epidemiologic data, despite careful construction of the risk equations. Rather than calibrating the equations to the new data, a special event was incorporated that triggered a change in the force of infection, increasing the chance of an outbreak. The sole purpose of the special *Outbreak Demon* entity was to trigger this event and thus instigate outbreaks of disease by changing the force of infection (Getsios et al. 2004). The frequency with which the *Outbreak Demon* triggered this special event could be set to produce any desired outbreak pattern.

3.5.4 Agents

In a standard DES, each entity moves through the model unaware of the other entities. It proceeds from one event to the next, accumulating the experience and whatever consequences that brings, but with no actions of its own nor any direct effect on other entities. This limitation can be relaxed by defining a set of behaviors that govern how the entity reacts when encountering elements of the model, including other entities. These additional features transform the entity into an *agent* and the DES is now known as an *agent-based simulation* (see Chapter 9) (Auchincloss and Diez Roux 2008; Borshchev and Filippov 2004; Djanatliev et al. 2014; Levy 2014). Unlike entities that are governed entirely by the model logic, the agents are *aware* of their surroundings, can make decisions, and, thus, have some capacity to alter their own course through the model. Few HTA problems today are addressed using agent-based simulation—the standard DES being sufficient to meet most of the decision makers' needs. The main exception to this is in the area of communicable diseases, where interactions among people can be a very important component (Meng et al. 2010).

3.6 Assigning Attributes

A newly created entity has only its type and identification number as characteristics. To properly reflect the target population representing a specific group of interest, each entity must be allocated the attributes needed for the model, and if appropriate at the time of creation, relevant values for each one must be assigned. These values might be set for the entire time horizon (e.g., the entity's sex) or they may be assigned initially, but allowed to be updated based on some function (e.g., age increasing with the passage of time), or with the occurrence of events (e.g., the number of strokes experienced so far). Indeed, they may initially have no value (e.g., the survival time accumulated in the model), and they may stay empty throughout the simulation (e.g., the number of cancer recurrences in someone who dies of other causes while in remission). There are several ways to make the initial assignment.

3.6.1 Using Summary Characteristics

One way to specify the target population is to describe its relevant characteristics using distributions. The distributions can take many forms (see Section 3.2). Typical ones are discrete distributions that specify the proportions of the population with particular characteristics (e.g., 45% males) or continuous ones such as the normal distribution given by a mean and standard deviation (e.g., age). Reports of randomized clinical trials, for example, usually include a *Table 1* that describes the study population in this way. From these aggregated measures, it is not possible to create entities that reflect the real population with 100% fidelity at the individual level. It is possible, however, to create a simulated population which, as a group, matches the real one. This is done by sampling the distributions (see Section 3.2.3) to obtain specific values and assigning those values to each individual entity's attributes. Specialized DES software typically provides a module that handles the selection of values and assigns them to the corresponding attributes. In general software, the instructions to sample the distributions are included in the loop that creates the entities and the values selected are written to the array that stores the entities and their attributes. In a spreadsheet, this is done by incorporating the formula defining the selection from the appropriate distribution in each cell of the column representing each attribute. For example, for the attribute *sex* with the specification that 45% are males, the formula would read *If random number (0,1) ≤ 0.45, then "Male," else "Female."* The precise syntax depends on the spreadsheet program, and for many standard distributions, built-in functions may be provided. Of note, it is important to make sure that the random number—and thus the attribute value—is not updated every time the spreadsheet is re-calculated.

In using distributions to assign attribute values, it is important to take into account any correlations among the data (see Section 3.2.3). Many characteristics of a population are correlated with sex and age. Indeed, sex and age are usually correlated as well. Implementing this requires that the value of one characteristic be established before the other. Although the order in which the assignments are made doesn't really matter (as long as one is consistent), it is more straightforward to sample the discrete distribution first and then address the continuous one conditional on the first value selected. For example, if the body weight distribution is different for males and females, as is typically so, then the simulation first establishes the sex of the entity and then samples from the respective distribution to obtain the weight. In specialized DES software, there are several ways to do this. A very transparent way is to add a branch point after the selection of the first value to separate the entities according to the value of sex obtained from the discrete distribution and then implement separate assign statements after each branch so that the continuous weight distribution is sampled conditionally (Figure 3.19). Alternatively, the correlation can be implemented directly by adding an outer *if* statement to the weight selection: *If male then weight = M(random number), else F(random number)*, where *M* and *F* stand for the male and female distributions of weight, respectively. How this more compact statement is written depends on the specific package. Care must be taken to ensure the order of the statements corresponds to how the conditional selection is done. Thus, in this example, the *sex* must be assigned first and then the *body weight* selection

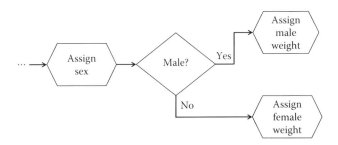

FIGURE 3.19
Assigning weight conditional on sex using a branch point. After the entity's sex is assigned, the logic splits up males and females so that their weight can be assigned dependent on sex.

follows. A similar approach can be taken when using general software by including this code in the *create* subroutine, instead of implementing the branching logic in DES software. In a spreadsheet, it is easiest to incorporate the nested *If* statements directly in the cells that store these values.

Using distributions to assign attribute values is handy and may be imperative if that is the only way the target population is described. It does present, however, a difficulty for creating realistic entities in the simulation. Since the distributions describe the population in the aggregate (e.g., the mean body weight and its standard deviation), it is possible to select attribute values that correctly reflect the separate specified distributions but that together describe an individual that is much less common than in reality. At the extreme, the specified individual may not exist in reality at all! For example, based on separate random selections, *sex* may be assigned as male, *age* as 45 years, and *weight* as 20 kilos. Although taking correlations among the data into account helps reduce this problem, the selection of values becomes less straightforward as the number and complexity of correlations increases, regardless of which type of software is used.

3.6.2 Using Real People

To avoid the risk of creating entities with combinations of attributes that would be extreme or non-existent in the real world—without getting tangled in sampling from multiple, complexly correlated distributions—it is better to use data from actual individuals to assign attribute values. Such a dataset can be assembled from, for example, the information collected at enrollment in a clinical trial or from a set of electronic medical records. Increasingly, countries carry out national surveys that document the occurrence of various characteristics, including disease biomarkers, across the population. In the United States, for example, the National Health and Nutrition Examination Survey (NHANES) periodically provides data of this type (National Center for Health Statistics 2005). Whatever the source, there are several ways to use the data from the source population, without resorting to aggregated measures.

The simplest approach is to use the data directly. In other words, for each real person in the source population, a corresponding entity is created, with each relevant attribute of the latter matching the value of the real person's characteristic. This ensures that every entity in the model reflects an actual person in the population. Implementation of this approach is straightforward. The source data are organized so that each person's information occupies one record and the characteristics are in the same order for everyone. When using a spreadsheet, these data are read into the entity worksheet page, taking care to ensure that the information for each attribute is entered in the appropriate column. If using general software, a loop is set up to read the data into the entity array. In DES software, the method for opening a file and reading data depends on the specific program, but once that is done, the assignment of attribute values proceeds by taking the set of values from one real person in the dataset that has been read in and setting those in the appropriate entity memory slots.

If the source population is, itself, the target population, then this one-to-one match is ideal and nothing further needs to be done, provided there are sufficient individuals to serve the simulation's purpose. In most cases, however, the target population does not necessarily match the source population, or the latter has too few people in it to allow for proper simulation. In either situation, the solution is to sample from the source population, assigning the characteristics of the real person to the attribute values of the simulated one. If the sampled person is left in the source dataset (i.e., sampling with replacement), then as many entities as needed can be created. To alter the characteristics of the simulated population to accord with its target description rather than with the source population, weighted sampling can be used, with the weights reflecting the desired mix (Figure 3.20). In a dataset like that provided by NHANES, each record bears a weight expressing how many real people it reflects, and that can be used in weighted sampling.

To implement sampling of the source population as a means to assigning the attribute values, the same steps to organizing and reading in the data are taken, except that values are not entered directly into the attributes. Instead, they are placed on a separate worksheet or array. That table is then sampled by randomly drawing a number from 1 to the maximum number of source people. This number is matched to the corresponding record and the values of that person are read into the attribute slots. This is repeated until the required number of entities is created. Any record can be picked multiple times, depending on the random draws. Implementing weighted sampling is somewhat more complicated because it requires identifying the features

Patient ID	Sex	Age	Score	Share (%)	Factor ...	Accumulated share (%)	
							Rnd = 0.106
1	M	52	3	15	...	15	$0 \leq$ Rnd < 0.15: Copy patient 1
2	M	48	2	18	...	33	
3	F	69	5	9	...	42	Rnd = 0.541
4	M	78	4	22	...	64	$0.42 \leq$ Rnd < 0.64: Copy patient 4
5	F	64	3	12	...	76	
6	F	76	2	8	...	84	Rnd = 0.941
7	M	59	4	16	...	100	$0.84 \leq$ Rnd < 1.0: Copy patient 7
...	...	...	...	...	...	...	

FIGURE 3.20
Weighted sampling from a file containing individual data on real patients. Each patient on this list represents a profile that occurs in the population with a frequency given in the accumulated share. Using a random number to select a profile is shown in the arrowed boxes on the right. Rnd, random number.

that weighting will be based on and then creating an additional field in the source records to tag them according to the profile they represent. Various methods for weighted sampling can then be applied. If the weights will not change (i.e., there is a fixed target population), then this can be simplified by copying records with the desired profiles until the enlarged source population reflects the desired mix.

Often, even when there are individual records available to use in assigning attribute values, some necessary characteristics may be missing. Rather than abandon the use of individual data, a hybrid approach can be used by applying the kind of technique that is used by statisticians for interpolating missing data in clinical studies (Enders 2010). For each real person, the record is supplemented by sampling from appropriate distributions to fill in the information gaps. This can be done directly in the source data. An even better approach is to derive the missing value in the simulation after the entity is created and its known attribute values have been assigned. This allows for appropriate representation of uncertainty in the interpolated value.

3.6.3 Duplication of Entities

One way to assign attribute values to entities is to create new ones by duplicating an existing one. These duplicated entities have the same values for all attributes at the time of duplication. This is useful if the objective is to enlarge the simulated population without adding any new profiles. Some DES software include a module that does this duplication (i.e., one entity enters the module and emerges along with one or more copies). In general programming languages, or in spreadsheets, the same can be accomplished by copying the entries in the entity array or worksheet. If the mix of profiles is to be kept the same, then care must be taken that all profiles are duplicated to the same extent. The modeler may take this opportunity, however, to create a different mix by selectively duplicating entities or altering the number of times some profiles are copied.

A situation where duplicating entities is highly desirable is when comparing interventions, a major objective of HTA. In this case, the efficiency of the comparison (see Chapter 7) is improved considerably if the interventions are compared among initially identical populations because this removes extraneous variability. To achieve this, one set of entities is created and values are assigned to all attributes except those that specify something about the technologies to be compared (e.g., type, dose, frequency of use, efficacy, and safety). Then, each entity is duplicated as many times as necessary to be able to assign all the interventions in the assessment, one to each copy. All attribute values are indistinguishable between the original and its duplicates, except the *entity identification*. This ensures that the populations compared are identical, at least at the beginning of the simulation. Each entity is then assigned the characteristics corresponding to the specific intervention allocated, and from there their course may differ, and almost certainly will.

In a spreadsheet, this is easy to do by simply copying the entire page containing the entities and their initial attribute values. One page is allocated to each intervention. Then, the columns on each page representing the intervention-specific attributes are filled in as appropriate. In general software, this is achieved by copying the entity array after it is filled in with the pre-intervention values. Each array (or new dimension, if using multi-dimensional arrays to hold the information) is then expanded to include the intervention-specific data. The module for duplicating entities is used with DES software. The original entity is then routed to a module where attributes specific to the first intervention are assigned. If there are only two interventions to be compared, the duplicate entity proceeds to its corresponding module for assignment of its intervention-specific values. If additional interventions are to be compared, then the duplicate undergoes further duplication until the desired number of identical entities is obtained (Figure 3.21).

3.6.4 Updating Values and Retaining History

The description so far has been about assigning attribute values when an entity is created or shortly thereafter when it is assigned an intervention. In HTA models, however, it is necessary to update values as the simulation progresses and to track what has happened to that person, much as is done in a person's medical record. Thus, it may be important to change values of things like biomarkers, to retain the history of different treatments taken, including when the switching occurred, the number of clinical events that

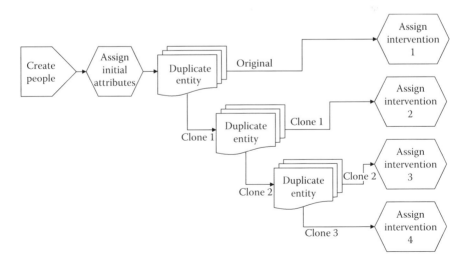

FIGURE 3.21
Duplicating an entity to create clones who are then each assigned a different treatment. The original entity is successively duplicated creating three additional copies. An intervention is assigned to each duplicate, ensuring that the technologies are compared in identical people.

have happened to that person and when, and when a new diagnosis was made. Many of these values may be of interest in their own right, but some are tracked because they affect what happens down the line or they alter the person's response to treatment, modify the costs of managing them, and so on. Assigning new values to existing attributes is done in the same way as for the initial ones, and additional attributes may be created to store the various elements that become part of an entity's history.

One particular use of attributes is to store the original values before they are altered by the simulation. For many characteristics, it is useful to retain the values at baseline because reporting or analyses will take these into account. For example, it may be necessary to report results by groups of age defined at entry into the model, despite also needing to update age as the simulation progresses. This can be done by representing age with two attributes: *original age* and *current age*. The latter can be easily maintained by adding current simulation time (if the entity started at time zero) to the original age.

3.6.5 Handling Quality of Life

Most HTA problems involve consideration of the impact of interventions on people's quality of life. To enable these calculations, it is very convenient to carry the required values in one or more attributes. When the conceptualization of quality of life is very straightforward, it may be sufficient to carry a single attribute *quality of life* and modify it as events happen, or if it is felt to be age-dependent, then adjusting it whenever age is updated. If time lived in the simulation is to be quality-adjusted (e.g., to compute QALYs), then the time interval since the previous event can be multiplied by the quality value and added to the ongoing accumulated amount stored in its own attribute or global variable: $QALY = QALY + (time\ interval \times quality\ of\ life)$. A discounted accumulation can be stored in the same way (see Section 3.9).

In many cases, the quality of life is computed based on multiple components and one or more of them might change over time. Often, a base value is set (possibly adjusted by age and sex) and events are assumed to decrease it (i.e., they pose a tariff) by a fixed amount (i.e., an *additive* assumption) or proportionately (i.e., a *multiplicative* assumption) for a period of time. There may also be positive effects (e.g., due to treatment). One approach to handling this complexity while retaining maximum flexibility is to define an attribute for each component (Figure 3.22). This way, each component value can be stored and changed whenever appropriate, and their impact can be combined under whatever assumptions are to be implemented.

Another useful layer of complexity may be added to handling quality of life when multiple scales measuring quality of life are considered in a DES. For example, the modeler may want to implement data obtained using different techniques, such as the time trade-off approach (Dolan et al. 1996), the visual analog scale (De Boer et al. 2004), or the three- versus five-level versions of the EuroQol 5-Dimension (EQ-5D) questionnaire (Janssen et al. 2008), or even a

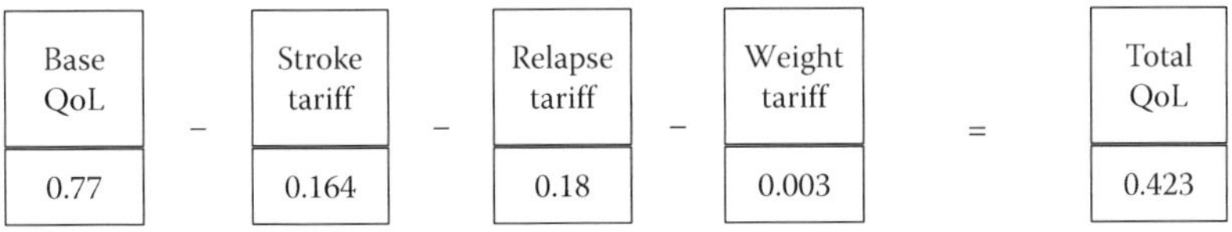

Base QoL		Stroke tariff		Relapse tariff		Weight tariff		Total QoL
0.77	−	0.164	−	0.18	−	0.003	=	0.423

FIGURE 3.22
Calculation of quality of life using a base value and tariffs stored in an entity's attributes.
The entity's quality of life is derived by adjusting the base value according to a series of tariffs imposed by the effects of a stroke, a disease relapse, and obesity. QoL, quality of life.

completely different measure such as the disability-adjusted life year (DALY) (Anand and Hanson 1997). For any number of these, the approach is generally the same: define as many attributes as necessary to carry the required information and update these values as needed. It is much better to carry all the information simultaneously, rather than rerunning the model multiple times for each variation. Incorporation of numerous attributes poses little challenge given modern computing capabilities and memory capacity.

3.6.6 Other Values

There are many other kinds of values that may be usefully stored as attributes. One example is the series of random numbers (or a subset) that are drawn for an individual. This may be done to ensure that any selections from distributions or logical checks that use random numbers can be replicated. A reason to implement this is to force duplicate entities to remain the same and follow the same course through the simulation beyond the selection of initial attribute values—except when the specific intervention dictates otherwise. This is a major way to reduce nuisance variance (see Chapter 7). Another example is to facilitate structural uncertainty analyses (see Chapter 5). For this purpose, the model may store in the attributes one or more indexes that reflect which equation the entity will use at relevant points in the simulation. For example, in a model of cardiovascular disease, duplicate entities with index 1 will use the Framingham cardiovascular risk equation (D'Agostino et al. 2008), those with index 2 will use an equation derived from the ARIC study (Chambless et al. 2003), and those with index 3 will use the QRISK equation (Collins and Altman 2012; Hippisley-Cox et al. 2008; Robson et al. 2012).

3.7 Handling Time

In a DES, a central component is the simulation time, which runs continuously from the start of the simulation until its end. Most DES use the time-to-event approach to advancing the simulation clock, but sometimes a simulation is set to check periodically (e.g., every month) to see what events have happened in that interval.

3.7.1 Advancing the Clock Using Time-to-Event

Implementation of the time-to-event approach involves use of an appropriate formulation for the distribution of event times to select a time of occurrence for each event. For example, if the information about the occurrence of death is that there is an annual mortality of 10%, then the modeler would need to convert that information into a suitable distribution of times until death. In this case, the use of the phrase *annual mortality* suggests that the hazard of death is constant over time, and that an exponential distribution of death times is applicable. With this assumption, Equation 3.6 can be used to derive the underlying constant hazard of death from the reported annual probability of death:

$$\text{Hazard} = \frac{-\ln(1 - \text{reported probability})}{\text{reporting interval}} \tag{3.6}$$

The formula inside the parentheses is 1 minus the probability because the value given was for death rather than survival, so the complement is required. Inserting 10% as the reported probability and 1 as the reporting interval yields a hazard of 0.105 per person-year. This hazard is the parameter, usually symbolized using the Greek lowercase letter λ, that describes the exponential distribution, and, thus, it can be applied to obtain a time of death for an individual using the equation:

$$\text{Time to event} = \frac{-\ln(\text{random number})}{\lambda} \tag{3.7}$$

where the random number ranges from 0 to 1, and the resulting time-to-event is in the same time units as the hazard (this is the same equation as Equation 3.2). For the example above with the hazard of 0.105, an individual drawing a random number of 0.23 would be expected to die at 2.49 years. Note that the median time to death could be obtained by using 0.5 instead of a random number, yielding 6.58 years until half of the individuals have died (any other centile of the distribution can be derived using the appropriate number).

Provided the distribution of event times can be specified and the resulting equation for its quantile function has a closed form, then implementation is straightforward. With DES software, the logic to implement this drawing of a random number and selection of an event time is simple (Figure 3.23). Many of the commonly used time-to-event distributions are built into DES software and combine the drawing of the random number with its application in the appropriate formula (the first two modules in Figure 3.23), directly yielding a selected time for the event. Some software also include functions

FIGURE 3.23
Time-to-event check for the event death. A random number is drawn and used to select a time of death. If this is the next event, then it will be the earliest on the event calendar and the simulation will delay until that time is reached and then process the death event.

that implement algorithms for distributions that do not have a closed form of the quantile function.

Using general software, the equation for the required quantile function is written into the code that computes event times. It is helpful to create a function that does this for each type of distribution and to call that function in the code. In spreadsheets, the formula (see Table 3.14) is written into the cells of the column where that event's times are calculated. If the chosen spreadsheet software allows for creating user-defined functions, then it is helpful to do so for each type of time-to-event equation that will be used. These user-defined functions can then be used in the appropriate cells, rather than writing out the full formula each time. With a very simple one, like that for the exponential distribution, the gain in efficiency is minor, but with some of the more complex ones it can be very useful. Some spreadsheet software have built-in functions that implement the quantile function for many of the standard theoretical distributions.

The modeler must be careful, however, to ensure that the time referent of these calculations is correctly used. The time selected using the random number in the quantile formula for the distribution of times refers to a zero time that is not necessarily the same as the zero time of the simulation. If no zero time has been specified for the selection, then that zero time is understood to be the moment that the selection is made. In other words, the selection yields a time that is measured from the point of selection (in whatever

TABLE 3.14

Quantile Function Formulas for Deriving Event Times Using Commonly Implemented Standard Distributions

Distribution	Formula
Exponential	$-\ln(\text{random number})/\lambda$
Weibull	$\left[-\ln(\text{random number})/\lambda\right]^{1/\gamma}$
Gompertz	$\left\{\ln\left[-\gamma\ln(\text{random number})\right]-\lambda\right\}/\gamma$
Log-logistic	$\left\{\ln\left[-\gamma\ln(\text{random number})+1\right]-\lambda\right\}/\gamma$
Log-normal	$e^{\left[\text{NomInv}(\text{random number})/\gamma\right]\cdot\lambda}$

units the quantile function was specified). To convert the selection to an absolute simulation time, it must be added to the model time when the calculation is made.

Another problem arises when age is used as a proxy for time in the quantile function. This may occur because the source data provides the number of events occurring at various ages. In this case, the time referent of the quantile function is zero age, that is, birth. Thus, the selected time is no longer measured from the moment it is calculated in the simulation, but from birth—it is the age at which the event will occur. To properly set the simulation time for the event, the difference between this event age and the current age must be used. This issue arises not only with age but whenever the time referent for the hazard function is a specific point in the entity's life. For example, in models of cancer, times to recurrence of the tumor are usually referenced to the time remission was attained; and in many diseases, times are referenced to when the illness was diagnosed.

When using age or other time proxies in the derivation of event times, an additional complication arises if the event time is not selected at the zero time referent of the quantile function. In most models, individuals do not enter at birth, but rather at some later age, often well into adulthood. In this situation, any selection using a quantile function specifying the distribution of event times in terms of age of occurrence must take into account that a substantial portion of the distribution is no longer applicable—the times corresponding to ages before the entity entered the model or before the sampling. For example, if the distribution specifies the times of death from birth onward, then using it to select a time of death for a 20-year-old individual cannot allow an age of death younger than 20. The formulae must be modified so that they reflect this: they must be conditional on the event not yet having occurred at the time the sampling is made. This is tantamount to reformulating the distribution so that all of it pertains to the remaining ages where the event is possible. For example, to calculate an event time (an age at which the event will occur) with a Weibull distribution that has parameters λ and γ, the quantile function is

$$\frac{-\ln(\text{random number})^{1/\gamma}}{\lambda} \tag{3.8}$$

With λ of 0.1 and γ of 0.9 and a random number draw of 0.58, the selected time is 5.1 years, if the sampling is made at birth. If the entity for whom that event time is sampled is older than that age already, then the time to that event must be beyond the current age. The formula that applies if the time sampling for that event is carried out at some time (i.e., age) T later than birth is

$$\frac{\left[(\lambda T)^{\gamma} - \ln(\text{random number})\right]^{1/\gamma}}{\lambda} \tag{3.9}$$

For example, if the time-to-event selection is made when the person is already 20 years old, then a random number draw of 0.58, with the same λ of 0.1 and γ of 0.9, yields an age for the event of 46.8 years (26.8 years beyond age 20)—vastly different from the 25.1 years that would be obtained if the sampling used the unconditional function.

3.7.2 Advancing the Clock in Fixed Steps

Instead of the time-to-event approach, a simulation may be set up to advance the clock in fixed steps and repeatedly check whether the event has happened in the period that has just passed. This requires a different formulation of the distribution of event times. Periodic checking to see if the event has happened involves computing the probability of it happening in a period, drawing a random number, and comparing it to that probability. If the random number is less than or equal to the probability, then the event is deemed to have occurred; otherwise, that event does not happen in that period.

In order to do this correctly, the probability has to refer to the period chosen as the time step. Often, the time step represented in the source data from which probability values are estimated is different from the time step in the simulation. For example, the probability of death is often cited in annual terms (e.g., 10%, or 1 in 10 individuals will die in a year), but the model is checking for death daily. In this case, the annual probability needs to be converted into a daily one. This is done by deriving the underlying hazard implied by the reported probability using Equation 3.6.

This hazard can then be converted to the probability for any desired time step for application, assuming it remains constant over that period, by using

$$\text{Period probability} = 1 - e^{-\text{hazard} \times \text{time step}} \tag{3.10}$$

with the time step defined in the same units as the hazard. These two steps can be combined into one using

$$\text{Period probability} = 1 - (1 - \text{reported probability})^{\text{time step}} \tag{3.11}$$

For a 10% annual probability, this formula yields a daily probability of death of 0.0003. If the random number drawn on a given day is equal to or lower than the period probability, death is deemed to have occurred on that day. The comparison of the random number with the period probability only indicates that the event happened in that specific period—not at what exact time in that period it occurs. For each period over which the probability applies, one random number draw is required for each entity, rather than the single calculation involved in the time-to-event approach.

In DES software, the logic for implementing the fixed step approach looks something like that in Figure 3.24. Depending on the specific package, the selection of the random number may be built into the branching module.

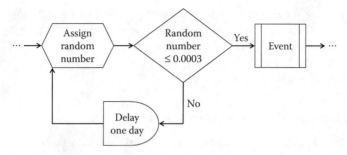

FIGURE 3.24
Implementing a daily check for the occurrence of death with a 10% annual probability of dying (equal to 3 per 10,000 per day). In this periodic checking approach, the occurrence of an event is checked after each cycle. If the random number indicates that the event has taken place, it is processed, but if not, the clock is advanced one period (here a day) and the check is repeated.

The fixed step approach is implemented in a spreadsheet by incorporating it in a statement that does the check: *If (random number (0,1) ≤ period probability, event occurs else event doesn't occur)*. In general software, this kind of statement is incorporated in the code that checks for event occurrence on a periodic basis.

3.7.3 Recording Times

It is often useful to store for each entity the times at which key events occur in the model. This may inform analyses of the timing of events across groups of entities. Such analyses may be outcomes needed for a particular HTA, or the timings may be recorded for use in validation or as calibration targets. For event times that are sampled from time-to-event distributions, the times of interest correspond to event times already contained in an entity's attributes, either an indexed attribute or a set of attributes. If there is a potentially recurring event represented, the timing of each event should be copied to another attribute, to avoid having it overwritten by further calculations of times to that event. If the time is for something that occurs as part of the logic but is not a scheduled event (e.g., a patient waiting for a transplantation is selected for an available organ), then the simulation clock must be read, and the time observed stored in an attribute. Of note, since most of the times pertain to people, the modeler may prefer to store the age at which events occur, by adding the current time to the *age at start*.

In DES software, the recording of times is straightforward as there is usually a built-in function that gives the current time on the simulation clock (e.g., in ARENA® it is *TNOW*) and the attribute value can be set to equal that. For example, if the modeler wants to record the time of a decision to increase dose, then the instruction would be *time dose increase = current time*. In general software, the modeler has to manage the simulation clock, so the approach to this instruction depends on how the clock is being handled. In a spreadsheet, the simulation clock is often implicit in the structuring of the worksheets and the event times that are stored in the cells. An explicit simulation clock can be implemented by storing the value of current time in a cell but the manner in which a spreadsheet calculates (i.e., all at once) means that handling such a clock inevitably requires using a macro language to force repeated sequential triggering of the spreadsheet.

3.7.4 Matching Actual Time

The simulation clock is typically set to begin at time zero and it runs until the end of the time horizon, without reference to specific actual calendar dates. This is a reasonable approach when the specific dates do not matter to the HTA. There are circumstances, however, when they might be important. For example, in a disease exhibiting seasonality, the simulation clock would need to be set to match calendar time, perhaps starting on the first of January and continuing through several years (although the exact year might not be relevant). If the time of day matters, then the starting time for the clock may have to be set appropriately. For example, in a model of the management of acute cerebrovascular accident, the availability of imaging machines will vary at nights and weekends because the personnel required to do the scans and interpret them may not be present at those times. Thus, the actual time of day of the stroke would be required. At the extreme, it may be necessary to set the simulation clock to actual times matching specific calendar dates. This might be the case in a model that involves prediction of epidemics and compares alternative approaches to stockpiling supplies, such as antibiotics and ventilators.

3.7.5 Delaying Time Zero

In some situations, for example, when modeling an emergency room, there is no clear time zero, though day of week and time of day are likely to be important. The facility is never empty in reality (except the minute it opened). If the simulation follows the standard HTA approach and begins introducing some entities into an empty facility at the starting point in time, from which results are accumulated, this will be unrealistic and the outcomes will be distorted. The problem can be avoided by running the simulation for a period of time before beginning to accumulate outcomes (Figure 3.25). During this *warm-up* period, the model is allowed to reach a realistic state with various entities and appropriate values already in place. That point is now set as time zero, and results accumulate from that point onward.

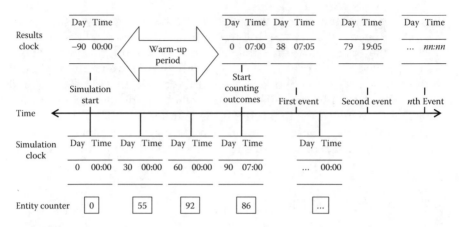

FIGURE 3.25

Diagram illustrating the use of a warm-up period. Continuous real time is represented by the double-headed solid arrow. The simulation starts at some day zero (arbitrary but sufficiently far back) and allows entities to accumulate in the facility, as indicated by the entity counter. This is the warm-up period, and what happens during that time is not considered in the model results (it is *negative* time as far as results are concerned). On day 90 at 7:00 AM, the model starts counting events and accumulating results, as indicated above the time arrow (the timescale differs from that below the arrow).

3.8 Applying Intervention Effects

HTA models must, of course, incorporate the effects of one or more technologies assessed relative to a reference *comparator*. Depending on the specifications for the HTA, the comparator may be a particular other technology reflecting the current standard of care or may be no intervention (e.g., in evaluations of screening technologies, where *no screening* is often the alternative). In some cases, the comparison may be with a mix of alternatives representing actual practice in the setting of interest. Interventions may have an impact on various outcome dimensions, including life expectancy, quality of life, productivity, and costs. Effects can include both the intended ones (i.e., what the intervention is supposed to do) and unintended ones. The latter are often thought of as adverse *side effects* but sometimes they can be beneficial as well (e.g., some treatments meant to reduce hyperglycemia in diabetics also induce weight loss). The extent to which unintended effects are incorporated in the simulation depends on the specifics of the evaluation, but there is nothing special about them from a simulation point of view. There may also be effects that extend beyond the direct ones on the person receiving the intervention. For example, the burden of caring for a patient may be affected differentially by alternative technologies. In models that represent resource constraints, certain technologies may be associated with greater or lesser use of specific resources, which may affect access to such resources by other patient groups.

The effects to be considered may involve a change in the distribution of a time to an event—or, more properly, a change in the hazard of occurrence of a given event. This is often expressed as an increase in survival or decrease in time to a desirable outcome (e.g., entering remission). Implementation of this kind of effect involves changing the distributions of time to the events in question for the entities who receive the intervention. The changes are applied by modifying the quantile function for the distribution so that it reflects the impact of the intervention. Care must be taken in selecting the new event time so that it does not move in the wrong direction. For example, if a treatment increases survival by an average of 6 months, then an entity who was expected to die on day 700 should now die some time beyond that (not necessarily exactly on day 880 because the average effect is not applied to individual entities). Thus, the modified death time needs to be selected conditionally on what the time was before the intervention effect is applied. If the effect of an intervention is applied at time zero, then the appropriate conditional selection can be handled by using the same random number for selecting the time to death for all clones of a particular entity. If an intervention effect is applied at a later time in the simulation, then the selection must take into account the time that has already elapsed.

Another type of effect is a change in a condition. For example, body weight may decrease, blood pressure may return to normal, hemoglobin may increase in a person with anemia, and severity of symptoms may diminish. Implementation of this type of effect is straightforward: it is applied in a DES by changing the relevant attribute values by an appropriate amount. For example, if the treatment leads to loss of, say, 10% of body weight over 26 weeks, then the value of each treated entity's *weight* attribute is decreased by that amount at the appropriate time. If the effect varies among individuals, then the impact on a particular entity's weight is chosen from a distribution describing that variation. As the effect is unlikely to happen all at once, it may be applied over time by expressing it as a rate of change in the attribute's value. An *updater* event can be instituted expressly for implementing these changes.

Sometimes, an effect that involves changes in a condition is expressed in terms of the proportion of people achieving a particular level over time. For example, the interest may be in the proportion who achieve a normal cholesterol level, or whose symptom severity drops to mild or none. This can be represented effectively as a time-to-event treatment effect, and the time when the level of interest will be reached can be computed from the rate of change in the attribute value.

Perhaps the simplest kind of intervention effect is one that involves a change in a probability. For example, in patients with atrial fibrillation who are receiving an anticoagulant to prevent cardiac emboli, the probability of dying increases if they suffer a stroke anyway (Caro et al. 1999). Implementation of this kind of effect involves altering the probability that is modified by treatment and applying the new value to the entities receiving the intervention at issue. Often, the source of this information reports the effect as a relative risk, and this can be stored as a global variable and applied as needed. Sometimes, the effect is computed using an equation (e.g., obtained from a logistic regression) that incorporates various patient characteristics as determinants. In this case, the new probability is computed for each patient based on his or her values for the attributes that correspond to those determinants.

Whether an intervention affects a probability, alters the time of an event, changes an attribute level, or affects some other aspect of the entities' pathways through the simulation, its implementation requires deriving and using an appropriate value that is specific to the intervention, in a form that is applicable in the simulation, be it an equation or an input parameter. This required effect is sometimes available and usable directly (e.g., a study of cardioversion provides an estimate of the probability of converting atrial fibrillation to normal sinus rhythm) and can be stored in a global variable. Much of the time, however, an equation is used to derive the value. Specificity to the intervention in this case may be a matter of adjusting a factor in the reference equation or use of a completely different equation. The techniques for collecting the data and estimating the effects—intended or unintended—are beyond the scope of this book and are well covered in textbooks such as Indrayan (2012).

To derive equations to represent intervention effects, such as time-to-event distributions (covered in Chapter 6), sources of data include randomized clinical trials of the competing technologies, meta-analyses of the information from various trials, and even observational studies such as registries. Sometimes, the modeler may be able to obtain the individual data from one or more clinical trials. Availability of the raw source data makes it possible to derive the required equations directly, considering variation of the effect over time and incorporating factors reflecting relevant individual and disease characteristics as appropriate, something which is rarely fully reported in publications. This provides much more flexibility, and the DES will tend to be better able to predict results than when having to-depend solely on summary measures. Implementation of intervention effects, once the equations are derived, follows the procedures for determining a time-to-event and triggering that event, or for modifying attribute levels, or for altering other components, as the case may be. The resulting effects may be quite individual (i.e., there are interactions between the intervention and other factors pertinent to the individual). If so, the resulting values may need to be stored in attributes. This is unnecessary if the effect is the same across individuals, in which case the values may be stored in global variables.

If the raw data are not available, the modeler has to depend on information reported by others. Occasionally, one is lucky to find equations reported that incorporate the intervention's effects over time. Usually, however, this is not the case, and the modeler must work with what is found in the relevant literature to deduce the equations. In this situation, the approach taken depends on how the information on effects is provided (e.g., as a summary measure, as a failure-time curve), to what the effects apply (e.g., an event occurrence, a condition such as pain, use of a resource such as hospitalization, or effects on another entity, like a caregiver), and how each aspect is modeled for the reference intervention. It is common in publications to present the results in terms that are relevant to answering the qualitative question about whether the technology *works*—the familiar hypothesis testing. This is less useful for simulation, where it is necessary to have quantitative estimates of the interventions' impact on the time course of the relevant disease aspects.

Regardless of how the data were obtained or what form the effect takes, it is important to be careful when extrapolating effects beyond the time period over which they were observed in the clinical studies, particularly if the period over which data were obtained is short relative to the full disease pathway. As an example, HTA models of hormonal therapy for early breast cancer have estimated event rates using meta-analyses for the comparator (tamoxifen), for which up to 20 years of data have been reported (Karnon 2007). Less data were available to inform the new intervention's effects. At the time of model development to inform funding decisions, most trials reported up to 5 years follow-up, coinciding with the common duration of hormonal therapy, but the data suggested the presence of carryover effects beyond the treatment period. In the absence of conclusive data within the time frame for which a funding decision is to be made, scenario analyses were presented to represent the effects of alternative carryover effect scenarios.

3.8.1 Effects on Event Times

3.8.1.1 Using Kaplan–Meier Curves

Published data from clinical trials or other primary studies are often summarized in terms of the times at which events occurred—in a Kaplan–Meier curve (Figure 3.26), for example (Bland and Altman 1998)—or as the proportion reaching a particular level of disease over time. This kind of information represents the quantile function (i.e., the complement of the cumulative distribution function). Thus, the distribution of event times can be taken directly from the curve or a standard parametric survival function can be fit to the reported data to smooth the distribution. The effect of an intervention can be incorporated into such a fit as one or more factors in the resulting equation. If the curves show that the underlying hazards are not proportional over time, then the intervention effect may need to be considered separately. One way to do this is to generate separate curves, with each fit corresponding to a different intervention. This can be problematic if different characteristics emerge as predictors in the separate equations. Another approach is to apply a time-varying intervention effect by including one or

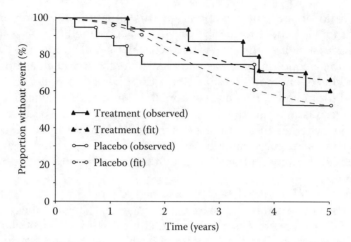

FIGURE 3.26
Example of a Kaplan–Meier curve typically reported in a trial. Two Kaplan–Meier curves
have been reported by the clinical trial, one for the treatment group and another for the placebo
arm. One set of fits for these is shown. It seems reasonable for the treated group but is a poor
reflection of the placebo experience.

more interaction terms with parameters of time. If a single parametric form
(e.g., Weibull or exponential) does not fit the observed data, then piece-meal
fitting may be necessary (see Chapter 6). Once the parametric curves have
been fit, then the quantile functions can be derived and the selection of event
times is implemented as detailed earlier in this chapter.

3.8.1.2 Using Summary Measures

Alternative, or supplementary, forms of reporting intervention effects are
measures that summarize the effect across the population and over time. For
events, this may be done using a relative technique, the most common and
useful being the hazard ratio, though others are also popular (e.g., odds ratio
and relative risk). The summary measure may be an absolute one instead
(e.g., risk difference). These summary measures may be fine for hypothesis
testing but are somewhat problematic for simulation as they do not include
the critical time dimension and, thus, require more effort and assumptions
to incorporate them in the model. For example, a relative risk reported based
on the events that have occurred by the end of one year of follow-up does
not contain information about how the effect accumulated over that year. It
almost certainly did not arise full-blown on day one, nor did it appear sud-
denly at the end of the year—most likely, it grew slowly over the year. That
one-year relative risk also says nothing about what will happen to the effect
subsequently. Will it remain constant? Will it wane? Or, will it continue to
grow?

To implement the summary measure in a time-to-event equation, it must be applied to the reference equation because the measure itself does not carry sufficient information to derive the intervention quantile function separately. This involves converting the summary measure into a form that can be applied to the reference quantile function. The required steps depend on the specifics of the reference equation (see Chapter 6).

For DES that check at fixed time steps to determine whether an event happened, summary information regarding risk or proportion achieving a result can be applied directly in the simulation by using it to adjust the corresponding reference probability of event occurrence. For example, the study may report that the ratio of mortality at one year is 0.75 indicating that the intervention decreases the probability of death over the year by 25% versus the comparator. If the measure is such a relative risk, implementation is straightforward: the probability for the comparator is multiplied by the relative risk of the new technology. If the measure is a relative risk reduction (or increase), or a change in the proportion achieving a threshold (e.g., 20% more patients achieved a normal level with the intervention compared to the reference), then the product is subtracted (or added) to the reference probability. If the measure provided is an odds ratio, then it needs to be converted to a relative risk using Equation 3.12.

$$\text{Relative risk} = \frac{\text{odds ratio}}{1 - \text{reference} + \text{reference} \times \text{odds ratio}} \quad (3.12)$$

where *reference* is the reference probability.

When an absolute measure of effect is available, the timescale of the intervention effect must be the same as that of the reference probability. If, for example, the reference is a monthly probability of an event, but the intervention is reported to reduce the annual probability of experiencing that event, then the intervention's effect must be transformed using Equation 3.11. The application of an absolute intervention effect in the DES is also straightforward: the intervention probability is derived by subtraction (or addition) from the reference probability or proportion. The appropriate formula is used to derive the resulting intervention probability, and it is stored in a global variable. This is then used for those entities receiving the intervention.

Sometimes, the impact of individual characteristics on event times is reported as derived from regression equations that consider various determinants of the event rate, such as Cox proportional hazards (Cox 1972). If this is available, then the effect can be computed for various profiles of interest, or even at the individual level. In that case, it may be worth storing the resulting probability in an attribute rather than in multiple global variables representing the possible profiles.

3.8.1.3 Dealing with Composite Outcomes

As noted in Section 3.4.5, many clinical trials define their primary endpoint as a composite of various events that reflect the impact of the disease and, thus, the effect of the intervention. For example, related hospitalizations, procedures that may prevent further complications, and death due to the disease may be included in the endpoint. If the effects of intervention are only presented with respect to this aggregate outcome, then the DES must resort to computing the effect of the intervention on the time to this composite event. This allows the simulation to determine when one of these events occurs, but such composite endpoints are generally not suitable for HTA because the implications of the constituent events in terms of results of interest, such as the costs and quality of life, can vary substantially. Thus, the event must be decomposed, which requires the application of a second distribution to represent the proportional contribution of the alternative event types to the composite endpoint.

3.8.2 Changes in a Level

The effects of an intervention may be to modify the level of a disease marker (e.g., bone mineral density in osteoporosis); or reduce the disease severity directly (e.g., breathlessness in asthma); or to improve the quality of life (e.g., by reduction of dysfunction in multiple sclerosis). All of these are aspects of an intervention's effect that are not naturally thought of as modification of the time to an event. Such measures of effect reflect a change in a quantity such as a score or a laboratory result, not a time-to-event distribution. The approach to implementation in a DES is, however, similar: define an equation to represent the intervention effect (e.g., as a relative or absolute change in the disease marker) and use it to derive the modified value for the entities receiving the intervention. As with the time-to-event effects, if the disease marker effects operate at the individual level, the modified values should be stored in attributes.

In some cases, publications may report the marker level reached after a certain period (e.g., mean difference and proportionate difference), the amount of change over a period of time (e.g., difference in mean change and percent increase or decrease in the mean change), whether a threshold is crossed (e.g., proportion achieving normal value), or other measures (e.g., area under a curve). These measures of effect are less useful for DES. The mean change achieved after a certain period, for example, does not address what the rate of change looks like over time. If nothing else is known about the effect, then the modeler may have no choice but to assume that the summary measure applies constantly over time, but every effort should be made to elucidate its time components. Some trials report longitudinal data or provide graphs with the means at study visits. This allows examining whether the effect is modified by time, and even deriving a simple equation of effect versus time to allow projection or application at different times.

3.8.3 Unintended Effects

Interventions often have one or more unintended effects, often considered adverse. It is important to model these in order to achieve a complete picture of the interventions' consequences. Typically, these kinds of outcomes are reported simply as the proportion of people exposed to an intervention who suffered the adverse effect, without regard to multiple events happening to one person, or any correlation among them. The modeler has little choice, in this situation, but to treat them as independent risks and expose the simulated people receiving a particular intervention to all the competing risks. This is easy to do in a DES as each probability is used in turn, in either a periodic check or to derive a time to each type of adverse event.

3.9 Recording Information

DES can provide copious amounts of information because it counts or accumulates things in very many places in the model and can report these at any arbitrary time interval, for any parameter that is of interest. Typical stratifications are by intervention, by stakeholder, and possibly according to various subgroups. This flexibility introduces the danger that critical outcomes are drowned in masses of other data and users can be overwhelmed. It is the modeler's responsibility, therefore, to define which outputs are relevant for the decision maker and at what intervals they should be reported.

3.9.1 What to Record

Generally, the results can be classified into four categories: counts, times, levels, and costs (Table 3.15).

TABLE 3.15

Types of Results

Type	Examples
Counts	The number of each relevant event that occur. These can include disease events, adverse events, drug doses consumed, treatment discontinuations, drug switches, deaths.
Times	Duration of various states, such as survival (e.g., in years), quality-adjusted time in the model (e.g., in QALYs), time on treatment, duration of symptoms.
Levels	Average level over time (e.g., mean glycemia), maximum level observed (e.g., peak 6 minute walk distance), highest change (e.g., average weight loss), proportion attaining threshold (e.g., percent normalizing blood pressure).
Costs	Of interventions, concomitant drugs, monitoring, disease events, adverse events, hospitalizations, productivity, caregiver time.

Many DES software provide for automatic recording of information, typically of two types. One has to do with observations of what is happening at the moment it happens (called *observational* statistics), and the other has to do with data that are accumulated over a time interval (called *time-persistent* statistics). Automatically collected observational statistics include equally weighted observations (e.g., wait times, flow times, and inter-arrival times), while time-persistent statistics are observations weighted by time (e.g., queue length, resource state, and number of entities in the system).

The automatic data collection tools are rarely useful for HTA because these models focus on the experience of the individual entities (often individuals) who exist during the entire time horizon or a large part of it. In the typical DES in other areas, the models focus on resource utilization, throughput, work in progress, and so on; the time horizon tends to be quite short and the entities are barely distinguished one from the other (e.g., parts in an assembly line). Therefore, in most DES platforms geared to that type of model, the statistical accumulators are built in to obtain information from entities as they use resources, wait in queues, and so on, so that the performance of the system can be characterized. These types of results are not so interesting to most HTAs, and, therefore, the modelers need to design and implement collection of the required information separately.

3.9.2 How to Record

Most DES software provide generic modules for recording information, and the modeler needs to only specify what should be obtained (e.g., QALYs lived) and when (e.g., at every disease event). An alternative is to store all the data for every entity and output it as a dataset for analysis after the end of the model run. This is commonly the approach taken with general software and with spreadsheets. Depending on the software used, this output can be stored within the model itself, be written to text files or spreadsheets, or be linked directly to other interfaces and displayed on screen.

The most straightforward way of accumulating results is to store them directly in global variables. There are two distinct benefits of using global variables instead of attributes for accruing the results. First, memory usage is much lower since there is only one variable per outcome compared to using one attribute per entity per outcome. Second, the values stored in global variables can be reported directly without further processing, whereas data stored in attributes need to be extracted and combined across entities. Moreover, global variables are constantly updated and ready to be reported at any time, not just at the end of the run. Such data can also be displayed in animation during the running of the DES, something which can be very helpful for model verification (debugging) purposes.

There are circumstances in which results should be recorded in attributes because the individual values are important to the analysis or may become so later on. For example, although it is possible to accumulate some measures of variance in global variables, different measures might be needed later, requiring access to the individual values. Another situation is a post-run request for analyses in a subgroup of the population different from the pre-planned ones or from the perspective of a different stakeholder. In these situations, recording individual values in attributes saves the modeler from having to repeat analyses in additional runs. Some results (e.g., event counts) that may be recorded in global variables, likely stratified by intervention, to provide ongoing access may also be stored in attributes because they are needed for use in some of the model equations, or if an outcome like the maximum number of events experienced by an individual is required (e.g., for validation).

For an HTA, the output of the results usually starts when the time horizon of the model has been reached. The timing of the writing or transfer must be carefully set so that no entities are inadvertently left somewhere in the simulation without having reported their last values (e.g., costs and life years) to wherever the data are being accrued. The purpose of the *model end* event is to make sure that the values of all entities will be duly processed, and it is necessary to make sure it is triggered before the simulation ends. If the end of time horizon is being used as a condition that terminates the simulation, then it may be necessary to increase it slightly to enable the *model end* event to happen. Depending on the research question and the time horizon, some models may also be set up to report the results at periodic intervals (e.g., at 5 and 10 years, or annually). This is fine, but it is important to remember that the volume of output increases quickly and can easily become overwhelming.

Given the stochastic nature of DES, a model run often consists of multiple replications, each one allowing a different set of random numbers to play out. If this is done (instead of one very large replication), results are accrued and reported for each replication. These have to be analyzed and summarized to provide the results for the complete run. All DES software have built in functionality for this, but it is also easy to do the analysis post-hoc either with general software, statistical packages, or even in spreadsheets. These results across replications provide a sense of how the outcomes vary due to the randomness in the model and can be used to judge how many replications would be appropriate for the final run of the model. Another use of replications is for probabilistic sensitivity analysis, where the results are generated in each replication for a different set of inputs selected to reflect parameter uncertainty (see Chapter 5). Recording of the results of each replication is essential in this case as they provide a picture of the impact of uncertainty. Although less common for HTA DES, a model run may comprise multiple replications in order to represent the experience over alternative time periods (e.g., the daily operation of a clinic or emergency department). In this case, the results of each replication are of interest, in and of themselves, but can be combined to represent the results for the facility as a whole.

3.9.3 Discounting

To reflect the common human preference for attaining good things sooner and postponing bad outcomes to later, HTAs usually require that costs, survival, QALYs, and any other outcomes of interest be valued according to when in time they accrue, with those happening later in time having less value than those happening sooner. Implementing this time-preference involves discounting each value back to some common analytic point— usually the start of the simulation. With a DES, this discounting can be implemented properly, without the need for approximations, since time is handled continuously and any value can be immediately discounted using the exact time point at which the event or recording occurs (Equation 3.13).

$$\text{Discounted value} = \frac{\text{value}}{\left(1 + \text{discount rate}\right)^{\text{time}}} \quad (3.13)$$

where *value* is the undiscounted value and *time* is the period from the event occurrence back to the model initiation or point of present value.

For example, if an event costing \$1,000 occurs at 3.4 years, then the cost discounted at 3% back to the model start is \$904.39 [1000/(1.03³·⁴)].

If the *value* is accrued over a period of time, then Equation 3.14 is used.

$$\text{Discounted value} = \frac{\text{value}}{r}\left(e^{-r \text{ time } A} - e^{-r \text{ time } B}\right) \quad (3.14)$$

where:
  r is the discount rate
  time A is the start of the valuation period
  time B is its end

The equation requires all time units to be the same: if cost is expressed in daily terms, then the times must be in days and the discount rate should also be the daily one. For example, if the daily cost of a treatment is \$10, and the treatment starts at 380 days into the model and ends by day 815, then the accumulated treatment cost discounted at 3% per year is \$4,141.76, with the discount rate converted to a daily value.

3.10 Reflecting Resource Use

In HTA models, it is important to consider the consumption of resources, particularly those that the health care system agrees to pay for. Most HTA models currently assume, however, that all resources have infinite capacity, and, thus, they impose no limitations: when an entity needs a resource, it can access it and use it immediately. This simplified situation is addressed in Section 3.10.1. A more realistic approach requires explicit modeling of health care resources, with capacities that reflect the actual context of a particular health care system. This is covered in Section 3.10.2.

3.10.1 Implicit Handling Using Cost Accumulators

If a model does not implement resource constraints, then there is no reason to include resources explicitly and no provision needs to be made for capacity, scheduling, or queues. In this situation, the only relevance of resources is that their consumption generates costs. These costs can be tracked by accumulating them whenever an event occurs that implies use of covered resources. For example, if an entity experiences an asthma exacerbation on day 165, then the cost of managing that episode is accrued. If it is sufficient to know the total costs for the population, then appropriate amounts can be added to the corresponding global variables, but if there is a reason to hold the information individually (e.g., it is necessary to understand the distribution of costs), the total is accumulated in the corresponding attributes. For costs that depend on the duration of use (e.g., the cost of a hospitalization given length of stay rather than case-based costing) or that pertain to expenses incurred between events (e.g., the daily cost of a treatment), the costing is handled by adding to cost accumulators the product of the cost per period and the duration of the period. Note that despite electing to have no resource constraints in the simulation, a modeler may still elect to implement resources explicitly for face validity reasons or to allow for structural analyses where capacity constraints are tested.

3.10.2 Explicit Representation

In a DES that limits the capacity of resources (i.e., a constrained resources model), each resource must be explicitly represented so that it can be assigned a capacity and entities can *compete* for it, as appropriate. The terminology for resources is very much influenced by models of industrial production facilities, where they are a critical part. There, resources are considered to be in one of four states: an *idle* machine is available and not in use; it is *busy* when it is processing one or more entities and thus, *working*; if it is turned off (i.e., its capacity is set to zero), it is considered *inactive*; and it has *failed* if it has broken down and cannot be used. In such production models, the focus and critical outcome is often the utilization of the different resources in relation to the production capacity and schedule. This is not usually considered important in HTA models but it can be if the assessment involves a physical facility. Even if this is not the case, explicit representation would be of greater interest if unconstrained models were not the norm. To implement the *constrained-resource* functionality, the required components are the resource and its queue.

When modeling a clinic or other health care facility, there can be many types of resources, and these may be required in different combinations or at varying times (Berg et al. 2010). For example, in a model to evaluate the effects of a new technology in a colonoscopy clinic, there were a fixed number of procedure rooms, each one only allowing a single individual at a time and containing a procedure bed and a monitoring unit. There was also a preparation area and a recovery unit, each with its own number of beds, more in the preparation area than in the recovery one. Nurses who might or might not be able to substitute for each other in the various units, gastro-enterologists, anesthesiologists, and technicians were also represented. The clinic had a limited number of colonoscopes, which could be out of service or in the process of sterilization. The individuals checked in at the reception and then waited in that area to be called for the preparation, after which they might have to wait again to get into the procedure room, and depending on the examination type, different resources were required in diverse sequences and for various durations. After completing the procedure, another wait could ensue if there were no beds available in the recovery room—this time blocking the procedure room even though none of the resources there were needed any longer. Each of these resources had to be explicitly defined, its capacity set and the behavior of the queue delineated.

Once the simulation is running, the model checks to see if the resource an entity needs next (e.g., recovery room bed) is available (i.e., idle). If the resource is busy (or inactive, or failed), the entity waits in place (e.g., stay in the procedure room) or takes some other action until the resource becomes available (e.g., move to the corridor to wait). If the resource is available, the entity occupies it (i.e., decreases its capacity by 1) for a particular amount of time and then releases it. At that point, the resource may be spent (if not reusable); may become inactive and, thus, remain unavailable for some time (e.g., for the colonoscope to be sterilized); or may become idle immediately and, thus, available for another entity to use it (i.e., its capacity increases by 1). For resources that have a capacity greater than one (e.g., a nurse in the recovery room), incoming entities can occupy a place if available or need to wait if they are all occupied (i.e., the nurse is busy). If there are multiple resources of a given type, then the model logic needs to define whether they are completely interchangeable (i.e., the entity takes the first available one) or they must use a specific one. The cost of using resources can be accrued based on a per use price or on the amount of time it is accessed, when it was used (e.g., during regular hours or when overtime is paid), or combinations of these.

TABLE 3.16

Steps in the Use of a Constrained Resource

Step	Instruction
1	Enter the queue in front of resource
2	Attempt to access the resource
3a	If unsuccessful, wait for resource of required type to become available
3b	If successful, decrease resource capacity by 1 and wait until duration of use passes
4	Accrue costs for using the resource
5	Release the resource and increase its capacity by 1 (if reusable, and no lag to reuse)

Implementation of constrained resources is best done using specialized DES software, although it can be programmed using general languages. It is difficult to do in a spreadsheet without additional programming. The steps to be defined for the use of one resource are given in Table 3.16.

If the task of the model is to compare two or more versions of the system (e.g., with one treatment vs. another), it is critical to duplicate the resources for each intervention, so that entities compete for a resource only within their intervention group and not across groups. In other words, each version of the system must be mutually exclusive with the others.

3.10.2.1 Capacity

The capacity of a resource is the number of entities it is able to serve simultaneously. For the procedure room in the colonoscopy clinic example, it was one, since only a single individual can be in there at any one time. The preparation area, on the other hand, was a resource with multiple capacity because it could accommodate several individuals at once. Resource capacity may change with time and the changes may be random or scheduled. For example, a breakdown of one of the recovery unit beds would temporarily and unexpectedly reduce the capacity of that area, while over a holiday weekend or during scheduled time for maintenance the lower capacity is planned. Failures can be set to happen at random time points by defining a mean time between failures, and when a failure has happened, the resource is unavailable for a mean time to repair.

The capacity of a resource is implemented by making it a feature of the resource. In specialized DES software, this is a standard capability available whenever a resource is created. With general software, the feature must be stored in a variable or array that is linked with the resource, and code must process that capacity each time the resource is used or released, fails, or is taken offline. With spreadsheets, it is more awkward to implement this aspect but the capacity can be stored in a cell of the array corresponding to the resource's features. Processing it requires use of either macros or formulas that adjust the capacity according to a use indicator stored in another cell.

3.10.2.2 Scheduling

The scheduling of resources to optimize production can be an important component of an assessment. This scheduling involves specifying which resources are available when and with what capacity, and it can be very simple (e.g., available daily between 9:00 and 17:00) or can be as complicated as may be needed to be sufficiently realistic (e.g., including breaks, lunch, and variation in these times by day). The specifications can be implemented using logic external to the resource or be carried as logic intrinsic to the resource. DES software offers components to create the schedules for resources. These may also be pulled from external tables. As with capacity, implementing scheduling is somewhat more difficult with general software and quite painful in a spreadsheet.

3.10.2.3 Utilization

Apart from capacity and its scheduling, there are several parameters that relate to the specification of constrained resources. A simple one is utilization, defined as the amount of time a resource is busy divided by the total time it is available. For example, a nurse who is busy 6 hours out of an 8-hour shift has a utilization of 75%. Utilization is an important concept in service industries. This information can be very useful in planning the number of resources to make available, their scheduling, and so on. For example, if in the colonoscopy unit waiting times are high and the utilization of recovery room nurses is consistently 100%, then the bottleneck has been identified and alternative scenarios can be specified and tested to see if waiting time goes down. While low levels of utilization may imply a mismatch between supply and demand, high levels can lead to breakdowns (*burnout* in health care personnel) and congested systems with attendant patient dissatisfaction and highly variable service times. Utilization is determined by how frequently a resource is called upon to perform a task and how fast that resource can provide the service.

Other parameters include the flow time and the wait time, but these are rarely used in DES for HTA.

3.10.2.4 Queues

When a constrained resource is busy and one or more entities need to use it, those entities must wait in a queue for it to become available. For example, in Figure 3.27 an individual arrives for a monitoring visit and enters the first queue (i.e., the waiting room) to wait for the nurse to be available. Once the nurse is available, the individual *accesses* this resource and starts to have some tests performed (e.g., take blood pressure). Then, without releasing the nurse (i.e., they remain together), the individual waits for the doctor to arrive. The doctor arrives and, with the nurse, attends to the individual. In this case, the entity is occupying two resources. Then, perhaps, the doctor leaves (i.e., is released and, thus, available to attend to other entities) while the nurse does some closing notations before being released as well.

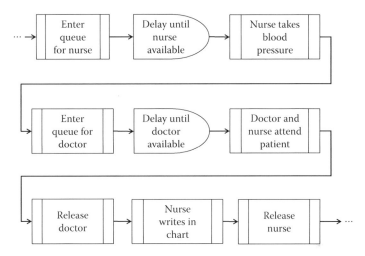

FIGURE 3.27
Logic sequence involving two resources (a nurse and a doctor) and their respective queues. The entity must wait first for the nurse to be available, then for the doctor to come in. The doctor leaves after attending to the patient while the nurse remains until the charting is complete.

A queue can function simply as a place to wait in an organized matter. A normal queue is one where the first entity to enter is also the first one to be served by the resource (i.e., first in, first out). For health care facilities, a very common type of queue is one where priority is accorded based on an attribute value or via an expression that makes it possible to rank individuals on the urgency of their need. Much less common are queues where last one in is the first one out. Apart from organizing the waiting and prioritizing who gets to use the resource, queues can also provide useful information, either as an output (e.g., how much time was spent waiting) or as an input to other decisions or calculations (e.g., waiting for a brain scan beyond a certain threshold precludes thrombolytic treatment in acute stroke).

To implement queues, there needs to be a mechanism for halting progress through the logic (though keeping track of values that may be accruing during the time spent in the queue), a process for sorting the priorities appropriately, and a means for maintaining and reporting the statistics pertaining to the queue. In DES software, a queue is automatically defined when a resource is created, and it provides various features which can be set by the modeler. It also reports statistics that can be used in other logic or reported as part of the results. In general software, the programmer must write code to achieve this, perhaps using an array to represent the queue and creating events for entities to *enter* that array or *leave* it. Another array must be created and maintained with the statistics of interest. The same can be attempted with a spreadsheet, although if there are many queues and some have complex rules, this rapidly becomes unwieldy.

4

A Simple Example

The previous chapters have introduced the concepts of discrete event simulation (DES) for health technology assessment (HTA) and the basics of implementation. To consolidate this understanding, it may be helpful to work through a simple, but complete, example. A suitable example which is reasonably straightforward is a model designed for an HTA of a new treatment for a given cancer. The HTA agency wishes to compare the consequences in patients who respond to treatment because the response rates are no different relative to standard therapy. Thus, the patients have been diagnosed with cancer and are currently in remission (the cancer is not active), and the concern is with the duration of that remission. The simulation contains the basic components common to all DES, but there are only two pathways through the model. The entities are the patients with cancer, and there are two main events: *recurrence* and *death*. The model starts with patients diagnosed with cancer and ends with the death of each patient.

In this chapter, this simple model of cancer is developed from its conceptualization through its implementation. Each of the main components of a DES (i.e., creating the entities, assigning them attribute values corresponding to their individual characteristics, duplicating them, implementing risk equations, deriving the times to each event, delaying until an event happens, triggering an event and processing its consequences, using distributions, accumulating results, handling global information, and ending the simulation via time horizon or death) is described in detail as it pertains to this cancer model. The explicit conceptualization of resources is not included, in accord with most HTA done today, and as a result, there are no queues. The chapter begins with the design of the model. This consists of the definition of the problem, the influence diagram that will guide model development, and a defense of the choice of technique. Given the selection of DES as the technique (not a big surprise), the flow diagram is developed. The next section provides the inputs that are obtained for the model, and then the equations that drive the model are described (their derivation is not addressed in this simple example, though they were borrowed from a real problem). The final section details each of the components of the DES.

4.1 Design

4.1.1 Problem

The HTA agency has received a request by the national health care system administrator to evaluate the health and economic impact of a new treatment, called *MetaMin*, for a particular cancer (kept unspecified to avoid getting into the clinical intricacies of a specific neoplasm) compared to the current standard of care (SoC). The cancer in question is incurable with overall survival from diagnosis typically ranging from 1 to 5 years, depending on the mode of treatment employed. Patients suffer from a range of debilitating symptoms, and while treatment frequently results in remission of disease, recurrence of the cancer is inevitable. For patients whose disease has progressed following initial treatment, prognosis is particularly poor, and survival is typically no more than 6 months.

MetaMin is an orally administered therapy for the treatment of patients whose disease has progressed following previous treatment. The key clinical evidence is taken from the phase III randomized controlled trials in which MetaMin was compared to standard treatment for patients whose disease had progressed following previous treatment. Its efficacy was measured in terms of its ability to prolong the duration of remission in patients who respond to the treatment. This interval is commonly known as progression-free survival (PFS). It was also documented that the treatment prolongs the time to death after the cancer has recurred. This is known as *post-progression survival* (PPS). Together, these two intervals add up to overall survival (OS). Both intervals are known to depend on age, sex, and the level of lactate dehydrogenase (LDH), a biomarker that reflects neoplastic activity (high values being worse). This construct is illustrated in Figure 4.1.

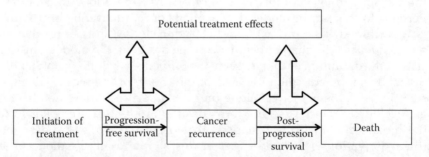

FIGURE 4.1
Schematic describing the key constructs for the example of an HTA of a new cancer treatment. The progression of cancer is summarized in three constructs for this simple HTA example. There is treatment initiation, recurrence of cancer after a period of PFS, and then death after some additional PPS. Treatment may have an effect on either of these periods, presumably lengthening them. HTA, health technology assessment; PFS, progression-free survival; PPS, post-progression survival.

TABLE 4.1

Characteristics of the Patients in the Clinical Trials and Main Results by Treatment

Characteristic	Standard of Care	MetaMin
Age (years)	60.3	60.1
Sex (% female)	61.0	62.0
LDH (units/liter)	209.8	210.1
PFS (median months)	13.6	22.9
OS (median months)	20.0	31.2

LDH, lactate dehydrogenase; PFS, progression-free survival; OS, overall survival.

In the randomized controlled trials, there was a highly statistically significant improvement in median time to progression (the primary endpoint of the clinical studies) among patients enrolled in the MetaMin arm, compared to in those randomized to the standard treatment arm. In fact, median time to progression was 22.9 months versus 13.6 months. The follow-up data also showed a statistically significant improvement in median PPS for MetaMin compared with those patients who were treated with SoC only (8.3 months vs. 6.4 months). This led to a gain in median OS of almost a year (31.2 months vs. 20.0 months). This information, along with the key patient characteristics, is summarized in Table 4.1.

MetaMin was also shown to produce minimal side effects. The most common adverse events ascribed to MetaMin were symptomatic in nature—principally headache and nausea—occurring at about the same rate as with SoC. Adverse events of this nature are familiar to, and well managed by, oncologists as part of routine care. The frequency of severe adverse events was very low and no different from SoC, illustrating the safety of this new treatment.

4.1.2 Influence Diagram

Early in the design process, an influence diagram is developed to map out all the relevant concepts that comprise the problem and their interrelations. This involves the modelers discussing the problem in an iterative series of brainstorming sessions, some of which may include expert clinicians and researchers knowledgeable about the new treatment. The initial diagram was developed by the modelers based on their understanding of the problem and the clinical trial publications (Figure 4.2).

Although the initial influence diagram may be quite complicated, the discussions will typically allow it to be pruned down to what is essential for the problem at hand. For example, during the development of the influence diagram for this problem, an issue that may have come up is whether

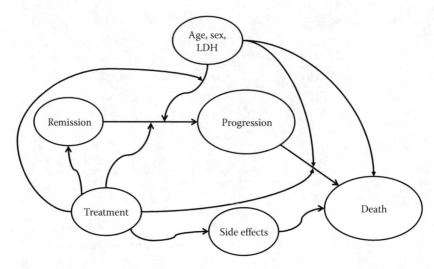

FIGURE 4.2
Initial, somewhat more complex, influence diagram developed by the modelers based on the problem statement. After some discussion of the problem posed by the health technology assessment, the modelers develop an initial influence diagram that comprises all the concepts they have come up with. LDH, Lactate dehydrogenase.

adverse effects of the treatments should be incorporated in the model. Here they were eventually deemed not to be severe enough or significantly different, so it was agreed that they would therefore not be modeled. Other connections were also deleted based on further analyses of the data and discussions with the experts (Figure 4.3).

4.1.3 Model Technique

After considering various approaches, the modelers chose to use DES. Although cohort Markov modeling was the most commonly used technique for HTA in their context, that approach does not have the flexibility required to address the heterogeneity in PFS and OS among patients with this cancer. DES avoids problems such as implementing all the parameters at their mean values rather than using the actual observed distributions. Instead of ignoring the impact of the patients' clinical history, which may have an effect on the future course of the disease for that particular individual, it can incorporate it without difficulty. Moreover, the problem statement characterized the illness in terms of the occurrence of recurrence of the disease and death, both of which are naturally thought of as events. A major benefit of the DES technique is that it can represent the management and the course of disease with higher accuracy by handling the events of interest exactly at the predicted times of these events. Therefore, appropriate effects of the events can be implemented at the time of the event, avoiding any

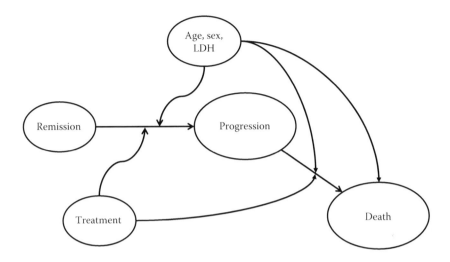

FIGURE 4.3
Influence diagram simplified after discussion with experts. The side effect concept has been deleted as the experts felt that it was unnecessary to model because the trials showed these to be mild and occurring at the same rate in the two treatment arms. The arrow from treatment to the one connecting patient characteristics to pre-progression survival was deleted because no evidence was found that treatment was a modifier. Similarly, no effect of treatment on PPS was detected so that arrow was deleted as well. LDH, lactate dehydrogenase.

artificial assumptions on the timing of the events and their consequences. The management of patients can be simulated in as much detail as necessary, while keeping the model logic transparent. All relevant aspects can be incorporated explicitly and efficiently, and the design can also be presented very clearly. These advantages (Table 4.2) led the researchers to select discrete event simulation to develop their cancer model.

4.1.4 Flow Diagram

The next step in the development of the model is to construct the flow diagram that will serve as the blueprint for programming the simulation (Figure 4.4). As the simulation is about treating patients who are diagnosed with cancer, the flowchart indicates that at the start. Patients' attributes are assigned and then, per the problem statement, they are in remission and will

TABLE 4.2

Reasons DES was Chosen for the Cancer Model

Need to include heterogeneity in the population
Need to keep medical history
The occurrences of interest are event based
Transparency

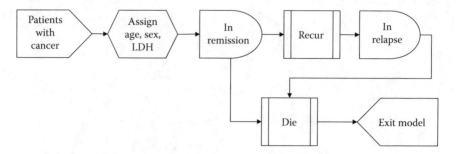

FIGURE 4.4
Flow diagram for the cancer DES without considering treatment. This is a sketch of the initial flow diagram the modelers have agreed on. It does not yet incorporate treatment but covers the major control logic. LDH, lactate dehydrogenase.

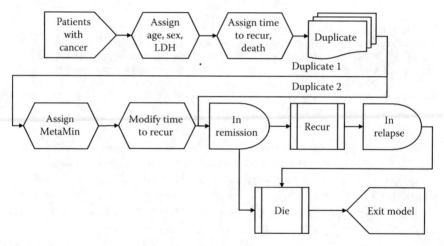

FIGURE 4.5
Modified flow diagram to incorporate treatment. With the addition of a duplicate module and assignment of the treatment effects, the flow diagram is complete. LDH, lactate dehydrogenase.

stay in that condition until an event happens. This could be recurrence of the cancer or death for some other reason. If cancer recurs, then the patient will be in a relapse until death.

This initial flow diagram does not have treatment in it, however. To carry out the HTA, treatment needs to be introduced, and this requires several modifications to the flow diagram (Figure 4.5). First, the assignment of event times is made explicit to indicate that the times under standard care are selected before treatment is assigned. Then, the entities are duplicated so that the duplicate will receive MetaMin. Since the new treatment delays recurrence, the time until the cancer recurs—already assigned before duplication—needs to be modified to incorporate the effect of the treatment. The other duplicate continues in the simulation as in Figure 4.4 with no modification.

4.2 Obtaining the Inputs

A critical step in building the model is the acquisition of data that supply all the required inputs for the simulation. Needless to say, the data should be suited for the task, meaning they should be collected in a population sufficiently similar to the one that will be simulated, and they should be in a form that enables DES, either by providing descriptions of the distributions or, preferably, consisting of the data for individual patients so that the modelers can derive all the required distributions and equations. The details of searching for suitable data, and even more so those of designing studies to collect data *de novo,* are beyond the scope of this book, but a typical process will be briefly described for this simple model.

Several categories of inputs are involved in constructing and analyzing this cancer DES. One type has to do with the characteristics of the patients involved. According to the problem specifications, these are age, sex, and LDH level. Another category of information pertains to the equations that specify the distributions of times until recurrence of the cancer and until death. An estimate of the effect of the new treatment will be required as well. Finally, information on the values, in terms of quality of life (QoL) and costs, to be assigned to the treatments, to the time before recurrence, and to the time afterward will also be needed. In a real HTA, there would be additional requirements (e.g., the discount rate).

4.2.1 Patient Characteristics, Costs, and Quality of Life

In this case, the modelers were able to obtain the individual patient data from one of the MetaMin trials. This dataset contains the required information in terms of age, sex, and LDH level at baseline. Table 4.3 shows an extract from this dataset covering the first 20 patients. For this simple example, it will be assumed that the full dataset is sufficient for purposes of the simulation, but, in reality, it is possible that the randomized controlled trial is small or failed to include some types of patients that are common in actual practice. If this is the case, then additional data might be obtained from another study like a disease registry. Also, for this simple model, it will be assumed that the distribution of patient characteristics as observed in the trial reasonably corresponds to the target population, but if it didn't then weighted sampling of the individuals could be implemented to achieve a different mix.

TABLE 4.3

Extract from the Clinical Trial Dataset
Containing the Individual Patients'
Data at Entry into the Clinical Trial

Patient ID	Sex	Age (Years)	LDH
1	Female	64	144
2	Female	61	210
3	Male	52	188
4	Female	59	207
5	Female	70	158
6	Male	58	179
7	Male	60	223
8	Male	68	172
9	Female	60	233
10	Male	55	229
11	Male	62	311
12	Female	48	187
13	Male	72	209
14	Female	48	161
15	Male	76	186
16	Female	49	186
17	Female	60	321
18	Male	68	195
19	Female	58	285
20	Male	55	202

LDH, lactate dehydrogenase.

Costs are one of the key outputs for the HTA. To compute these, the simulation will need to record them as they are accrued by patients. In a full HTA, even with this very simple model, there would be many categories of costs (e.g., doctor visits, monitoring tests, diagnostic tests, admissions to hospital, management of side effects, concomitant medications, and lost productivity). Their consideration, and, possibly, unit values would depend on the perspectives considered in the analyses. In this simple example, however, only the daily cost of the treatment is included. In the jurisdiction of the HTA, it has been determined in another study that the SoC costs $3 per day, on average. Using MetaMin instead will cost $24 per day (Table 4.4). Based on the conduct of the clinical trials, it will be assumed that the treatment is continued until the cancer recurs.

In order to compute QALYs, the effects of events on the QoL need to be estimated as well. Many events, including treatment itself, its side effects, and diagnostic tests, could have an impact on QoL, but for this simple model

TABLE 4.4

Costs and Quality-of-Life Adjustments

Item	Value
Cost	
Standard of care	$3/day
MetaMin	$24/day
Quality of Life	
In remission	0.83
After recurrence	0.61

it will be assumed that only cancer recurrence does, changing the QoL from a higher level while the patient is in remission to a lower one once the cancer has progressed. The modelers identified a study that published estimates of the decreases in QoL that ensue with this cancer. After progression, patients rate their QoL as 39% lower than without cancer, on average. Even in remission, they feel that it is still 17% lower. Thus, the modelers assign the progression-free time a QoL adjustment of 0.83. In other words, one year of life in remission is equivalent to 0.83 years in full QoL. After recurrence, the adjustment is 0.61.

4.2.2 Equations

The DES needs equations that specify the distributions of the time until recurrence and the time from recurrence until death for both standard care and MetaMin, so that appropriate times can be selected for each patient. If only the published results of the clinical trial were available, then fitting of these would need to be done, for example, using the published Kaplan–Meier curves. Chapter 6 provides more detail on how to do this. In this case, the modelers were able to obtain the individual patient data from the clinical trial, so they can carry out statistical analyses to derive the equations directly. They found that for standard care, a Weibull distribution fits the observed data well, and the known effects of sex, age, and LDH could be represented using a Cox proportional hazards component. Thus, the time-to-event equations for both PFS and PPS will have the form given in Table 3.14, but with the addition of the Cox proportional hazards component. This is given by Equation 4.1. (Of note, an alternative would have been to fit a single parametric equation that incorporated all the factors, rather than fit a Weibull to the overall times and do a separate Cox proportional hazard analysis for the risk factors.)

$$t = \left[\frac{-\ln(\text{random number})}{\lambda_0 PI_j} \right]^{1/\gamma} \tag{4.1}$$

where:

λ_0 is the Weibull scale parameter for the average patient

PI_j is the prognostic index for the jth patient, given by the Cox proportional hazards

γ is the Weibull shape parameter

These parameters and the coefficients from the Weibull fits are given in Table 4.5.

For example, the first patient in the data extract (Table 4.3) is a female aged 64 years with an LDH of 144 units per liter. A patient like her receiving SoC treatment would be expected to have a median time until recurrence (i.e., PFS) of 15.8 months. This is given by using the random number 0.5 (i.e., the median) in the PFS equation (Equation 4.2).

$$PFS = \left[\frac{-\ln(0.5)}{0.000916 \times e^{1 \times -0.458 + 64 \times 0.032 + 144 \times 0.003}} \right]^{1/1.67} \tag{4.2}$$

A male of the same age and with the same LDH value could expect a shorter median PFS of 12.0 months, while a female a decade older could still expect a median PFS of 13.0 months. A much higher LDH at baseline of 350 units per liter would reduce the expected median PFS to 10.9 months for the original female.

For the PPS, the equation would be implemented the same way, using the corresponding coefficients (Equation 4.3).

$$PPS = \left[\frac{-\ln(\text{random number})}{0.00019 \times e^{(\text{if female}, 1 \times -0.188) + \text{age} \times 0.086 + \text{LDH} \times 0.004}} \right]^{1/1.24} \tag{4.3}$$

TABLE 4.5

Weibull Parameters (Scale, Shape) and Cox Proportional Hazards Coefficients for Age, Sex, and LDH Obtained for Standard Care

Parameter	PFS	PPS
Weibull Fit		
Scale	0.000916	0.00019
Shape	1.67	1.24
Cox Equation		
Sex	−0.458	−0.188
Age	0.032	0.086
LDH	0.003	0.004

LDH, lactate dehydrogenase; PFS, progression-free survival; PPS, post-progression survival.

The next step is to assess the effect of MetaMin. According to the influence diagram, it is expected that it will affect both PFS and PPS, and, indeed, the clinical trials showed this to be so. The modelers could have reanalyzed the individual patient data to obtain separate equations for MetaMin, but they decided instead to accept the published result which was given in terms of hazard ratios (HR). For PFS, this was reported to be 0.42, indicating a 58% reduction in the hazard of recurrence, while for PPS, the HR was reported to be 0.73, indicating a 27% reduction in the hazard of death following progression of the cancer. To apply these values to individual patients, the equations need to be modified slightly to incorporate HR (Equation 4.4).

$$t = \left[\frac{-\ln(\text{random number})}{\text{HR} \times \lambda_0 \text{PI}_j} \right]^{1/\gamma} \tag{4.4}$$

Thus, for a patient of the same age, sex, and baseline LDH level, who draws the same random number (i.e., who is a duplicate of the patient receiving SoC treatment), the time to the corresponding event will increase by a factor (F) given by Equation 4.5. For PFS then, this turns out to be 1.681.

$$F = \left(\frac{1}{\text{HR}} \right)^{1/\gamma} \tag{4.5}$$

For example, the 64-year-old female with an LDH of 144 units per liter, who is the first patient in the data extract (Table 4.3), could expect her median PFS to increase to 26.5 months by receiving MetaMin treatment. Of course, with different random numbers, the results would change but the relation would stay the same.

Before implementing these estimates of efficacy in the DES, the modelers decide to check how well the estimated values for MetaMin correspond to the observed results for PFS, under the assumption that the impact of age, sex, and LDH is the same as with SoC and that the Weibull shape also holds. To carry out this check, they estimate the entire PFS distribution using the mean values of age and LDH and the proportion of females in the MetaMin arm of the clinical trial (Table 4.1). Then, they compare the resulting time-to-event curve (often referred to as a *survival* curve) to the observed results from the clinical trial (Figure 4.6).

They found that the predicted curve matched reasonably well with the Kaplan–Meier one derived from the observed clinical trial data, at least for the overall population. This provided enough confidence in the assumption that the Cox proportional hazards coefficients applied to the MetaMin population as well, and, thus, the modelers implemented the approach as described. It should be noted that estimating the entire PFS distribution using the mean values can yield poor results because of the non-linearity of the equations. A better way would be to estimate PFS for each individual and

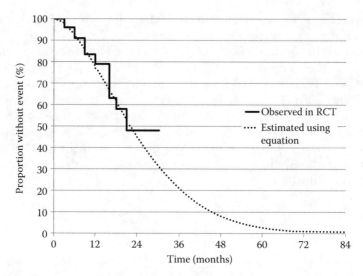

FIGURE 4.6
Comparison of PFS predicted by equation and observed clinical trial results. The solid stepped line is the Kaplan–Meier curve reported by the clinical trial for the MetaMin arm. It ends after 30 months as that was the maximum follow-up in the clinical trial, and it does not extend all the way down to the horizontal axis because nearly half the patients were still alive when the trial ended. The dotted line is the prediction made by applying the reported hazard ratio to the Weibull equation obtained from the standard-of-care arm of the trial. PFS, progression-free survival; RCT, randomized controlled trial.

derive the simulated PFS curve that way, but the simpler approach was taken here for didactic purposes.

In the flow diagram (Figure 4.5), there is a connection between *in remission* and *die*, indicating that a patient may die without a cancer recurrence. This corresponds to deaths due to causes other than the cancer itself (e.g., cardiovascular disease). To inform this branch of the simulation, the modelers need an equation that gives the distribution of deaths due to causes other than cancer in patients who are in remission with this particular neoplasm. One possibility would be to use the life tables reported by the national authority for the country where the HTA takes place. These life tables typically provide the proportion of people alive at each age from birth until 100 years or more, usually separately for females and males. They often also provide estimates of the survival function and hazards, but, if these are not given, they are fairly easy to derive. These distributions of time until death are for all causes, however. They are not specific for causes other than the neoplasm at issue. The latter are what is required for the model, and these can be estimated by subtracting the hazard of death due to the neoplasm from the all-cause hazard. The resulting estimated hazards of death due to other causes can then be fit with a known function (e.g., Weibull and Gompertz) providing the required equation.

This equation may still not reflect the population of interest, however, because it is obtained from a broad cross section of the population, not just those people who are in remission from this neoplasm. Since the patients in question may suffer increased mortality from other causes (e.g., because of cardiac toxicity of the treatments already received) or they may, in fact, have decreased mortality compared to their contemporaries (e.g., because they are closely followed clinically and receive optimal care), it is uncertain how well the equations based on the life table will apply to the DES.

Seeking an alternative, the modelers go back to the clinical trial data to see if there are enough deaths due to other causes to enable estimation of the equations directly from the observed results. At this point, they discover that all the deaths observed in the clinical trial were attributed to the cancer, regardless of whether a recurrence had already been diagnosed prior to the death. The clinical trial protocol explicitly stated that this would be done because the diagnosis of recurrence may not happen before death, and, in common with most other clinical trials in this area, this assumption avoided difficult subjective attribution of the cause of death and possible biases that may arise.

This leads the modelers to check back with the experts who agree that in actual clinical practice deaths in patients with this advanced cancer would also be attributed to the neoplasm, and that it is reasonable to assume, therefore, that the mortality due to other causes is extremely small and could be safely ignored, even if re Individual Patients currence has not been diagnosed. Consequently, the modelers modify the influence diagram (Figure 4.7) and flow diagram (Figure 4.8) accordingly and stop looking for an equation for this component.

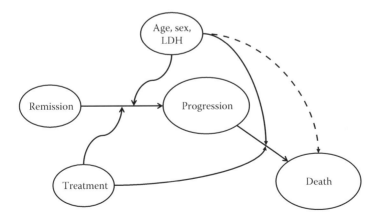

FIGURE 4.7
Revised influence diagram after checking with the experts regarding mortality due to causes other than the cancer in question. The dashed line indicating that age and sex separately influence mortality will now be deleted. LDH, lactate dehydrogenase.

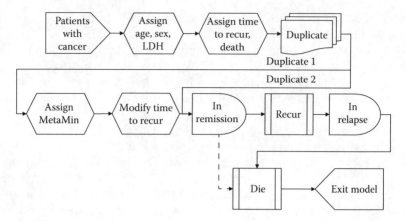

FIGURE 4.8
Modified flow diagram based on the experts' opinion that deaths due to other causes than recurrence can be safely ignored. The pathway indicated by the dashed line will now be deleted from the flow diagram. LDH, lactate dehydrogenase.

4.3 Structuring the Model

Once the inputs and equations required for the model have been obtained, the modelers can proceed to construct the DES. The exact steps will depend on the choice of software and, to some extent, on the modelers' preferences and experience. There are many ways of arriving at the same DES. In this section, the example is constructed step by step following the blueprint provided by the flow diagram. Each step is shown for a spreadsheet such as Microsoft Excel®, but pointers are given for the use of specialized DES software and for general programming software as well. To keep the display of all the steps manageable, the DES will be limited to the first 10 patients. In reality, there would be many more patients simulated.

4.3.1 Creating the Entities

The first step in the construction of the DES is to create the entities that will be simulated. For this example, the entities are patients with the cancer in question who have achieved remission. Creating the entities is very simple. In a worksheet, a table can be created with the first column containing the entity identification (ID). These entity identifiers are assigned sequentially using integers starting with 1 and ranging until the maximum number of entities in the simulation. Each row will contain the information identifying a particular individual. It is convenient, therefore, to include columns for each of the characteristics that will distinguish the entities—the values of their

	A	B	C	D
1	Entity ID	Sex	Age (Years)	LDH (Units/l)
2				
...				
Max				

FIGURE 4.9
Structure of the Worksheet Containing the Entities. LDH, lactate dehydrogenase.

attributes at the start of the simulation. For this cancer model, the worksheet will look something like Figure 4.9.

In specialized DES software, a module is typically provided for creating the entities. This module usually provides for declaring the type of entity that will be created and how many to create at that time. The module may also provide other options, such as setting times for the first creation of entities and for additional creations (e.g., for simulating incidence). An example using Arena® is displayed in Figure 4.10.

4.3.2 Assigning Initial Attribute Values

The next step is to assign characteristics to the entities. These will be the values that will be stored initially in their attributes. As the modelers were able to obtain actual patient data from the clinical trial and they feel that this population is reasonably representative of the target population for the HTA, they decide to use these data to assign the initial attribute values in their simulation. Nevertheless, to enable modeling of populations that differ, at least somewhat, from the actual one enrolled in the clinical trial, they implement a sampling mechanism that allows any particular replication of the model to select a different set of patients on whom to base the initial attribute values for the entities.

To provide the source information for the sampling, the dataset containing all the patients from the clinical trial, regardless of which treatment arm they were randomized to, is copied into a worksheet. That worksheet, call it *SourcePatients*, will contain columns labeled *Patient ID*, *Sex*, *Age*, and *LDH*. Each row will contain the information for one patient from the clinical trial. The actual treatment assignment is not copied in as it will not be used in the simulation—the concern is only with the baseline information as it pertains to a population of interest. The data in the worksheet will look like the extract displayed in Table 4.3.

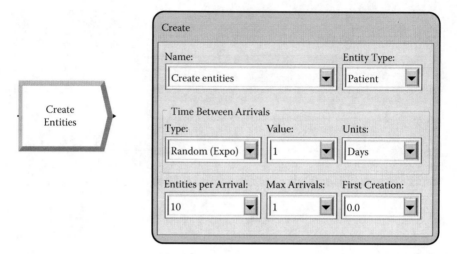

FIGURE 4.10

Example of a *create* module from Arena®. The pentagon on the left is the depiction of the module in the graphical user interface, and the form on the right provides fields where all the information defining the creation of entities can be entered. For the simple cancer model, it suffices to declare the *Entity Type* as *Patient* and the *Entities per Arrival* to the maximum number to be simulated (here limited to 10). As all the entities will be created at the start, *Max Arrivals* is limited to 1 and the *First Creation* is set for time 0.0. This disables the sub-form labeled *Time Between Arrivals*.

For a given replication of the model, the patients listed in *SourcePatients* will be sampled and their data will be assigned to the corresponding attributes of the entities that are created at the start of the simulation. To do this, a method of random sampling with replacement is required. This is easily implemented by drawing random numbers, one for each entity, and using them to select a patient from the clinical trial. Random number generators yield numerals from the set of real numbers, typically ranging from 0 to 1 (including 0 but not 1). This generator can be called in Microsoft Excel® using the formula RAND(). As the purpose of the random numbers here is to identify which patients will serve as the basis for the entities, the sequence of random numbers is transformed into integers ranging from 1 to the maximum number of patients in *SourcePatients*. This function is available in all general programming languages, in all bespoke simulation software, as well as in Microsoft Excel®. In the latter, it is a simple formula: INT(RAND()*maxpatients) + 1. By multiplying by the maximum number of patients in the source data (i.e., *maxpatients*), the formula sets the upper limit of the selection range and the INT component converts the real number that emerges into an integer. Adding 1 changes the sequence from zero-based to one-based, to correspond to the patient IDs.

To assign attribute values to the first 10 entities, suppose the formula gives the following random sequence of 10 integers: 10, 19, 18, 17, 12, 13, 2, 13, 19, and 11. It should be noted that both number 13 and 19 appear twice because

TABLE 4.6

First 10 Entities Created by Randomly Selecting
Patient Records from the Clinical Trial Population

Entity ID	Patient ID	Sex	Age (Years)	LDH (Units/l)
1	10	Male	55	229
2	19	Female	58	285
3	18	Male	68	195
4	17	Female	60	321
5	12	Female	48	187
6	13	Male	72	209
7	2	Female	61	210
8	13	Male	72	209
9	19	Female	58	285
10	11	Male	62	311

LDH, lactate dehydrogenase.

the random draws are independent. This is tantamount to sampling with replacement in that any given patient from the source data can be selected more than once. The attribute values for the first 10 entities, based on this random number sequence, are shown in Table 4.6 (the ID number of the patient from the source population is included for reference).

If the modelers did not have access to individual patient data, then the attribute values for the entities would be obtained by sampling from distributions specifying the population. The implementation of this type of sampling is described in Chapter 3.

Since the HTA will compare MetaMin to SoC treatment, two sets of patients will be needed, one to receive each of the treatments. As described in Chapter 3, it is a good idea to have the same patients experience each of the treatments. This minimizes nuisance variance (see Chapter 7) by ensuring that the only thing that differs between the groups is the treatment itself and its effects. In a spreadsheet, this is easily accomplished by creating a copy of the entity worksheet once all the attribute values have been filled in. The initial one can be named *SoC entities* and the copied worksheet *MetaMin entities*.

Implementation of sampling with replacement from a set of patient records follows the same process in a specialized package like Arena®. First, the source dataset containing the patient data from the clinical trial is read into an array, named say *iPatData* (of note, it helps to insert an initial lowercase letter before the name of an item to indicate the type of item for debugging purposes—here *i* for *input*, see Chapter 9). Like the worksheet *SourcePatients*, the array will contain four columns (for *Patient ID*, *Sex*, *Age*, *LDH*) and as many rows as are necessary to include all the patients' records. Assignment of the attribute values then takes advantage of the module designed for that

Assignments			
	Type	Attribute name	Value
1	Attribute	aPatNum	AINT(RA*20)+1
2	Attribute	aSex	iPatData(aPatNum,2)
3	Attribute	aAge	iPatData(aPatNum,3)
4	Attribute	aLDH	iPatData(aPatNum,4)
5	Attribute	aAgeOrig	aAge

FIGURE 4.11

Example of an *assign* module from Arena®. The octagonal icon for the *assign* module is shown on the left and its data form is shown on the right. The first instruction in the data form assigns the randomly selected integer (here between 1 and 20) to the temporary attribute *aPatNum*. This integer is then used to find the corresponding record in the array *iPatData*, and the values in that record from the corresponding columns are assigned to each of the attributes *aSex*, *aAge*, and *aLDH*. Each entity triggers these instructions in sequence as it passes this module.

purpose. This module assigns the required random integer to a temporary attribute, say *aPatNum* (*a* for attribute), using the same formula as in the spreadsheet. In Arena®, the formula is written as AINT(RA**maxpatients*) + 1, where RA is the shorthand for drawing a random number (Figure 4.11). The integer stored in *aPatNum* is then used to find the required record in the array that contains the clinical trial patient data. Then, the value in that row for each column is copied into the appropriate attribute for the entity. If there was a reason to retain the age at baseline (e.g., because results will be reported by subgroups defined by the age at which MetaMin is started), then it would be stored by assigning *aAge* to *aAgeOrig*. This *assign* module would be placed immediately following the *create* module so that it is the first thing the entities trigger after entering the simulation.

To duplicate the entities, specialized DES software usually offers a dedicated module. In Arena®, this module is known as a *separate* module. It takes each entity as it comes in and duplicates it so that two copies emerge that are identical in the sense that they have the same attribute values (but with different entity IDs). Before implementing this module, however, it is better to assign the event times that will pertain to SoC.

4.3.3 Assigning Recurrence Event Time

Given the decision not to include deaths due to other causes, the model will simulate two events in sequence. The first one will be recurrence of the cancer, and its time of occurrence will be given by the distribution specified by Equation 4.1 and the values given for PFS in Table 4.5. For a particular entity, the values stored in the attributes for the three determinants (age, sex, and LDH) will determine the applicable distribution of event times.

4.3.3.1 For Standard of Care

The simulation must select a specific time from that distribution and store it in the corresponding attribute, say *PFS*, for that entity. In a complex model it may be useful to have a separate worksheet where all event times are stored. In a simple model like this cancer example, this can be done in the same worksheet, *SoC entities*, where the initial attribute values are stored. A new column is labeled *PFS*, and the formula for selecting the event time is entered there (Figure 4.12).

Even for simple equations like those used in this example, however, the formula can become quite large, with many nested parentheses to indicate the order of calculation. For example, the formula for PFS for the first entity in Figure 4.12 would be as follows: (–LN(RAND())/(scalePFS*(EXP (B2*BetaSexPFS + C2*BetaAgePFS + D2*BetaLDHPFS))))^(1/shapePFS). In this formula, advantage has been taken of the Microsoft Excel® feature that allows naming cells. Thus, the values for the equation have been stored in a worksheet *Inputs* in a table (like Table 4.5) and named *scalePFS*, *shapePFS*, *BetaSexPFS*, *BetaAgePFS*, and *BetaLDHPFS*. It can be much easier to write, inspect, and verify the formulas if they are broken up into logical components. For example, this formula could be written as follows: (–LN(RAND())/ (scalePFS*E2))^(1/shapePFS), which is more readily recognizable as the PFS equation, by adding a new column E that contains the prognostic index for each entity: EXP(B2*BetaSexPFS + C2*BetaAgePFS + D2*BetaLDHPFS) (Figure 4.13).

It may also be helpful to extract the random number component to a separate column, reducing the formula to (–LN(F2)/(scalePFS*E2))^(1/shapePFS). Care should be taken to make sure that the random number is not redrawn each time the worksheet calculates. Otherwise, the PFS time for each patient will keep changing (as will other event times) even if a new replication of the simulation has not been initiated. Figure 4.14 shows how the *SoC entities*

	A	B	C	D	E
1	Entity ID	Sex	Age (Years)	LDH (Units/l)	PFS
2					
...					
Max					

FIGURE 4.12
Including PFS Values in the SoC Entities Worksheet. Max, maximum; PFS, progression-free survival; SoC, standard of care.

	A	B	C	D	E	F
1	Entity ID	Sex	Age (Years)	LDH (Units/l)	Prognostic Index PFS	PFS
2						
...						
Max						

FIGURE 4.13
Disaggregating the Formula for PFS in the SoC Entities Worksheet. Max, maximum; PFS, progression-free survival; SoC, standard of care.

	A	B	C	D	E	F	G
1	Entity ID	Sex	Age (Years)	LDH (Units/l)	Prognostic Index PFS	Rnd Num PFS	PFS (m)
2	1	Male	55	229	11.55	0.9048	3.8
3	2	Female	58	285	9.52	0.8242	6.4
4	3	Male	68	195	15.82	0.3939	12.1
5	4	Female	60	321	11.30	0.0149	36.5
6	5	Female	48	187	5.15	0.2329	31.0
7	6	Male	72	209	18.75	0.6974	6.2
8	7	Female	61	210	8.36	0.8835	5.3
9	8	Male	72	209	18.75	0.0348	23.6
10	9	Female	58	285	9.52	0.2300	21.5
11	10	Male	62	311	18.49	0.2218	14.7

FIGURE 4.14
Extracting the Random Number into a Separate Column in the SoC Entities Worksheet. SoC, standard of care.

worksheet would look like at this point. Each value in the PFS column indicates a time until recurrence for each entity in the simulation.

If specialized DES software is used, then the time until cancer recurrence is estimated and stored in each entity's attribute *aPFS* using an *assign* module. As with the spreadsheet implementation, it can be a good idea to decompose

Variable—Basic process							
	Name	Rows	Columns	Data type	Clear option	Initial values	Report statistics
1	iBetaSexPFS			Real	System	1 rows	☐
2	iBetaAgePFS			Real	System	1 rows	☐
3	iBetaLDHPFS			Real	System	1 rows	☐
4	iScalePFS			Real	System	1 rows	☐
5	iShapePFS			Real	System	1 rows	☐

Double-click here to add a new row.

FIGURE 4.15
Use of global variables to store the parameters of the PFS equation. The panel shows the declaration of the variables for the PFS equation, their names, and other information required by the software. The values of each parameter to be inserted by clicking on *Initial Values* are those displayed in Table 4.1. PFS, progression-free survival.

the formula into its components and store these in *aPIPFS* for the prognostic index, *aRndNumPFS* for the random number. The three coefficients for the equation can be stored in global variables *iBetaSexPFS*, *iBetaAgePFS*, and *iBetaLDHPFS* (Figure 4.15). The Weibull parameters would be stored in global variables as well: *iScalePFS* and *iShapePFS*.

The formula EP(aSex*iBetaSexPFS + aAge*iBetaAgePFS + aLDH*iBetaLD HPFS) would be used with the values in the three attributes for each entity and the three coefficients stored in the global variables to give the value of *aPIPFS*. The value of *aRndNumPFS* would be drawn using the function *RA*. Then, the value of the attribute *aPFS* for each entity is obtained using the formula (–LN(aRndNumPFS)/(iScalePFS*aPIPFS))**(1/iShapePFS), where ** is the operator for *raise to the power of*. This is assigned in another *assign* module (Figure 4.16). The statements assigning the event time could also be entered as an additional row in the existing *assign* module, but a *separate* module is used here to emphasize the different aspects that are at issue.

At this point, the patients receiving standard care could begin their simulation since their first event, cancer recurrence, has been scheduled to happen after a delay equal to PFS. One would wait until the cancer recurs before assigning the time of the next event, death. It is better, however, to do one more assignment at this point so that the event times can be properly assigned to the duplicate patients who will receive MetaMin. This is the assignment of the random number that will be used to determine the PPS when the cancer recurs. A column labeled *Rnd Num PPS* is created and a new random number is drawn for each entity. A separate random number is needed, otherwise the two time selections will be linked, as if the PPS were completely dependent on the PFS, something which has not been specified in the model design. In other words, using the same random number would imply that an entity would select the exact same point in the two event-time distributions, rather than the correct approach of being allowed to have a

	Type	Attribute name	Value
Assignments			
1	Attribute	aPIPFS	EP(aSex*iBetaSexPFS+aAge*iBetaAgePFS+aLDH*iBetaLDHPFS)
2	Attribute	aRndNumPFS	RA
3	Attribute	aPFS	(−LN(aRndNumPFS)/(iScalePFS*aPIPFS))**(1/iShapePFS)

FIGURE 4.16
Additional *Assign* module for the SoC PFS time. The module icon on the top represents the module where the time of progression will be selected for each patient when SoC treatment is used. The panel shows the assignment statements for each component. PFS, progression-free survival; SoC, standard of care.

	A	B	C	D	E	F	G	H
1	Entity ID	Sex	Age (Years)	LDH (Units/l)	Prognostic Index PFS	Rnd Num PFS	PFS(m)	Rnd Num PPS
2	1	Male	55	229	11.55	0.9048	3.8	0.8924
3	2	Female	58	285	9.52	0.8242	6.4	0.5664
4	3	Male	68	195	15.82	0.3939	12.1	0.4827
5	4	Female	60	321	11.30	0.0149	36.5	0.5783
6	5	Female	48	187	5.15	0.2329	31.0	0.9444
7	6	Male	72	209	18.75	0.6974	6.2	0.4846
8	7	Female	61	210	8.36	0.8835	5.3	0.9408
9	8	Male	72	209	18.75	0.0348	23.6	0.2758
10	9	Female	58	285	9.52	0.2300	21.5	0.5872
11	10	Male	62	311	18.49	0.2218	14.7	0.8493

FIGURE 4.17
Assigning a New Random Number into a Separate Column in the SoC Entities Worksheet. SoC, standard of care.

better or worse selection for each one. With these additions, the *SoC entities* worksheet now looks like Figure 4.17.

The random number for deriving each entity's PPS is easily assigned in the specialized DES software using the function *RA* and storing the value in *aRndNumPPS*. This assignment can be entered as another statement in the *Assign SoC PFS* module. At this point, the model in Arena® has three linked modules and a global variable worksheet containing the parameters for the equations (Figure 4.18).

4.3.3.2 *Applying the Treatment Effect*

In an earlier step, the *MetaMin entities* worksheet was created by copying the *SoC entities* worksheet containing the *Entity ID*, *Sex*, *Age*, and *LDH* attribute values for each entity. This ensured that the MetaMin treatment

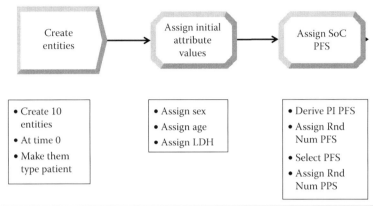

FIGURE 4.18

Model in Arena® up to this point. The linked modules address the creation of the entities, assignment of their initial characteristics, and selection of a PFS time for each one. PFS, progression-free survival.

would be applied to the patients who are identical to those receiving the SoC. To derive the time until recurrence for each entity, the same steps could be followed as for SoC. This is unnecessary, however, since the treatment effect is given by the hazard ratio, and this can be applied directly to the time of recurrence already estimated for each entity. Since the effect is the same for all entities receiving MetaMin, the factor F can be computed once using Equation 4.5 and stored in its own cell, perhaps named *FPFS*. A column labeled *PFS* is added in the MetaMin entities worksheet and in the cell E2, the formula 'SoC entities'!G2*FPFS is entered (Figure 4.19) and copied into the rows for all the entities. This calculation adjusts the PFS that was estimated under SoC to what it is expected to be with MetaMin, for each patient.

The random number that was stored for use later when the time to death is calculated can now be copied into its own column F (Figure 4.19). This is somewhat redundant since any formulas could refer to the values already stored in *SoC entities* but it is done to make the process clear. Rather than pasting the values directly, it is better to enter a cell reference so that if the

The table within the figure:

Variable—Basic process							
	Name	Rows	Columns	Data type	Clear option	Initial values	Report statistics
1	iBetaSexPFS			Real	System	1 rows	☐
2	iBetaAgePFS			Real	System	1 rows	☐
3	iBetaLDHPFS			Real	System	1 rows	☐
4	iScalePFS			Real	System	1 rows	☐
5	iShapePFS			Real	System	1 rows	☐

Double-click here to add a new row.

	A	B	C	D	E	F
1	Entity ID	Sex	Age (Years)	LDH (Units/l)	PFS	Rnd Num PPS
2						
...						
Max						

FIGURE 4.19
Including PFS Values in the MetaMin Entities Worksheet. Max, maximum entities; PFS, progression-free survival.

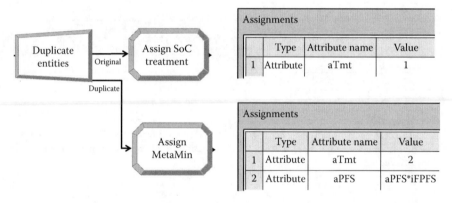

FIGURE 4.20
Duplications of entities and assignment of treatment. The first module creates an identical copy of the original entity, with the only difference being the entity ID. The duplicate is sent out of the module via a different path and this permits the assignment of treatment-specific aspects—in this case, the treatment identifies a modified PFS time for the duplicate receiving MetaMin. PFS, progression-free survival.

random number changes for the entity receiving SoC, it will remain identical for the entity receiving MetaMin.

To assign the treatment effect in the specialized DES software, the entities must first be duplicated. This is readily accomplished using the *separate* module mentioned earlier (Figure 4.20). This module creates an identical copy of each entity that passes through. Since all values assigned before the separation are duplicated, the cloned entities have identical copies of the attribute set at this point. In this case, the values of *aPatNum, aSex, aAge, aLDH, aRndNumPFS, aPIPFS,* and *aPFS* are identical. The value of *aRndNumPPS* is also copied over, hence its assignment before duplication despite it not being needed yet.

After duplication, the module conveniently splits the duplicate pair into separate paths. That way, any assignments that are specific to a given treatment can be applied to only those entities that are to receive that treatment before the entities are brought back together to continue in the simulation logic. The original entity in each pair is sent to a new *assign* module that labels the entity as a patient who will receive SoC. This is done by assigning a new attribute *aTmt* and storing a value that indicates which treatment the entity receives. For SoC, it is convenient to use 1.

Any other items that are specific to treatment would be assigned at this point before allowing the entity to continue unimpeded in the logic. It might be thought that the cost of SoC should be assigned here, but since it is common to all entities in this group, it is inefficient to do so. It is better to store it in a global variable available to all entities. Since there will be a cost for each treatment, a single global variable, *iCostTmt*, can be defined with two rows, the first row for treatment 1, which is SoC, and the second row for MetaMin, which will be treatment 2 (Figure 4.21). At this point, the effect of treatment on PFS needs to be stored in another global variable (since it applies to all the entities receiving MetaMin). This will be called *iHRPFS*. Using *iHRPFS* the factor *F* can be computed using Equation 4.5 and its value is stored in *iFPFS*. Although not yet needed, the two values for QoL can be stored now in their global variables, *iQOLRemission* and *iQOLProgress*, as well.

The entity emerging via the *duplicate* exit of the *separate* module will be a patient who receives MetaMin. Thus, an appropriate *assign* module is created and, in it, the attribute *aTmt* is assigned the value 2 (Figure 4.20). The efficacy of MetaMin on PFS can now be applied. A statement is added that reassigns *aPFS* using *iFPFS*aPFS*. The software uses the previous value (i.e., the one assigned under SoC) in the right-hand side of the equation and adjusts it

Variable—Basic process							
	Name	Rows	Columns	Data type	Clear option	Initial values	Report statistics
6	iCostTmt	2		Real	System	2 rows	☐
7	iHRPFS			Real	System	1 rows	☐
8	iFPFS			Real	System	1 rows	☐
9	iQoLRemission			Real	System	1 rows	☐
10	iQoLProgress			Real	System	1 rows	☐

Double-click here to add a new row.

FIGURE 4.21
Assignment of additional global values. The global variable table now contains a sixth row for the cost of treatment (rows 1–5 were shown in Figure 4.15). This item is defined as having two rows because it will contain the costs for SoC in the first cell and for MetaMin in the second cell. In addition, the hazard ratio for PFS (iHRPFS) and the two QoL values have been stored as well. The values to be inserted in the *Initial Values* are those in Table 4.5. PFS, progression-free survival; QoL, quality of life; SoC, standard of care.

by the factor *iFPFS*. Now, one entity in a pair will have a PFS determined by SoC, and its duplicate will have the PFS attained with MetaMin.

4.3.4 Accruing Values Until Recurrence

After treatment has been assigned, each entity will live until the cancer recurs. By definition, no other events happen during this time. The only aspect that needs to be accounted for is the value of that time itself. In this HTA, the time is valued in two ways. One is based on the cost of living—here simplified to only the cost of treatment. The other value is given by the QoL adjustment over that period. All three dimensions (the accrued cost, the survival time in life years, LY, and the quality-adjusted life years, QALY) need to be accrued. In addition, for most HTA, it is customary to also provide discounted estimates of these values to take into account time preference (Krahn and Gafni 1993). These accruals can be conveniently taken care of when the recurrence event occurs.

In the spreadsheet, a new worksheet is inserted and named *Accruals* (Figure 4.22). The *Entity ID* is entered in the first column (bearing in mind that the same number will be used for each duplicate in a pair) and the subsequent columns are labeled *LY*, *QALY*, and *Costs*. The next three columns can be labeled *dLY*, *dQALY*, and *dCosts* to store the discounted values. Since

	A	B	C	D	E	F	G
1		Standard of Care					
2	Entity ID	LY (y)	QALY (y)	Cost ($)	dLY (y)	dQALY (y)	dCost ($)
3	1	0.3	0.26	329	0.45	0.35	349
4	2	0.5	0.44	548	0.92	0.68	579
5	3	1.0	0.82	1,096	1.23	0.97	1,088
6	4	3.0	2.41	3,287	3.12	2.54	3,185
7	5	2.6	2.06	2,849	2.62	2.14	2,721
8	6	0.5	0.42	548	0.70	0.54	561
9	7	0.4	0.36	438	0.50	0.40	480
10	8	2.0	1.58	2,192	2.18	1.75	2,089
11	9	1.8	1.45	1,972	2.04	1.63	1,916
12	10	1.2	1.00	1,315	1.28	1.04	1,317

FIGURE 4.22
Accruing Values for SoC in Worksheet Accruals.

all of these accruals will first be for the entities receiving SoC, the group of columns is labeled accordingly. The *LY* column is simply the PFS, which was estimated in months, divided by 12, so the formula is 'SoC entities'!G2/12 for the first entity, and this is copied down for the others. For the QALY accrual, the LY need to be multiplied by the QoL adjustment for time in remission. If this has been stored in a cell named *Qrem* in the *Inputs* worksheet, then the formula is B3*Qrem for the first entity, and this is also copied down for the others. Finally, for the cost accrual, the annual cost of SoC (3*365.25) is stored in a cell named *CostSoC* in the *Inputs* worksheet and the formula is B3*CostSoC for the first entity. If discounting is to be applied, the discounted columns can then be filled in using the analysts, discounting formula of choice (see Chapter 3 for options) and the applicable discount rate, stored in *Drate* in the *Inputs* worksheet (3% was used in Figure 4.22).

For the patients receiving MetaMin, a new set of six columns is created under the title MetaMin, and these columns are filled in in the same way (Figure 4.23), except, of course, using the PFS estimates for MetaMin and its specific annual cost, stored in *CostMM* in the *Inputs* worksheet ($8,766).

It may be useful for the analyst to keep a count of the number of recurrence events that are happening in each treatment group. In the spreadsheet, this can be done by starting a new worksheet called *Results* and using the formula = COUNT('SoC entities'!G2:Gmaxrows), to count the events for SoC, and = COUNT('MetaMin entities'!E2:Emaxrows), where *maxrows* stands for

		H	I	J	K	L	M
1					MetaMin		
2	...	LY (y)	QALY (y)	Cost ($)	dLY (y)	dQALY (y)	dCost ($)
3	...	0.5	0.45	4,748	0.53	0.44	585
4	...	0.9	0.75	7,889	0.88	0.73	969
5	...	1.7	1.40	14,829	1.65	1.37	1,811
6	...	5.1	4.24	44,780	4.74	3.93	5,193
7	...	4.3	3.60	38,059	4.07	3.38	4,457
8	...	0.9	0.72	7,597	0.86	0.71	938
9	...	0.7	0.62	6,501	0.73	0.61	804
10	...	3.3	2.74	28,928	3.14	2.61	3,442
11	...	3.0	2.50	26,444	2.89	2.40	3,163
12	...	2.1	1.71	18,043	2.00	1.66	2,187

FIGURE 4.23
Accruing Values for MetaMin in Worksheet Accruals.

the highest row number covering the set of entities in the model (here 12). In this simple model, this is not really necessary as every row in the tables will be a recurrence event and all the entities will suffer a recurrence, so the number of these events is known. It is done here to illustrate how such an event count might be generated.

In specialized DES software, it is necessary to prompt the passage of time until the recurrence event before implementing the accruals. This is done by bringing the separated entities back together into the same logic (there is no reason to keep separate logic after the treatments and their characteristics are assigned) and, then, having them go into a *delay* module. This module simply holds each entity for the specified delay time (until the next event on its event list is also the first event on the event calendar). In the context of this cancer model, this amounts to waiting for the lowest PFS to pass before advancing that entity to the recurrence event and triggering its consequences. In Arena®, the delay time is specified in the module, labeled *PFS*, together with the units for this value, here stated to be in days (Figure 4.24).

As each entity culminates its PFS time, it will exit the *delay* module. In the next part of the logic, the recurrence event will happen and the relevant accruals will take place. Since there are several steps to the occurrence of this event, it is convenient to group them into a *submodel*, which can be labeled *recurrence* (Figure 4.24). This aggregation is purely for display purposes—it does not affect the logic at all—but it clearly identifies the modules that comprise the execution of the event.

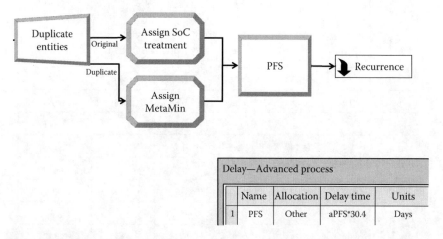

FIGURE 4.24
Addition of a *delay* module and a submodel. The module labeled PFS introduces a delay, equal to the duration of PFS, that each entity must wait for before continuing on to the submodel labeled recurrence. The *delay* module does not otherwise affect the logic. Its form specifies that the delay time for each entity should be obtained from the attribute *aPFS* (which is stored in months) multiplied by 30.4 days, the units this module is operating in. The allocation to *Other* is not used in this model. PFS, progression-free survival.

Inside the *recurrence* submodel, there will need to be a means of accruing the amounts corresponding to the survival in life years, the QALYs, and the costs. Before doing that, however, the recurrence events should be counted. To do this, a *record* module is inserted. This module, as its type indicates, registers something about the entities that trigger it. In this case, it will be set to count the number passing by as all entities going through this submodel are experiencing a recurrence event. Since the analyst wants to have the counts separately for patients receiving SoC and those treated with MetaMin, the counts are recorded into a set (i.e., a group of counters) called *recurrence*, with the entity's value in the attribute *aTmt* determining whether to register the count in the first (i.e., SoC) or second (i.e., MetaMin) counter (Figure 4.25). The specific counters need to be named in the *members* table that opens up when the *recurrence* set is created. Here, they are called *Recur SoC* and *Recur MetaMin*.

Once the event has been recorded, the simulation can proceed to the accruals. These are taken care of by inserting an *assign* module that will keep track of the amounts accruing to each category (Figure 4.26). Here the analyst has a choice. As was done in the worksheet *Accruals*, these amounts can be tracked in the corresponding attributes: *aLY, aQALY, aCost*, and *adLY, adQALY, adCost*. An alternative, however, would be to record them into a set of global variables *vLY, vQALY, vCost*, and *vdLY, vdQALY, vdCost*. The choice would depend on whether the analysts want to have the information at the patient level (e.g., because the range, variance, and median are of interest) or it suffices to

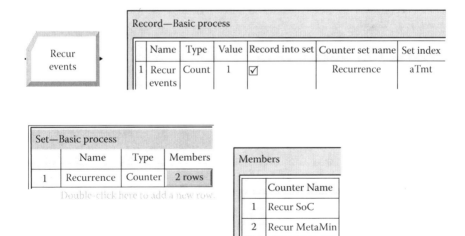

FIGURE 4.25
The *record* **module that registers counts of the recurrence events.** As each entity passes by, this module records the occurrence of a recurrence event. This recording of type *Count* is made into a set that groups two counters, distinguished by the value of the entity's *aTmt* attribute. Entities with aTmt = 1, which indicates SoC will use the first counter in the set, *Recur SoC*, while those with aTmt = 2, indicating use of MetaMin, will use the second counter, *Recur MetaMin*.

Assignments			
	Type	Attribute name	Value
1	Attribute	aLY	aPFS/12 + aLY
2	Attribute	aQALY	aPFS/12*iQOL remission + aQALY
3	Attribute	aCost	aPFS/12*iCostTmt(aTmt) + aCost
4	Attribute	aTime1	0
5	Attribute	aTime2	aPFS/12
6	Attribute	adLY	Discount*aLY + adLY
7	Attribute	adQALY	Discount*aQALY + adQALY
8	Attribute	adCost	Discount*aCost + adCost

FIGURE 4.26

Assign **module inserted in submodel to handle accruals.** This module assigns the values of life years, QALYs, and costs to each entity's corresponding attributes. The formulas add the current amount to the ongoing total in order to keep a running accrual (of course, at this point, the previous amounts are zero). The modeler has also implemented an *expression* called *Discount* that contains the analysts, chosen formula for computing the discount factor based on the values of *aTime1* and *aTime2* and then uses this to discount the amount and add it to the appropriate running total. QALY, quality-adjusted life year.

know the overall result. In this case, following what was done in the spreadsheet, the amounts will be accrued in attributes.

4.3.5 Updating Attribute Values

At the time that the cancer recurs, the patient's time to death (i.e., PPS) needs to be estimated. When the modelers go to do this, however, they realize that they need some additional information. In order to derive the PPS for each entity, the values of the prognostic factors have to be updated. While the sex of an entity will not have changed since the start of the simulation, the age and LDH levels do change with time. Deriving the new age is simple, of course. The PFS (divided by 12 so it is in years) is added to the baseline age and this gives the age at the time of recurrence. For LDH, however, it is necessary to have an estimate of the rate of change. The modelers go back to the clinical experts who tell them that during remission the LDH should have stayed stable but, with recurrence, it will increase by 5%–30%. With further probing, the experts agree that the distribution of values in this range can be summarized in terms of the proportions in each 5% range: 5%–9%, 10%–14%, 15%–19%, 20%–24%, and 25%–29%, and within each range the distribution can be taken to be uniform. In other words, any value between 5% and 9% is equally likely. They also state that the distribution does not depend on any other factors, except that it is different for men and women (Table 4.7).

TABLE 4.7

Distribution of Percent Change in LDH at Recurrence

Interval (% Increase)			
Lower Border	Upper Border	Females (%)	Males (%)
0	5	0	0
5	10	14	7
10	15	31	18
15	20	26	31
20	25	20	26
25	30	9	18

Note: Upper border value is included in next interval.
LDH, lactate dehydrogenase.

TABLE 4.8

Cumulative Distribution of Percent Change in LDH at Recurrence

Interval (% Increase)			
Lower Border	Upper Border	Females	Males
0	5	0.00	0.00
5	10	0.14	0.07
10	15	0.45	0.25
15	20	0.71	0.56
20	25	0.91	0.82
25	30	1.00	1.00

Note: Upper border value is included in next interval.
LDH, lactate dehydrogenase.

With this additional information, the LDH increase at recurrence can now be computed for each entity. First, the distributions given in Table 4.7 are entered in the *Inputs* worksheet and then converted to the cumulative by adding the total for each interval to the preceding total (entered as decimals in Table 4.8). The cells containing the cumulative distribution for the females are collectively named *fLDH*, and those for the males, *mLDH*.

Then, a new random number will be selected for each entity and stored in a column labeled *Rldh*. This random number will be used to find the interval of LDH increase that applies to each entity (Figure 4.27). For example, if the random number turns out to be 0.4, then for a female entity this indicates that the increase in LDH should be drawn from interval 2 (10%–15% increase), while for a male entity, this would be drawn from interval 3 (15%–20% increase). To find the appropriate interval for the first entity, the formula = MATCH(J2, IF(B2,fLDH, mLDH)) can be entered in column K of the *SoC entities* worksheet, assuming that cell J2 contains that entity's *Rldh* and

	A	B	...	I	J	K
1	Entity ID	Sex	...	Age (Years)	Rldh	LDH Interval
2	1	0	...	55.3	0.5637	4
3	2	1	...	58.5	0.0216	1
4	3	0	...	69.0	0.7065	4
5	4	1	...	63.0	0.2204	2
6	5	1	...	50.6	0.3380	2
7	6	0	...	72.5	0.3791	3
8	7	1	...	61.4	0.7238	4
9	8	0	...	74.0	0.0207	1
10	9	1	...	59.8	0.6265	3
11	10	0	...	63.2	0.2943	3

FIGURE 4.27
Finding the Interval of LDH Increase in the SoC Entities Worksheet. LDH, lactate dehydrogenase.

cell B2 contains that entity's indicator for sex (1 = female, 0 = male), with *fLDH* and *mLDH* containing the two cumulative distributions, for women and men, respectively.

To find the precise value for the increase in LDH, the interval must be interpolated according to the place of the random number *Rldh*. This is given by the distance between the random number and the cumulative percent of the distribution at the lower border. For example, for entity number 1, the distance between his random number and the lower border of the male cumulative LDH increase distribution is 0.0037 (0.5637–0.56). This result can be obtained using the formula = J2-INDEX(IF(B2,fLDH, mLDH),K2), with the INDEX function finding the lower border of the corresponding cumulative distribution.

For a linear interpolation of the interval, the slope of the increase in the interval is required, and to facilitate the calculation, this can be added to the LDH input table and named *sFem* and *sMal* (Table 4.9). The required slope is found using the formula = INDEX(IF(B2,sFem, sMal),K2) for the first entity. Multiplying this by the distance of the random number from the lower border of the cumulative distribution, obtained before, gives the precise point in the interval, and the increase to be applied to LDH can be derived by adding this to the lower border of the interval in a single formula = 5*K2 + (J2-INDEX(IF(B2,fLDH, mLDH),K2))*INDEX(IF(B2,sFem, sMal),K2) that is placed in a column labeled *ChgLDH*. The LDH value can

TABLE 4.9

Cumulative Distribution of Percent Change in LDH at Recurrence

Interval (% Increase)		Females			Males		
Lower Border	Upper Border	Prop	Slope	Cum	Prop	Slope	Cum
0	5	0.00		0.00	0.00		0.00
5	10	0.14	35.71	0.14	0.07	71.43	0.07
10	15	0.31	16.13	0.45	0.18	27.78	0.25
15	20	0.26	19.23	0.71	0.31	16.13	0.56
20	25	0.20	25.00	0.91	0.26	19.23	0.82
25	30	0.09	55.56	1.00	0.18	27.78	1.00

Note: Upper border value is included in next interval.
Cum, cumulative; Prop, proportion.

	A	B		I	J	K	L	M
1	Entity ID	Sex	...	Age (Years)	Rldh	LDH Interval	Chg LDH	LDH (Units/l)
2	1	0	...	55.3	0.5637	4	20.07	275
3	2	1	...	58.5	0.0216	1	5.77	301
4	3	0	...	69.0	0.7065	4	22.82	239
5	4	1	...	63.0	0.2204	2	11.30	357
6	5	1	...	50.6	0.3380	2	13.19	212
7	6	0	...	72.5	0.3791	3	17.08	245
8	7	1	...	61.4	0.7238	4	20.35	253
9	8	0	...	74.0	0.0207	1	6.48	223
10	9	1	...	59.8	0.6265	3	18.39	337
11	10	0	...	63.2	0.2943	3	15.71	360

FIGURE 4.28
Updating the Value of LDH in the SoC Entities Worksheet. LDH, lactate dehydrogenase; SoC, standard of care.

now be updated (Figure 4.28) by applying this percent increase to the original value, using the formula = (1 + L2/100)*D2.

Updating the LDH value for the entities that were treated with MetaMin takes the same approach but a decision needs to be made whether the change in LDH should be the same for the duplicate in each pair or whether this increase can diverge. Uncertain, the modelers consult the experts again and find that they are not entirely sure either. After some discussion, it is agreed that the treatment may alter the severity of the recurrence, and thus the change in LDH but considerable uncertainty about this remains, so the modelers make a note to include this aspect in their structural uncertainty analyses. Age is also updated for the MetaMin entities, and the worksheet now has all the information required for assigning the time of death (Figure 4.29).

In the specialized DES software, the process is considerably less laborious because Arena® offers a built-in function for interpolating continuous distributions specified by interval. Inside the *recurrence* submodel (since this update is part of the *consequences* of the cancer recur event), an *assign* module is inserted after the one that performed the accruals. This module can be named *update values*. In it, the entity's age is updated by adding *LY* to the stored value of *aAgeOrig*. LDH is updated by drawing a value for the increase from an array *iLDH* where the two cumulative distributions are stored, one for males and the other for females. This is done using the built-in function CONT(cp1,ub1, cp2,ub2 …), where cp is the cumulative probability for each interval and ub is the upper border of the interval. The number drawn is the proportional increase in LDH, so it is added to 1 before multiplying by the original *aLDH* (Figure 4.30).

	A	B		G	H	I	J	K
1	Entity ID	Sex	...	Age (Years)	Rldh	LDH Interval	Chg LDH	LDH (Units/l)
2	1	0	...	55.5	0.0601	1	9.30	250
3	2	1	...	58.9	0.7271	4	20.43	343
4	3	0	...	69.7	0.5608	4	20.02	234
5	4	1	...	65.1	0.8518	4	23.55	397
6	5	1	...	52.3	0.5671	3	17.25	219
7	6	0	...	72.9	0.9198	5	27.77	267
8	7	1	...	61.7	0.0696	1	7.49	226
9	8	0	...	75.3	0.6921	4	22.54	256
10	9	1	...	61.0	0.2709	2	12.11	320
11	10	0	...	64.1	0.3485	3	16.59	363

FIGURE 4.29
Updating the Value of LDH in the MetaMin Entities Worksheet. LDH, lactate dehydrogenase.

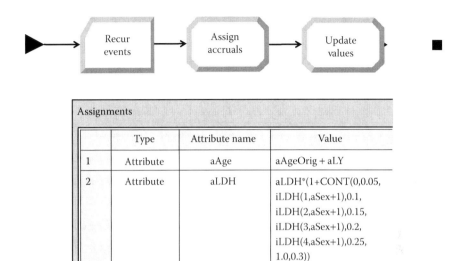

FIGURE 4.30

Inclusion of an *update* module in the recurrence submodel. Inside the recurrence submodel (indicated by the triangular entry point at left and square exit point at right), an *assign* module is added where age and LDH values are updated. For age, this involves simply adding the PFS time in years. For LDH, an increase is drawn for each entity from an array called *iLDH* where the cumulative distributions are stored. The array is indexed by interval, with column 1 for males and 2 for females. The CONT function starts with zero frequency to set the floor of the distribution at 5% (otherwise any value less than 0.1 could be selected). LDH, lactate dehydrogenase; PFS, progression-free survival.

4.3.6 Assigning the Time of Death

Once age and LDH values have been updated, the time for death can be selected for each entity. A similar process is followed for the death event as was done for recurrence. In the *SoC entities* worksheet, a column is added to compute the prognostic index for each entity, given the updated values of age and LDH with the formula = EXP(B2*BetaSexPPS + I2*BetaAgePPS + M2*BetaLDHPPS). The random number for PPS was already computed, so the next column implements the formula for the PPS itself = (-LN(H2)/(scalePPS*N2))^(1/shapePPS). With these additions, the worksheet now looks like Figure 4.31. Overall survival can be computed by adding PFS and PPS for each entity.

For entities receiving MetaMin, the PPS cannot be computed from the SoC entity's PPS using the handy factor that was employed in deriving PFS because their prognostic index is no longer the same, since the recurrence takes place at a different time (i.e., age is different) and the increase in LDH was selected using a separate random number. Thus, the full computation has to be implemented, following the same steps as for SoC, but using the hazard ratio for PPS to adjust the hazard of death with the formula = (-LN(F2))/(HRPPS*

	A	B		H	I		M	N	O
1	Entity ID	Sex	...	Rnd Mum PPS	Age (Years)	...	LDH (Units/l)	Prognostic Index PPS	PPS (m)
2	1	0	...	0.8924	55.3	...	275	349.3	1.55
3	2	1	...	0.5664	58.5	...	301	422.8	4.84
4	3	0	...	0.4827	69.0	...	239	982.4	3.00
5	4	1	...	0.5783	63.0	...	357	779.0	2.87
6	5	1	...	0.9444	50.6	...	212	150.1	1.75
7	6	0	...	0.4846	72.5	...	245	1359.7	2.30
8	7	1	...	0.9408	61.4	...	253	447.8	0.76
9	8	0	...	0.2758	74.0	...	223	1416.6	3.53
10	9	1	...	0.5872	59.8	...	337	546.1	3.74
11	10	0	...	0.8493	63.2	...	360	968.0	0.91

FIGURE 4.31
The SoC Entities Worksheet with the Calculation of PPS. m, month; PPS, post-progression survival.

(scalePPS*L2)))^(1/shapePPS). The random number is the same, however, as it was copied over when the attributes were initially filled in. With these additions, the *MetaMin entities* worksheet looks like Figure 4.32.

In specialized DES software, assigning the time of death to each entity follows the same process as for PFS. First, the three coefficients for the equation are stored in global variables *iBetaSexPPS*, *iBetaAgePPS*, and *iBetaLDHPPS*, as are the Weibull parameters *iScalePFS* and *iShapePFS*, and the effect of MetaMin in *aHRPPS* (Figure 4.33).

The survival after cancer recurrence is estimated and stored in each entity's *aPPS* attribute. Since this calculation takes place at the time of recurrence, the additional set of statements can be implemented in another *assign* module that is included in the *recurrence* submodel. As before, the calculation is disaggregated into its components and *aPIPPS* is computed using *aSex*, the updated *aAge* and *aLDH* values and the three coefficients stored in the global variables in the formula EP(aSex*iBetaSexPPS + aAge*iBetaAgePPS + aLDH*iBetaLDHPPS). Before proceeding to derive the value of *aPPS*, however, the hazard ratio appropriate for each entity's treatment must be determined and stored in *aHRPPS*. This could be done by separating the entities according to *aTmt* using a *decide* module in order to assign the appropriate hazard ratio before bringing the entities back together. A more concise way of accomplishing the same thing is to use a conditional statement: (aTmt == 2)

	A	B		F	G		K	L	M
1	Entity ID	Sex	...	Rnd Mum PPS	Age (Years)	...	LDH (Units/l)	Prognostic Index PPS	PPS (m)
2	1	0	...	0.8924	55.5	...	250	322.7	2.12
3	2	1	...	0.5664	58.9	...	343	517.7	5.30
4	3	0	...	0.4827	69.7	...	234	1022.0	3.74
5	4	1	...	0.5783	65.1	...	397	1095.9	2.81
6	5	1	...	0.9444	52.3	...	219	179.4	1.96
7	6	0	...	0.4846	72.9	...	267	1532.3	2.69
8	7	1	...	0.9408	61.7	...	226	414.0	1.05
9	8	0	...	0.2758	75.3	...	256	1807.7	3.74
10	9	1	...	0.5872	61.0	...	320	566.5	4.68
11	10	0	...	0.8493	64.1	...	363	1054.7	1.09

FIGURE 4.32
The MetaMin Entities Worksheet with the Calculation of PPS. m, month; PPS, post-progression survival.

Variable—Basic process							
	Name	Rows	Columns	Data type	Clear option	Initial values	Report statistics
11	iLDH	4	2	Real	System	8 rows	☐
12	iBetaSexPPS			Real	System	1 rows	☐
13	iBetaAgePPS			Real	System	1 rows	☐
14	iBetaLDHPPS			Real	System	1 rows	☐
15	iScalePPS			Real	System	1 rows	☐
16	iShapePPS			Real	System	1 rows	☐
17	iHRPPS			Real	System	1 rows	☐

Double-click here to add a new row.

FIGURE 4.33
Additional values added to the variable table. The coefficients for *Sex*, *Age*, and *LDH* are stored globally as are the scale and shape parameters and the hazard ratio for MetaMin, all corresponding to PPS. The figure also shows the specification for the array containing *iLDH*, the distribution of increases in LDH at recurrence, used earlier. LDH, lactate dehydrogenase; PPS, post-progression survival.

*iHRPPS + (aTmt == 1). Arena® evaluates the equivalences (with two equal signs in the parentheses) as true or false statements, assigning the value 1 to true and 0 to false. When an entity that has received MetaMin passes through here, the first equivalence is true and the value stored in *iHRPPS* will be assigned as the hazard ratio, since the second equivalence evaluates as false and, thus, adds nothing. Conversely, when an entity on standard care comes through, the first equivalence is false, so that term becomes 0 and the second one is true, so the value 1 is assigned as the hazard ratio (i.e., no effect).

Using the formula (-LN(aRndNumPPS)/(aHRPPS*iScalePPS*aPIPPS))**(1/iShapePPS), the value of PPS is obtained for each entity (Figure 4.34). Of note, the random number is common to the duplicates as it was stored in *aRndNumPPS* before duplication.

4.3.7 Accruals at the Second Event

After recurrence has been processed and the PPS time has been computed, the values pertaining to this time can be accrued. All three dimensions (the additional cost, the PPS time in life years or LY, and its quality-adjusted contribution to the QALY) have to be accrued, along with the discounted estimates of these values. The appropriate amounts can be added in the worksheet *Accruals* (previously displayed in Figure 4.22). The formula in the *LY* column is modified to add the PPS for the first entity = ('SoC

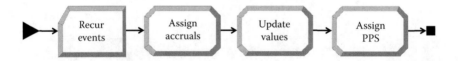

Assignments			
	Type	Attribute name	Value
1	Attribute	aPIPPS	EP(aSex*iBetaSexPPS+aAge*iBetaAgePPS+aLDH*iBeta LDHPPS)
2	Attribute	aHRPPS	(aTmt == 2)*iHRPPS + (aTmt == 1)
3	Attribute	aPPS	(−LN(aRndNumPPS)/(aHRPPS*iScalePPS*aPIPPS))**(1/iShapePPS)

FIGURE 4.34
Completed recurrence event submodel. The final module of the submodel assigns the time until death (PPS) to each entity. This is done by first computing the prognostic index *aPIPPS* using each entity's updated values of age and LDH. Then, the hazard ratio corresponding to the entity's treatment is stored in *aHRPPS*, with the value 1 given to entities treated with SoC and *iHRPPS* assigned to those receiving MetaMin. This hazard ratio is then used in the Weibull formula for PPS. LDH, lactate dehydrogenase; PPS, post-progression survival; SoC, standard of care.

entities'!G2 + 'SoC entities'!O2)/12, before dividing by 12 since it was also estimated in months, and this is copied down for the others. For the QALY accrual, the formula needs to be modified because the PPS added to LY does not have the same QoL adjustment as for time in remission—each interval has its own QoL. The formula becomes = ('SoC entities'!G2*Qrem + 'SoC entities'!O2*Qpro)/12 for the first entity, and this is also copied down for the others. For the cost accrual, no change is required in the formula since treatment is continued until death. Thus, B3*CostSoC for the first entity is correct. The formulas for discounting, if it is to be applied, need to take into account that the values should be discounted not just over the PPS interval, but back to the starting point as well. The worksheet now looks like Figure 4.35.

For the patients receiving MetaMin, the formulas are updated in the same way, except, of course, using the PPS estimates for MetaMin (Figure 4.36).

In the specialized software, the PPS time must pass before processing the second event (Figure 4.37). This is taken care of by introducing a second *delay* module named *PPS* and setting the delay time to each entity's PPS stored in *aPPS* (suitably converted to days). After the entity lives through this time, it enters a new submodel that will process the death event. As the required modules will be quite similar to those in the *Recurrence* submodel, that entire submodel can be copied and the copy is linked to the new *delay* module. Since Arena® does not allow duplicate module names, these must be changed to unique values. The copied submodel is renamed *Death*, the *record* module is renamed *Death events,* and the *assign* module is renamed *Accruals PPS* (Figure 4.37). The other two modules can be deleted as they are not necessary given that this will be the final event for the entities.

In the *record* module's form, the counter set name is set to *Deaths,* and the items in that counter set need to be named as was done before for

	A	B	C	D	E	F	G
1		Standard of care					
2	Entity ID	LY (y)	QALY (y)	Cost ($)	dLY (y)	dQALY (y)	dCost ($)
3	1	0.5	0.3	501	0.45	0.35	498
4	2	0.9	0.7	1,026	0.92	0.68	1,011
5	3	1.3	1.0	1,379	1.23	0.97	1,353
6	4	3.3	2.7	3,594	3.12	2.54	3,423

FIGURE 4.35
Updated Values for Standard of Care in Worksheet Accruals. y, year.

		H	I	J	K	L	M
1		MetaMin						
2	...	LY (y)	QALY (y)	Cost ($)	dLY (y)	dQALY (y)	dCost ($)
3	...	0.7	0.55	6,265	0.7	0.55	6,199
4	...	1.3	1.01	11,732	1.3	0.99	11,499
5	...	2.0	1.60	17,594	1.9	1.55	17,075
6	...	5.3	4.39	46,866	4.9	4.06	43,301
7	...	4.5	3.70	39,459	4.2	3.46	36,911
8	...	1.1	0.86	9,564	1.1	0.84	9,409
9	...	0.8	0.67	7,268	0.8	0.66	7,178
10	...	3.6	2.93	31,657	3.4	2.78	30,003
11	...	3.4	2.74	29,882	3.2	2.61	28,405
12	...	2.1	1.76	18,843	2.1	1.71	18,248

FIGURE 4.36
Updated Values for MetaMin in Worksheet Accruals. y, year.

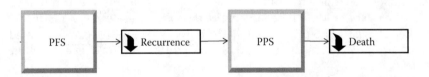

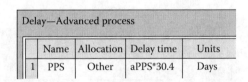

Delay—Advanced process			
	Name	Allocation	Delay time	Units
1	PPS	Other	aPPS*30.4	Days

FIGURE 4.37
Introduction of a delay for PPS and a submodel to process the death event. The *delay* module holds each entity until the PPS time (converted to days) has passed and then allows the entity to pass into the *death* submodel. The allocation to *Other* is not used in this model. post-progression survival.

the recurrence set. No other change is needed. In the *assign* module, the accrual formulas need only changing PFS to PPS, as they already add the new amounts to the previous accumulation. The *assign* module can then be linked to the square exit point and the submodel is complete (Figure 4.38).

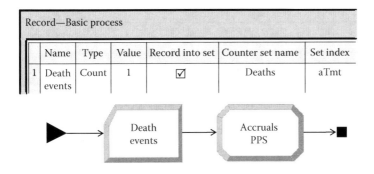

FIGURE 4.38

Submodel to process deaths. The death events are recorded upon entry of the entities into this submodel. This involves setting up another set to store the counts separately according to treatment. Next, the PPS is added to the various accruals. If discounting is implemented, then the appropriate discounted accruals would be included as well, taking care to redefine *aTime1* and *aTime2*. Then, the entity exits the submodel. PPS, post-progression survival.

4.4 Obtaining Results

In the worksheet *Results*, formulas are added to count the deaths as was done for the recurrence events. Then, the three dimensions of value, and their discounted amounts, are tallied. In HTA, the mean across entities is taken, so the function AVERAGE is applied to the appropriate column in the worksheet *Accruals*. The incremental cost-effectiveness ratios can then be calculated using the standard formula of increase in costs divided by gain in QALYs. The *Results* worksheet now looks like Table 4.10.

In the specialized DES software, the entities exiting the *Death* submodel will have ended their simulation. This is indicated in Arena® by adding a *dispose* module whose only function is to remove the entities from the simulation. The complete DES structure in Arena® looks like Figure 4.39.

The counts of recurrence events and deaths will be reported by Arena® automatically once the model completes an execution. The remaining results can be obtained in several ways. One is to add a series of *record*

TABLE 4.10

Results

	SoC	MetaMin
Recurrence	10	10
Deaths	10	10
LY	1.55	2.50
QALY	1.24	2.02
Costs	$1,702	$21,913
ICER		25,868
dLY	1.51	2.38
dQALY	1.21	1.92
dCosts	$1,650	$20,823
dICER		26,752

ICER, incremental cost-effectiveness ratio; LY, life year; QALY, quality-adjusted life year; SoC, standard of care.

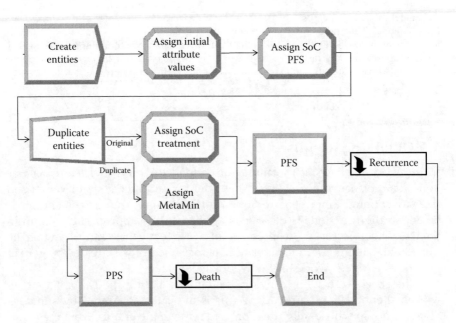

FIGURE 4.39
Completed structure of cancer DES. The linked modules show the full structure of the DES for the cancer example. The two submodels indicated by the black downward arrows contain additional modules that process the two events: cancer recurrence and death. DES, discrete event simulation.

	Name	Rows	Columns	Data type	Clear option	Initial values	Report statistics
13	vLY	2		Real	System	2 rows	☐
14	vQALY	2		Real	System	2 rows	☐
15	vCost	2		Real	System	2 rows	☐
16	vdLY	2		Real	System	2 rows	☐
17	vdQALY	2		Real	System	2 rows	☐
18	vdCost	2		Real	System	2 rows	☐

Variable—Basic process

Double-click here to add a new row.

FIGURE 4.40
Addition of variables to track overall results for SoC. The six global variables used to accrue overall results for each dimension and its discounted values are displayed. Each one has two cells to accommodate the results for SoC and MetaMin. An alternative would be to keep each one to one row, add six more variables to store the results for MetaMin, and check the box Report statistics for all. SoC, standard of care.

modules between the *Death* submodel and the *end* module. Each of these can register only one item so there would need to be six of them to capture LY, QALYs, and costs plus the three discounted amounts. These recorded tallies would be reported by Arena® at the end of a model run. Rather than adding six *record* modules, the totals can be accrued in six global variables (Figure 4.40). The statements to do this can be added to the *Accruals PPS* module, but these results will not be automatically reported by Arena®. To ensure that they are, the Report statistics box in the *variable* table needs to be checked for each one (and there would need to be separate variables for each of SoC and MetaMin).

The easiest way to obtain the necessary results, however, is to take advantage of the software's *Statistics* function, where it can be specified which quantities are to be tracked and reported (Figure 4.41). The two-level variables are added as *output* to the *Statistics*, where it is also possible to add calculated output like the ICER.

	Name	Type	Expression	Report label	Output file
Statistic—Advanced process					
1	Statistic 1	Output	vLY(1)	LY SoC	
2	Statistic 2	Output	vQALY(1)	QALY SoC	
3	Statistic 3	Output	vCost(1)	Cost SoC	
4	Statistic 4	Output	vdLY(1)	Disc LY SoC	
5	Statistic 5	Output	vdQALY(1)	Disc QALY SoC	
6	Statistic 6	Output	vdCost(1)	Disc Cost SoC	
7	Statistic 7	Output	vLY(2)	LY MM	
8	Statistic 8	Output	vQALY(2)	QALY MM	
9	Statistic 9	Output	vCost(2)	Cost MM	
10	Statistic 10	Output	vdLY(2)	Disc LY MM	
11	Statistic 11	Output	vdQALY(2)	Disc QALY MM	
12	Statistic 12	Output	vdCost(2)	Disc Cost MM	
13	Statistic 13	Output	(vCost(2)-vCost(1))/(vQALY(2)-vQALY(1))	ICER	

Double-click here to add a new row.

FIGURE 4.41
Values are to be written out to the default Arena® report. Each statistic needs to be provided with a report label to identify it in the report. The calculation of the ICER has also been added so that it is reported as well. ICER, incremental cost-effectiveness ratio.

5

Analyses

An analysis using a discrete event simulation (DES) involves several steps. First, all the input values are read in and stored in their appropriate locations (e.g., global variables). Next, the entities are created and they are introduced into the simulation where they experience the events specified in the model logic, accruing consequences such as costs, survival time, and quality-adjusted life years (QALYs). This happens either for the specified time horizon or for a fixed number of individuals (i.e., until the last entity dies or otherwise leaves the model). The outcomes of interest are recorded and all the results required to inform the health technology assessment (HTA) are transferred to a suitable medium (e.g., a spreadsheet), if necessary. There may be some additional processing of that information required to produce the final results in the form desired by the decision makers (e.g., an incremental cost-effectiveness ratio, ICER) (Rutter et al. 2011).

As a DES incorporates chance at many junctures (e.g., selecting from distributions, applying risk equations, and making decisions), there is stochastic uncertainty, and this must be dealt with in the analyses. In addition, many of the input values are subject to uncertainty because they are estimated from limited data, and this parameter uncertainty must also be considered at the analytic stage. During the design and construction of a typical DES, the modeler also makes many assumptions and decisions about the applicable logic. Since other choices could also be justifiable, there is always structural uncertainty to take into account. Finally, to address the heterogeneity in the population and variations of interest in other inputs (e.g., the time horizon), the modeler should analyze a range of scenarios.

The individual steps that need to be taken in a DES analysis have already been described in Chapter 3. In this chapter, the focus is on the overall process, particularly the additional processing required for an HTA. First, the base case is defined and the analyses required to produce it are detailed. The handling of stochastic parameter and structural uncertainty is described. Finally, the assessment of sensitivity to changes in input values that reflect heterogeneity and other scenario features is addressed.

5.1 Base Case

For HTA, the analysis of a DES typically involves generating a set of relevant outputs (e.g., costs, rate of clinical events, survival, and QALYs) for each assessed technology. These outputs are generally derived by using what are thought to be the most relevant estimates of the true values of all the inputs. This most-relevant analysis is known in HTA as the *base case*.

To produce the base case, the DES collects information on the experience of each entity and applies valuations to that experience (e.g., the costs and the quality-of-life adjustments). Although the experience is individual, the interest, ultimately, is in the aggregate values across the population (e.g., total cost and total QALYs lived). These aggregate values can be obtained by accumulating the pertinent quantities in global variables, separately by technology. At the end of the simulation, these aggregated values are reported and used to compute the differences between interventions, and these are used to derive additional measures like the ICER. Alternatively, the DES may accumulate the values at the individual level in attributes and carry out the aggregation separately at the end of the simulation, or even export the data for aggregation separately in other software. Although this consumes more computer memory and time, it provides for a fuller description of the results as measures of variation can also be derived (e.g., upper and lower quartiles, and standard deviation) as they would be for any dataset containing individual values.

Either way, the specific approach to carrying out the analyses depends on the software used. In a spreadsheet, all of the calculations can be implemented directly using the built-in functions that are typically available in such programs. If a general programming language is used, then subroutines must be developed to carry out the necessary calculations, and these are called when the simulation has concluded. In DES-specific software, there are often built-in functions that can track these outcomes and produce the required statistics. Many analysts, however, prefer to output the results of the simulation to a spreadsheet and carry out the final calculations there. Regardless of the approach taken, it is important to turn off all animation and other calculation-intensive processes not needed for the runs to ensure that the computations are executed as fast as possible.

5.1.1 Dealing with Chance (Stochastic Uncertainty)

Although the inputs in a simulation are common to all the entities, the results will vary across individuals because the pathways taken through the model are determined by the sequences of random numbers sampled for each entity during the simulation. For example, the time of death for one individual is assigned as 6 months into the simulation because the random number drawn for that individual sampled the twentieth percentile of the time-to-death

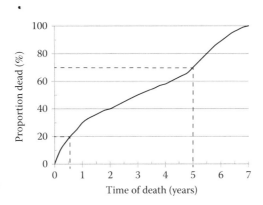

FIGURE 5.1

Example of identical individuals having very different experience in a simulation. The figure shows the cumulative fatality curve for all individuals with the same characteristics and treatment in a simulation. One individual dies at 6 months into the simulation because he was assigned a random number of 0.2 for this event, while another one dies at 5 years was given a random number assignment of 0.7.

cumulative distribution; the sampling for the next individual with identical characteristics yields the seventieth percentile of the distribution, and, therefore, that entity is assigned a time of death of 5 years (Figure 5.1). Across separate analyses of the model, these two individuals who were identical at the start will sample different percentiles of the relevant cumulative distributions each time and, thus, be assigned different times to those events. Hence, even if the analysis is repeated without varying the inputs, the results may still vary since the random numbers used will be different each time (unless steps are taken to force the use of the same set of random numbers) and the number of entities that are simulated is finite. This type of variability due to the play of chance is known as *stochastic uncertainty*.

The set of random numbers used in a particular analysis defines one *replication* of the model (also known as one *trial*). This situation is analogous to that of a randomized controlled trial. Even if the clinical study is repeated with an identical protocol, the second study will produce different results. It is another *replication*—another trial—trying to estimate the same effects. To reduce stochastic uncertainty, a modeler may carry out many replications. For each one, the model is analyzed using the same set of input values, but with different random number sequences for each replication. This is typically accomplished by using different sets for the random number generators in each replication. This leads to the sampling of different values from the input parameter distributions and, consequently, variation in the pathways taken through the model across the replications, despite the use of the same input values. The outputs are reset to zero between replications so that each replication provides a fresh collection of results. By averaging the results across the series of replications (akin to what is done in meta-analysis across a group of randomized controlled trials), the effect of stochastic variation (i.e., the risk of basing a decision on the results of a single, potentially unrepresentative replication) is reduced. An analysis like this, based on a given set of input values, is considered a *model run*. Since stochastic uncertainty cannot be completely eliminated, no matter how large the model run, it is important to quantify it by expressing the degree of variation observed (Briggs et al. 2012), much as is done when analyzing clinical trials.

In DES that represent the operation of physical facilities, such as an outpatient clinic or emergency department, relevant outputs might include estimates of mean waiting times and measures of the distribution of waiting times to inform the likelihood of patients incurring very long delays (Stahl et al. 2004). The runs for such models are generally defined with respect to the time frame of interest (e.g., one day). In order to remain realistic, each replication is then restricted to the number of individuals presenting at the facility within that time frame. Thus, generating stable estimates of the outputs requires undertaking multiple replications within each run.

By contrast, HTA models that are not addressing specific physical facilities are not required to reflect realistic numbers of patients, nor are they restricted to finite time periods over which individuals can enter the model, and thus, a model run can include as many individuals as are required to achieve stable outputs. Instead of multiple replications, the modeler can increase the number of entities until the results stabilize sufficiently. In the context of clinical trials, a similar decision looms: carry out a single very large trial, or engage in a series of smaller trials and aggregate the results across the series afterward. In the real-life situation of clinical trials, the decision is often driven by regulatory (e.g., a minimum of two separate trials required) and practical considerations. Given there are no limits to the number of entities that can be included in a replication (other than whatever is imposed by the hardware available), a DES model run can comprise a single large model replication.

As, due to the stochastic nature of DES, multiple model runs using the same input values will produce different model outputs, it is desirable to ensure that multiple model runs using the same input values generate reasonably stable output values for the parameters of interest to the decision maker (e.g., mean total cost by intervention and QALYs obtained with each intervention). There is no consensus on the definition of stable model outputs. Stability is observed when the difference in results produced by an additional model replication using the same input values but alternative random number sequences is considered insignificant (Figure 5.2). What constitutes an insignificant difference may vary according to the decision problem being addressed and the analyst making the assessment. A rule of thumb used by many is that differences of less than 1% across model runs are insignificant. Since HTAs are carried out to inform decisions, the threshold for decision making may be used as a guide to the degree of stability required. If the variation does not alter the decision, then the results can be considered sufficiently precise. Although the base case analyses may be

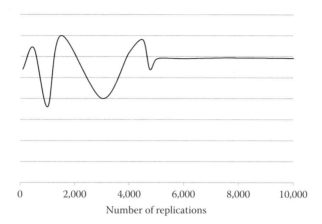

FIGURE 5.2
Control chart used to select the number of replications. This simulation was run for a varying number of replications, and it was found that variation in the value of an outcome of interest became minimal after about 5000 replications.

driven to a high stability (i.e., a low threshold of change), this tolerance may be relaxed when conducting the full uncertainty analysis in order to stay within feasible running times while still reflecting uncertainty.

An alternative approach to attaining sufficiently stable results is to determine the standard error of the model outputs that is considered tolerable and use this to estimate the number of entities to include in a model run. For example, in a published analysis of osteoporosis treatments using a DES (O'Hagan et al. 2007), a mean incremental net benefit of £1,308 was estimated in one model run using the base case input values, with an individual-level variance of about 2.4×10^9. A run size of 15,000 individuals across model runs produced a standard error of 400 (the square root of $2.4 \times 10^9/15,000$), and this provided sufficient confidence that the true mean incremental net benefit was positive.

5.1.2 Reflecting Uncertainty in Parameter Inputs

In addition to controlling for stochastic variation that arises from the random sampling of the pathways taken through the model for each entity, it is generally important to represent the effects of uncertainty around the parameter values used as inputs to the model. This is known as parameter uncertainty (parameter here is used as a synonym for estimated input). This kind of uncertainty arises because the true value of each model input parameter is unknown and the value used as an input is only an estimate. For example, the effects of a new technology may be estimated from a clinical trial. Those estimates, regardless of how large and well designed the trial may be, are uncertain, and this is typically reflected in a 95% confidence interval or other such statistical measure. In a model using that clinical trial result as an input, it would be important to incorporate the parameter uncertainty, generally based on the reported 95% confidence interval. In some cases, the input value may be known to be uncertain, but there may not be sufficient empirical data to quantify that uncertainty adequately. This poses a quandary for the analyst as ignoring variability in this parameter estimate is tantamount to assuming it is known with certainty. Applying an arbitrary range (e.g., ±50%)—although often done—is not an appropriate way to address this uncertainty, but such analyses can be useful to determine if the model outputs are sensitive to variation in these input parameters. If the outputs are judged to be sensitive to those changes, then additional effort should be expended to identify or collect relevant data to better inform the uncertainty. If there is insufficient time or funding to do this, then it may be necessary to elicit ranges for such parameters from experts in the clinical area.

Not all input values are uncertain, however. For example, the price of the technology, the recommended dose of a medication, and the time horizon are known with certitude, and, thus, there is no parameter uncertainty to consider (though there may be interest in how *sensitive* the model results are to alternative values for these inputs—see Section 5.2). Thus, for these kinds of inputs, it is inappropriate to consider variation in their values as part of parameter uncertainty.

The representation of parameter uncertainty is usually an important input to decision making, and so it is not appropriate to report only the results of the model run defined as the base case. Reporting the uncertainty around the base case results (mistakenly, but commonly, known as *sensitivity analysis*) is an essential component of all HTA analyses. Analyses of parameter uncertainty describe the variability among model runs, where the input values are allowed to change to take into account their uncertainty. These analyses can be either deterministic or probabilistic.

5.1.2.1 Deterministic Uncertainty Analyses

Deterministic uncertainty analysis involves the use of particular sets of input values other than those implemented in the base case analysis. These substitute values are chosen so that they are consistent with the estimated variance around the base case (mean) input parameter values. This type of analysis is called *deterministic* because the analyst determines which other specific sets of values to use in exploring the uncertainty. These alternative sets may be assembled by either changing selected values one at a time (i.e., *univariate* analyses) or creating new combinations of values (i.e., *multivariate* analyses). Typically, the alternate values are chosen so that they reflect the underlying range of uncertainty. For example, if a 95% confidence interval has been calculated for an input value, then the DES may be run using the lower border of the interval and again using the upper border. This provides a range of results that addresses uncertainty (but, important to note, the outputs do not

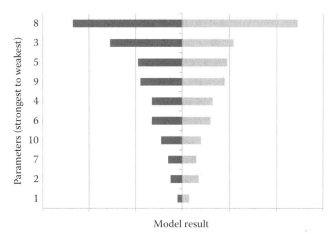

FIGURE 5.3
Display of a deterministic uncertainty analysis. This is a typical *tornado* diagram displaying the variation in an important model result with changes in the values of 10 input parameters. The base case result is at the center, and the bars give the extent of the change in results across the range of the input parameter values. Uncertainty in parameter #8 has the strongest effect on the results, with parameter #3 closely behind. Uncertainty in other parameters has much less impact.

constitute a 95% confidence interval). Results of the deterministic uncertainty analyses are often displayed using a tornado diagram (Figure 5.3).

Implementing this in a DES is very straightforward. The substitute inputs sets are entered in the simulation one at a time, and model runs are undertaken for each one. Depending on the number of inputs to be assessed, around 10 to 20 separate deterministic analyses are typically carried out for an HTA. As model running times tend not to be a major concern for such a limited set of analyses, each run should include the same number of individuals as in the base case. Of course, stochastic uncertainty is still an issue, but reporting of the results of the deterministic analyses focuses on the parameter uncertainty.

5.1.2.2 Probabilistic Uncertainty Analyses

Deterministic uncertainty analysis provides an indication of the parameters to which the base case results are most sensitive but does not provide an estimate of the overall uncertainty in the estimates due to all the components that have variance. To generate confidence intervals around ICERs or other measures (such as net monetary benefit), a probabilistic approach to sampling the sets of alternative input values is often used. This type of analysis, although about uncertainty, is most commonly known as a *probabilistic sensitivity analysis*, or by its acronym PSA (Baio and Dawid 2011; Claxton et al. 2005; Doubilet et al. 1984). In a PSA, the sets of alternate input values are chosen by the model rather than the analyst. This involves sampling input values from distributions that describe the uncertainty around the true value of each input. Each sampled set of input values is run through the model, just as in the deterministic analyses. The results of each run are stored, all the outputs are reset to zero, and another set of input values is sampled and run. This produces a series of results, which the analyst hopes adequately reflects the underlying parameter uncertainty (Figure 5.4).

Many (often 1,000 or more) sets of inputs are generated and replications are run for each one. To facilitate the process, it is usually automated. Once this is completed, interpretation of the large number of results can be assisted by ordering the values according to a decision-relevant outcome (typically the ICER) and then deriving the cumulative distribution of these results (Figure 5.5). This is commonly known as a *cost-effectiveness acceptability curve* (CEAC) and is interpreted as providing the probability that the technology meets any given threshold criterion for cost-effectiveness (Fenwick et al. 2001; Groot Koerkamp et al. 2007).

A probabilistic uncertainty analysis is readily done in a DES. The analyst begins by specifying the distribution (see Chapter 3) for each input that has parameter uncertainty. Inputs whose values are known with certitude, such as

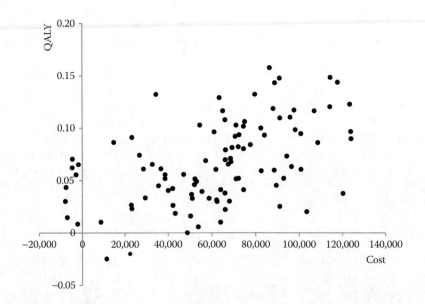

FIGURE 5.4
Scatter plot of results from a probabilistic uncertainty analysis. In this analysis, 100 replications were run, comparing two technologies, allowing the parameter inputs to change according to their uncertainty. The results of each replication, in terms of the net costs and QALYs gained, are plotted, showing the scatter produced by the uncertainty. Some results even reverse direction in the sense that they provide either cost savings or QALY losses. QALY, quality-adjusted life year.

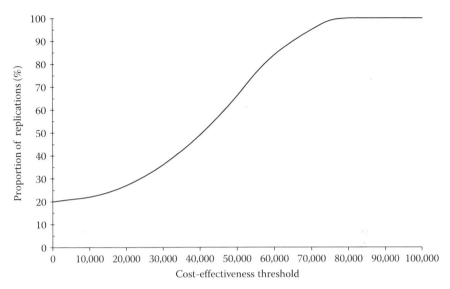

FIGURE 5.5
Cumulative distribution of cost-effectiveness ratios. A probabilistic uncertainty analysis was run and the results of the replications were ordered according to the ICER, from lowest to highest. The cumulative frequency was then computed and is displayed here. It indicates, for example, that two-thirds of the replications yielded a ratio below 50,000 and 75% were below 55,000. ICER, incremental cost-effectiveness ratio.

the discount rate, are not varied in this analysis, even if the analyst is interested in how sensitive the results are to variations in their values (see Section 5.2). There are two ways to implement sampling of the distributions that reflect parameter uncertainty. One is to create an event to do the sampling and store the selected input values in the appropriate global information places. This event has to happen immediately at the start of the simulation so that input values are available as needed. When using specialized DES software, this approach works well. It can also be implemented as an initial subroutine in a general programming language. In a spreadsheet, the values can be sampled directly into the cells that store them for the simulation. This process can be facilitated by using an add-in that has built-in distributions and tools for handling them. See, for example, an excellent recent review by Vose (2014). An alternative approach is to conduct the sampling externally to the simulation and supply a new set of inputs for each model run. Depending on the software used, special programming may be required to automate the process. Either way, the outputs are collected across all the replications and then they are analyzed to generate the required format for the presentation of the uncertainty results. A related form of analysis concerns the value of collecting additional data to reduce uncertainty (value of information analyses) (Claxton and Sculpher 2006), and this type of analysis is implemented in a DES in the same way as is PSA.

Given that many additional model runs are required for a PSA, model running times can increase significantly. This can be eased somewhat, as noted above, by reducing the threshold for achieving stability in the outputs. An alternative approach is to use analysis of variance to inform the combination of model runs and the number of entities per replication that are required to achieve defined levels of accuracy with respect to the mean of the estimate of interest and the variance estimates around that mean (O'Hagan et al. 2005). Alternatively, if a fixed time period for the analysis of a model is available (e.g., 100 hours), the analysis of variance method can inform the most appropriate combination of model runs and the number of entities per replication.

5.1.3 Structural Uncertainty Analyses

Structural uncertainty arises from the many decisions and assumptions that are made during the design and implementation of the model (Haji Ali Afzali and Karnon 2015). If alternative assumptions are made, then the results will likely change, and it is important for the decision maker to understand the degree and direction of those deviations. This structural uncertainty is taken into account by restructuring the model varying the assumptions made for the base case. Each alternative structure reflects one set of assumptions. As there are numerous aspects about which structural assumptions are made, and there may be many reasonable variations for each aspect, the volume of structural alternatives can be considerable. Needless to say, each restructured version of the model is still subject to both parameter and stochastic uncertainty, and the need to address those factors still applies. Thus, full accounting for structural uncertainty is quite challenging. It is not too surprising, then, that uncertainty around the structure of an HTA model is often overlooked by analysts! This should be of considerable concern to the HTA community as it has been shown that structural uncertainty tends to be very significant (Jackson et al. 2011).

DES facilitates the implementation of more complex model structures, which can represent disease processes and interventions more accurately, but also generates more opportunity for structural uncertainty. The more components a model has, the greater the number of alternative structures that may be considered reasonable possibilities. Fortunately, the DES framework also makes it easier to address structural uncertainty, particularly when using some of the specialized software packages.

The process for assessing structural uncertainty begins with careful documentation of the design, the assumptions made and implementation decisions taken, their basis, how they relate to each other, and any credible alternatives. During the implementation, provision can be made in the model structure for the more likely of these alternatives. For example, there may be some doubt about the form of one of the controlling equations—perhaps both a Gamma and a Weibull distribution fit the observed data equally well,

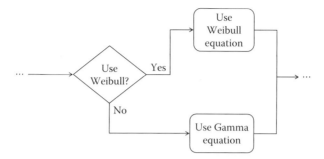

FIGURE 5.6
Implementation of a branching structure to enable structural uncertainty analyses. Here, the modeler sets a flag to indicate whether the Weibull equation should be used instead of a Gamma. Results of runs with the flag set one way can be compared to those with it set the other way to determine if the choice of distribution makes a difference in the information provided to the HTA.

and there is no *a priori* knowledge to help the selection. Having noted this, the modeler can then implement both forms of the equation and define a flag that tells the simulation which equation to use in any given model run (Figure 5.6). During execution, the flag is first set one way for one run and then reset the other way for another run. The results are then compared to determine the significance of the uncertainty around that particular structural factor. If it turns out that it matters (e.g., it changes the direction of the HTA decision), then at a minimum, the decision maker must be given this information. Ideally, further work is done to determine which structure makes more sense, perhaps by finding another dataset against which to validate the predictions, or even by collecting additional data.

Sometimes, alternative design decisions may involve different sets of events and connections. For example, one possibility in modeling the handling of an acute myocardial infarction may be to assume that everyone presenting to the emergency department is sent to the cardiac catheterization lab for angiography. An alternative might consider that some individuals are sent to the coronary care unit first. Ideally, this would be investigated further, but if that is impossible, then a structural uncertainty analysis should be carried out. There could easily be many more components affected by these choices. Consideration of the resulting structural uncertainty is facilitated by incorporating all the reasonable alternatives into the model framework and controlling which one is operative in any one model run by opening or closing one or more *gates*. Thus, in one run the gate allowing the *transfer to coronary care unit* logic is closed and in another it is opened (Figure 5.7).

The results from each model structure tested may be presented to the decision maker as alternative cases, analogous to the reporting of deterministic uncertainty analyses. This does not provide any information on the likelihood that each case reflects reality and may simply overwhelm the decision

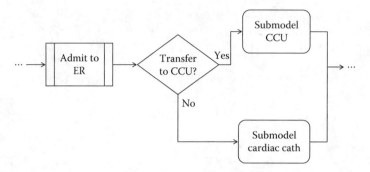

FIGURE 5.7
Use of a gate to implement very different structures reflecting diverse practices. Depending on how the gate is set, patients presenting to the emergency room with chest pain may be sent to the coronary care unit or to the cardiac catheterization suite for an angiogram. cath, catheterization; CCU, coronary care unit; ER, emergency room.

maker with figures that are difficult to interpret. Another possibility is to parameterize the structural uncertainty by specifying probability distributions for the flags and gate openings. For example, a structural alternative that is judged possible but very unlikely might have a controlling gate with an opening probability of 1%. Thus, 99% of the runs would not consider that structure but 1% would. For structural options where this can be done, the structural uncertainty is thus converted into parameter uncertainty and can now be analyzed using the deterministic, and even probabilistic, techniques described in Section 5.1.2. Although this approach of parameterizing structural uncertainty is conceptually simple, there are substantial practical hurdles. The likelihood of each option must be assessed quantitatively and the number of possible combinations may grow rapidly, with the likelihoods possibly correlated. These problems are not specific to DES, but DES makes it easier to address them by facilitating the incorporation of structural options, their controlling gates, and presenting the consequent alternative results.

In some cases, it may be too difficult to parameterize alternative structural assumptions, but it may be possible to compare the validity of alternative model structures by comparing the outputs of model to observed data for those outputs (see Chapter 8 for more details on validation). The simplest approach is to select the model that achieves the best validation results, though such an approach does not represent the effects of structural uncertainty. A model averaging approach generates probability weights that represent the accuracy with which each model predicts the observed data. The probability weights comprise an empirical distribution of the probabilities that each of the tested model structures is the most valid model structure. Combined with the probability distributions specified for each input parameter, parameter and structural uncertainty can be represented jointly (Claeskens and Hjort 2008).

To implement the joint modeling of parameter and structural uncertainty, a model structure may be sampled at the start of each run, and the relevant gates are opened and closed to implement the sampled model structure. Next, a set of input parameter values is sampled, and the model run completed and results stored, and the process repeated for the necessary number of model runs. In some cases, the analyst may decide to implement two or more separate models to represent the alternatively specified model structures. Here, values are sampled from the structural probability weights distribution to match the aggregate number of model runs to be generated. Each of the separate models is run according to the number of times each model was sampled, with alternative sets of input parameter values sampled for each model run. The outputs from each model run across the multiple models are then combined and analyzed to jointly represent parameter and structural uncertainty.

5.2 Exploring Sensitivity to Input Values

While uncertainty refers to imprecision in the outputs resulting from imprecision in the input values and doubts about the design and implementation decisions, *sensitivity* refers to the model's responsiveness to changes in the input values. For example, the price of each of the technologies modeled may be known exactly (i.e., there is no uncertainty), but it may still be of interest to assess the degree of change in the outputs if there were to be a particular adjustment in price (Figure 5.8). In this case, the model runs will show how sensitive the results are to changes in the prices, but price variation does not introduce additional uncertainty.

HTAs are fundamentally concerned with the expected performance of a technology in the HTA's jurisdiction. In that local context, additional specificity is required to address the characteristics of the people for whom the intervention is deemed appropriate. Indeed, the model may be used to help

Price	Result
10	23,250
20	26,500
30	28,750
40	31,000

FIGURE 5.8
Sensitivity analysis looking at the impact of changing prices. In this analysis, the modeler successively changed the price of some input and obtained the output for a result of interest. The amount of variation in the result reflects how sensitive this is to variation in this price.

determine in which people the intervention is most cost-effective. Thus, one major sensitivity concern has to do with the population modeled: do the results change substantially if a different *subgroup* is considered?

5.2.1 Subgroups

For many HTAs, it is feasible to make separate decisions regarding the provision of the evaluated technology for different subgroups. In such cases, decision makers may be interested not only in the value of the intervention for the whole population but also in the degree to which an intervention is cost-effective in different subgroups of the population.

Analyses to address subgroup differences can be undertaken as if they were separate problems. Thus, a full set of base case, parameter, and structural uncertainty analyses is undertaken for each subgroup. Fortunately, when using a DES, it is not necessary to actually run the analyses separately. Instead, the analyst identifies the subgroups of interest (e.g., by specifying ranges of age, gender composition, disease severity, and extent of prior treatment). An additional attribute is defined to reflect which subgroup an entity belongs to, or if it is possible to be in several subgroups, then an attribute is specified for each one. During the simulation, each entity's membership in one or more subgroups is stored in the respective attributes set up for that purpose. As each replication proceeds, the outputs are stored in indexed global variables corresponding to the subgroups. At the end of the model run, the outputs are processed by subgroup. This provides the decision maker with the required information more efficiently than running each subgroup separately. An increase in the number of entities simulated is still required, however, to allow for stochastic uncertainty within each subgroup. If parameter uncertainty differs by subgroup, then it may get too complicated to try to run all subgroups in the same analysis. Similar issues arise if structural uncertainty is not common across subgroups.

5.2.2 Other Sensitivities of Interest

This idea of simultaneously considering various subgroups can be extended to exploring the sensitivity of results to changing the values of other model characteristics, even if they bear no uncertainty. These other inputs may not be specific to entities, but instead broadly affect the entire model. One characteristic that is of particular interest in many HTA contexts is the *perspective* of the analyses. Perspective refers to the point of view taken in the analyses, usually having to do with who pays for what. For example, intervention costs will be different from the perspective of the insurer to that of an individual covering co-pays and yet again to that of a hospital provider concerned only about admissions. To consider the impact of these different perspectives, analysts typically run separate analyses for each perspective. Using a DES, however, these various perspectives can be accommodated in a single run by defining at the start what costs are pertinent to which perspectives and storing the outputs accordingly during each replication. Since much of the work of the simulation is processing the events, and there is much less effort involved in applying costs to any consequences, this simultaneous running of multiple perspectives is much more efficient than rerunning the simulation separately for each perspective.

The sensitivity of a DES to changes in the values of many types of inputs can be incorporated in a simultaneous analysis. For example, the analyst may be interested in the impact of different discount rates or in the degree to which the length of the time horizon matters. Just as with the perspective, these scenarios can be set up by specifying appropriate global variables to store the outputs. In the case of the discount rate, values discounted at different rates would be stored in separate variables. Since the discounting takes place when the related event occurs, the timing is known precisely, and any number of rates can be applied with no need to rerun the simulation. For different time horizons, it is simply a matter of reporting the outputs at the desired model times, but not terminating the simulation until the longest one has elapsed.

5.2.3 Threshold Values

While subgroup analyses address the variation in results in defined portions of the population, another sensitivity question can be about the value of a particular input that alters the results in such a way that the decision maker might reach a different conclusion. For example, if a technology is judged to be unattractive because its benefits do not justify its costs, the decision maker may be interested to identify the price at which the intervention becomes cost effective. This is known as a *threshold* analysis because it is trying to determine the value of an input that leads to the results crossing from one side of an HTA decision to the other.

If the relationship between the results and the input of interest is linear and easily derived, a threshold analysis can be addressed analytically outside the DES. In most cases, however, this is not possible because the connection is not linear or it cannot be readily deduced. In this situation, a threshold analysis requires running the model multiple times with different values of the input and using the results across runs to find the threshold value. This can proceed in the same way as a deterministic uncertainty analysis, varying the input value until the threshold is found. A more efficient alternative is to automate the process by letting the model select the value from a range and narrowing that range progressively until the threshold is found. Many specialized software packages now include functions or modules that facilitate this process.

6

Formulating the Required Equations

A discrete event simulation (DES) is a mathematical framework for addressing a problem—for health technology assessment (HTA), this has to do with the health and economic implications of employing a new intervention in the health care system. At the core of this mathematical framework is a set of equations that expresses numerically the relationships among the various factors. Although the number and type of equations in a DES vary based on the nature and complexity of the disease or outcomes modeled, the approach to building those equations is the same (Table 6.1). First, you must find appropriate sources of data. Often this is the most arduous part of the process. Then, you must determine which distribution will be applied to these data and what predictors should be included. Finally, the resulting equations have to be validated.

In this chapter, the approaches to formulating the required equations are presented. Throughout, it has been assumed that readers are familiar with basic statistical concepts (such as random variables, statistical distributions, and regression) at the level of an introductory course in statistics. Although equations are provided throughout, it is not necessary to understand these in any detail to be able to follow the presentation.

6.1 Requirements for the Equations

The disease process modeled in an HTA DES can be represented by an influence diagram showing how the various components involved in the process relate to each other (see Chapter 3). Figure 6.1 presents a simple influence diagram that will be used to illustrate explanations provided in this chapter. The diagram depicts the presumed relationships required for the model. A biomarker of the disease affects the risk of death. Treatment is expected to change this biomarker and, by so doing, lower the risk of death. Previous values of the biomarker influence subsequent ones. Treatment can also lead to adverse events, and these, in turn, increase the risk of discontinuing treatment. Discontinuation of treatment will affect the biomarker. Finally, patients' characteristics, such as risk factors and disease history, are determinants of all four components. Some relationships captured in the influence diagram are indirect. For instance, treatment (and, hence, discontinuation)

TABLE 6.1

Steps to Building Equations for a DES

	Steps
1.	Select appropriate data sources
2.	Determine the appropriate statistical distribution to use
3.	Select the predictors that will be included in the equation
4.	Validate the equation

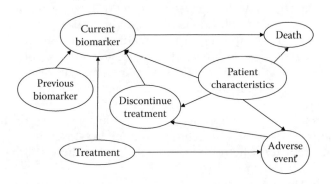

FIGURE 6.1

Simple influence diagram depicting relationships to be modelled. The diagram illustrates the relationships between variables involved in the simulation. Treatment affects the chances of a patient suffering an adverse event, which in turn affects the chances of the patient discontinuing treatment. Treatment and having discontinued treatment both affect the biomarker. The latter is also affected by its previous values and in turn affects the patients' chances of dying. The characteristics of the patient impact all components except treatment as this is controlled in the simulation.

is related to the risk of death even though there is no arrow linking them directly. The relationship occurs via their effect on the biomarker. There is no separate mechanism through which treatment lowers the risk of death.

Simulation of the processes displayed in the diagram requires equations that capture the stipulated relationships. Each elliptical node represents a condition or outcome for which an equation is required, and arrows represent predictors that should appear in the equation. For instance, an equation is needed to predict how the biomarker will change over time. (Note that time is not explicitly represented in the influence diagram as it is assumed that this is inherently part of the simulation and must be reflected in the equations.) This equation is expected to include relevant individual characteristics, previous values of the biomarker, as well as whether treatment is in use or has been discontinued. The various equations combined together provide a mathematical representation of the disease process simulated.

To serve the aims of the DES, the equations must fulfill certain requirements (Table 6.2). They should relate the occurrence of conditions to relevant individual characteristics. This ensures realistic predictions when different

TABLE 6.2

Key Requirements for Model Equations

	Requirement
1.	Relate conditions to characteristics of the people
2.	Link conditions to each other as appropriate
3.	Fit valid equations to represent the defined relationships

population profiles are simulated. Otherwise, the simulation would yield the same result regardless of which population is modeled. Equations must also properly link conditions per the specifications in the influence diagram. This ensures that predictions for associated conditions are logically consistent. For instance, if treatment leads to improvement in the biomarker, failing to also link treatment discontinuation to the biomarker would lead to overestimation of improvements after discontinuation. That is, for simulated individuals who were assigned initially to treatment but who are no longer using it, biomarker values would continue improving regardless!

The third requirement is that equations be clinically plausible (i.e., have face validity), fit the data used in the analysis well, and realistically predict outside the source data. The latter two aspects come under a variety of terms according to the field employing them. In accordance with the most recent modeling guidelines, fitting the source data will be called *dependent validity*, while predictions of outcomes observed in other data will be called *independent validity*.

Face validity is addressed by reference to expert knowledge in the corresponding field (see Chapter 8). Dependent validity is addressed by comparing the predictions made by the equations to what is observed in the source data (see Chapter 8). Independent validity—particularly important in the context of simulation, as the purpose of the model is to estimate results and explore situations that have not been studied—requires comparison of the predictions with observations made in data sources not used in constructing the equation (see Chapter 8). Short of this ideal, some sense of the independent validity can be obtained by running scenarios that deviate from the source data and assessing the credibility of those results.

6.2 Selecting Data Sources

Ideally, the equations used in a DES are derived from analyses of individual data obtained in a properly designed and conducted study. These data need not necessarily all come from the same source; data from multiple studies may be analyzed to create the equations. These studies can be interventional (e.g., randomized controlled trials), or they may be observational, either specifically designed to collect the required data (e.g., a prospective registry) or leveraging existing mechanisms of data collection (e.g., administrative databases or electronic medical records). When none of these sources are directly available, it is much more difficult to derive proper equations. It may be necessary to rely on information reported by others (e.g., in a publication) or even by asking experts for their considered opinion. Needless to say, the farther the modeler is from the actual data, the more questionable the reliability of the equations.

6.2.1 Interventional Data Sources

Randomized controlled trials intervene in patient care by assigning subjects to specific treatment strategies and controlling what happens via various design features and a detailed protocol. They are a rich source of data, particularly on the efficacy of the interventions of interest. Often, the clinical trial data can be obtained from the sponsors of the trial or the investigators who conducted it. For DES developed for commercial purposes, the manufacturer is typically the sponsor and may have access to the entire database. Data from trials funded by government or other organizations without a commercial interest may also be available, either directly via public access or with special permission. It is common to have access to the data on individual subjects from the trials for some interventions (typically those produced by the sponsor of the simulation work), but not for others. This forces the modeler to combine information at the individual level for some interventions with published aggregate results for other interventions. For interventions that have been on the market for some time, existing observational datasets may provide usable information as well, though these are prone to bias when estimating efficacy (discussed further below).

Data from randomized controlled trials are best for estimating the interventions' effects on the various events in the DES, as well as variations of these effects over time and according to subject characteristics (Figure 6.2). Given this preference for clinical trial data, it might be appealing to use the same source to derive all the equations for the model. This would provide consistency in the data collection methodology, context, definitions, and so

Pros	Cons
• Randomization reduces bias.	• Selected population.
• Data quality is high.	• Protocol alters clinical practice.
• Detailed clinical data.	• Higher adherence than normal.
• Preferred by many decision makers.	• Measurement tools not used in practice.

FIGURE 6.2
Advantages and disadvantages of using data from a randomized controlled trial for DES.

on, across all the information recorded in the study. Unfortunately, data obtained in clinical trials are affected by the protocol of the study, which is designed to promote high adherence to the interventions and optimal performance of all clinical procedures; to control the environment (e.g., use of concomitant medications), frequency of visits, testing, and so on; and to standardize assessments, often using tools that are uncommon in actual practice. This may not be an issue if the purpose of the simulation is to model the clinical trial itself in order to inform design and execution of other trials, where similar constraints are expected to apply. The aim of HTA DES, however, is to model the real-world context, and, thus, insisting on the use of clinical trial data may produce results that are not very relevant for the decision makers. Those results would reflect poorly the heterogeneity and variability that exist in real-world settings. Thus, the choice of data source for each equation must strike a balance between validity of the information and its relevance considering the purpose of the DES.

6.2.2 Observational Data Sources

The alternative to interventional data sources are observational studies that try to obtain data without intervening in patient care. This is done by trying to observe what happens to subjects in as unobtrusive a manner as possible. If the data are collected as part of routine care for reasons having nothing to do with a study—for administrative purposes or to maintain patient records—this low-profile ideal can be achieved. If this is not the case, then the study must set up procedures for obtaining the data and incur the problems of the Hawthorne effect (Roethlisberger et al. 1939).

6.2.2.1 Routinely Collected Data

In many jurisdictions, especially when health insurance of some sort pays for much of the health care provided, databases are maintained for administrative purposes. Naturally, these databases collect information that is required for adjudicating payment claims, not for research purposes, and much less to enable the estimation of equations for a DES (Figure 6.3). Nevertheless, they often store large volumes of data covering a broad (often representative) spectrum of people, with several years of follow-up, reflecting real-world practice patterns, use of medical resources, prescription and dispensing patterns, and various characteristics such as age, sex, and diagnosis. With some deductive thinking, it may be possible to determine the onset of clinical events, their duration, and outcomes. Access to these datasets is controlled by the administrative entities concerned, and, thus, depends on their regulations and procedures. Some of these have established processes that permit relatively quick retrieval (sometimes within weeks). In most cases, however, there is a price to pay, and this is often a significant component of the cost of building the DES.

Pros	Cons
• May be more representative of target population.	• Susceptible to bias due to confounding.
• Reflects routine prescribing patterns and usage (adherence) levels.	• Irregular frequency of data collection and inconsistent quality.
• Large samples of patients with long follow-up readily accessible at lower cost.	• Clinical data and information on treatment usage may be limited or unavailable.
• Loss to follow-up limited to death, change of coverage.	• Details limited to diagnosis or procedure codes.

FIGURE 6.3
Advantages and disadvantages of using data routinely collected for other purposes in a DES.

An important problem with routinely collected data, especially those obtained for administrative purposes, is that records are maintained on people only while they are covered by the insurer that is interested in the data. In settings where there is little or no movement among insurers (ideally where there is only a single payer), this is not a problem. When people change their insurer with some frequency, the records may be very incomplete. Indeed, a single person may be represented multiple times but with no link between one record and the next. To mitigate this problem, criteria requiring a minimum duration in the database are usually imposed (Berger et al. 2009; Cox et al. 2009).

Another type of routinely collected data is the information recorded clinically for purposes of patient care. These medical records are increasingly kept electronically (i.e., electronic medical records or EMR) and, thus, become a potentially convenient and informative data source, relative to the paper charts that have been the mainstay until recently. Their major advantage is that they record the very details that administrative databases are missing (like laboratory test results and other biomarkers, clinical history, and patient behaviors). As these EMRs contain much information that is considered deeply private in many jurisdictions, access is highly controlled and subject to many regulations. Moreover, the data are not recorded in a manner that is readily suited for statistical analyses, so they have to be processed—often extensively—before they can be used to derive equations for a DES. They are also subject to the vagaries of patient care, where any one record may reflect only a small portion of what has happened to that person. Nevertheless, the level of clinical detail available makes EMRs a highly attractive data source. Where EMRs are not available, abstraction of information from paper charts is a possibility.

6.2.2.2 Prospectively Collected Data

In many cases, routinely collected data are not available for the jurisdiction of interest, or information that is essential for the DES equations is not regularly recorded either administratively or clinically (e.g., patient-reported

Pros	Cons
• May be more representative of target population.	• Susceptible to bias due to confounding.
• Reflects routine prescribing patterns and usage (adherence) levels.	• Data may be missing, incomplete or inconsistent in quality.
• Necessary data can be obtained by design, even if not routinely collected.	• Selective loss to follow-up (e.g., due to side effect) will introduce bias.
• Less expensive than randomized clinical trial.	• Hawthorne effect may alter natural behavior.

FIGURE 6.4
Advantages and disadvantages of prospectively collecting observational data for DES.

outcomes such as a quality-of-life score). In situations like this, it is necessary to obtain data by designing and implementing epidemiological studies (e.g., registries) that prospectively collect the data (Figure 6.4). These studies can overcome the limitation of routinely collected data, as the study designers can target the data to be collected (e.g., laboratory test results, levels of other biomarkers, and patient-reported outcomes), as well as the desired level of detail, and can collect information directly from patients.

Epidemiological studies can be used to assess effects of marketed products by enrolling individuals receiving interventions of interest. Unlike in clinical trials, however, these interventions are not assigned randomly, but rather based on whatever criteria clinicians use to determine what they recommend for whom. Presumably, these criteria have to do with the expected benefit-risk balance for each individual, and this poses a problem because individuals receiving one intervention may be at a different risk than individuals receiving another. Thus, comparative evidence from these studies is confounded by indication—the estimates of the relative effectiveness of interventions are distorted by the differences in prognosis between groups. Analyses of data from observational studies must carefully adjust for these imbalances to minimize this bias.

Like trials, prospective observational studies involve design and execution that may require significantly longer time than the use of routinely collected data. They are also more costly to conduct (Berger et al. 2014).

6.2.3 Considerations When Using Multiple Sources of Data

In most HTA DES, data must come from various sources to achieve a realistic simulation. For instance, in the simple example (Figure 6.1) the modeler would be reluctant to simulate changes in the biomarker based on the rates observed in the trial because these may not be representative of what might occur in actual practice where there may be greater discontinuation of treatment. EMR data may be sought to derive an equation for changes in the biomarker, but these would not include the effect of treatment. The benefit of treatment would be incorporated by applying the treatment effect estimated from the clinical trial data to the predictions made by the EMR-based equation. Similarly, an administrative database might be used to create an equation for real-world discontinuation patterns as a function of adverse events occurring and other risk factors.

The metric used to quantify treatment effects should be determined prior to building equations, as this can have an influence on how the equations are parameterized. For instance, the effect of treatment on the biomarker might be measured as incremental change from baseline, or, alternatively, as the relative probability (or odds ratio) of achieving a minimum reduction. In the former case, the equation for the biomarker can be based on a linear regression for change, while in the latter case, it would be more convenient to analyze the probability of response with a logistic regression.

Combining elements from different sources can affect the credibility of the DES. To be able to enhance it, however, these disparate inputs must be closely aligned. This includes the specific factors that are included in the equations and how these are defined. Factors for which information is available in one source but not another can limit the ability of the DES to reliably model various populations and may mar the links among conditions. This undesirable consequence results from the predictions being more disparate than they should be. For example, if the natural deterioration in a biomarker (e.g., weight) is known to be different in women but the effect of treatment is only obtained without reference to sex, then the predictions applying the weighted average weight loss will be an underestimate for one sex and an overestimate for the other. Differences in the definition of factors can also distort predictions. For example, if diabetes in one equation is based on patients' self-report, but in another equation it is defined based on results of an oral glucose tolerance test, then the two equations will look like they are using the same factor, whereas in fact it is quite different. Risk factors will tend to be more precisely defined and ascertained in clinical trials or prospective epidemiological studies but have to be inferred from recorded diagnoses and even from medication usage in routinely collected data.

6.3 Taxonomy of Equation Types Commonly Used in DES

The equations required for a DES are best created by analyzing data on individual subjects using regression techniques. Although the equations may also be derived from published data (e.g., in the context of meta-analyses) by relating aggregate population characteristics and study features to outcomes, these types of analyses are not covered in this chapter.

Regression analysis involves relating the values of one quantity to those of another. To do this, a statistical distribution is specified for the values of a random response variable (i.e., a factor that is subject to chance variation, like body weight). A statistical distribution is a function that describes the probability of different values for a response variable, with the probabilities across all possible values of the response variable adding to a total of 100% (see Chapter 2). Then, a relationship is sought between a parameter of the specified distribution (e.g., its mean) and the values of factors that might explain the variability across individuals.

For example, body weight in the population of interest may be assumed to follow a normal distribution with some mean μ. The distribution for male and female individuals would be expected to have a different mean, μ_M and μ_F. This difference can be captured with an equation that relates the mean body weight to the individual's sex as follows:

$$\mu = \beta_F + \beta_1 X_M \tag{6.1}$$

In this equation, β_F represents the mean for women, β_1 represents the difference in mean for men and women, and X_M is an indicator of sex (1 = male, 0 = female). This is illustrated in Figure 6.5.

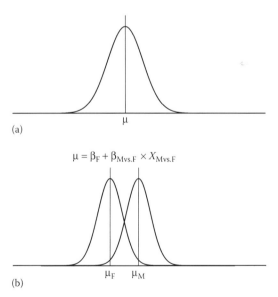

FIGURE 6.5
Relating parameters of a distribution to characteristics of people. (a) The top panel represents the distribution of body weight in the entire population; this is shown to have a normal distribution centered at mean μ. (b) The bottom panel represents distributions within male and female members of the population; weights in each of these subgroups also follow a normal distribution each centered at its own mean value and with smaller variances than in the overall population. The relationship between the mean body weight and patients' sex can be represented with a regression equation.

A type of regression that is familiar to most researchers is linear regression for a continuous (i.e., can take any value within a range) response variable. Linear regression implies that a given change in the value of the predictive factor always yields the same change in the value of the response variable, regardless of what the starting values are. For example, an equation might specify that after age 50 years, weight increases by a mean of 1.5 kg every year. Individuals are assumed to have independent normal distributions for the values of the random variable Y, and this is indicated with a subscript (Y_i). Each distribution has an expected value, or mean (μ_i) that can vary based on individuals' characteristics ($X_{1i}, \ldots, X_{ki}$), and (usually) a common variance (σ^2). This is mathematically written as

$$Y_i \sim N\left(\mu_i, \sigma^2\right) \tag{6.2}$$

$$\mu_i = \beta_0 + \beta_1 X_{1i} + \ldots + \beta_k X_{ki} \tag{6.3}$$

noting that there is no correlation between the values of different individuals

$$\text{Correlation}\left(Y_i, Y_j\right) = 0, \text{if } i \neq j \tag{6.4}$$

The analysis produces estimates of the coefficients ($\beta_0, \ldots, \beta_k$) which reflect the size and direction of the effect of the corresponding predictors ($X_{1i}, \ldots, X_{ki}$). The estimates are obtained by finding values of those coefficients that maximize the likelihood (probability) of the observed data. For example, consider a simple, intercept only model (i.e., $\mu_i = \beta_0$) for change in weight, where we assume a normal distribution for observed changes ($y_1, \ldots, y_n$) in a sample of individuals. (Note, the use of lower-case for y is meant to indicate that these are actual observed values rather than random variables.) The likelihood of the data, assuming observations were independent and arose from distributions with common variance, would be given by the product ($\prod$) of the probability of each observation:

$$L\left(\beta_0 \mid y_1, \ldots, y_n\right) = \prod_{i=1}^{n} P\left(Y_i = y_i\right) = \prod_{i=1}^{n} N\left(y_i; \mu_i = \beta_0, \sigma^2\right) \tag{6.5}$$

Plugging in the formula for the normal distribution, and taking the log of the likelihood to simplify its form, we obtain

$$LL\left(\beta_0\right) = \log\left[L\left(\beta_0 \mid y_1, \ldots, y_n\right)\right] = -\frac{n}{2}\left[\log\left(2\pi\right) + \log\left(\sigma^2\right)\right] - \frac{1}{2\sigma^2}\sum_{i=1}^{n}\left(y_i - \beta_0\right) \tag{6.6}$$

If we assume σ^2 from prior information, and since the y_i are observed values, the only unknown in this equation is β_0. Since the function represents the likelihood of having observed the data on hand, the *best* estimate for

β_0 is taken to be the value that maximizes the log-likelihood. In this simple case, the maximum can be easily seen by graphing the likelihood function, as illustrated in Figure 6.6. In typical cases where the model involves multiple parameters, the set of values that together maximize the likelihood is taken as the best estimate. These values are found mathematically (either in closed form solutions) or using numerical procedures that search for the maximum. The likelihood approach is broadly used due to the properties of the estimates obtained this way; in particular, with large enough samples, the parameter estimates can be assumed to have normal sampling distributions, allowing simple calculation of confidence intervals and p values.

Standard errors and p values reflecting the statistical significance of the coefficients are also produced to help determine the importance of predictors (i.e., predictors whose significance is too small to influence the response may be ignored). In addition, the analyses also produce statistics to assess the fit of the equation to the data. These include the log-likelihood ratio evaluated with the estimates of the coefficients (Wilks 1938), as well as measures like Akaike's information criterion (AIC) (Akaike 1974) or Bayesian information criterion (BIC) (Schwarz 1978). The AIC and BIC are derived from the log-likelihood, penalizing this to reflect the complexity of the equation (e.g., based on number of predictors).

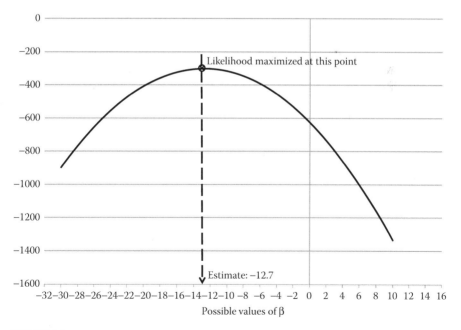

FIGURE 6.6
The log-likelihood function evaluated at various possible values of β. This figure illustrates values of the log-likelihood function (y-axis) evaluated at different values of the parameter β. The best estimate of β is chosen to be the value at which the likelihood function reaches its peak, which in this case occurs when $\beta = 12.7$.

Linear regressions are commonly used and are very useful but can only be applied to continuous response variables. This approach would not be appropriate, for example, for a dichotomous outcome (e.g., success/failure), nor for counts of occurrences of an outcome (e.g., number of hospitalizations in a year). Even for continuous response variables, the normal distribution may be inadequate if the values have a skewed (i.e., asymmetric) distribution (e.g., costs which typically cluster at some value but have many outliers with enormously higher values) or is subject to censoring (e.g., for event times in subjects who did not have the event in the timeframe of the study). These situations require other statistical distributions that are better suited for the type of outcome analyzed. For instance, a dichotomous outcome requires a binomial distribution, while counts are best modeled with a Poisson or negative binomial distribution.

Fortunately, regression analyses can be used with a broad range of distributions. These can be grouped into two broad classes: failure-time methods designed for time-to-event outcomes and generalized linear models (GLM) for other types of variables. Failure-time models are chosen because times to an event tend to have skewed distributions and have censored values for some individuals in the source study, while the GLM approach can accommodate continuous (both symmetric and skewed), dichotomous, polychotomous (i.e., multi-category outcome), and count data. This classification is illustrated in Figure 6.7. Particularities of these distributions and considerations in their selection are presented next. The approach to choosing which predictors to include and validation of the resulting equations follow similar principles regardless of the type of distribution used as the basis for the equation.

6.3.1 GLM-Based Equations

6.3.1.1 Continuous Response Variables

Equations for continuous response variables, like weight or blood pressure, can be based on various types of distributions. Two features of the response variable must be considered in determining the appropriate choice: whether it is bounded (e.g., negative values are not allowed so 0 is a lower bound) and whether the distribution of values is symmetric or is skewed. For unbounded and (reasonably) symmetric response variables, normal distribution regression (a special case of GLM) provides a reliable basis for developing equations.

This normal framework can also handle skewed data, especially if the response variable is transformed so that it is more symmetrical. A typical transformation is to take its natural logarithm. This dampens skewness, but is considered a crude approach because the interest is in the untransformed value and reverting the predicted variable to its original scale (e.g., by exponentiating the mean predicted from the equation for the log-transformed response variable) does not yield the correct value. The normal approach

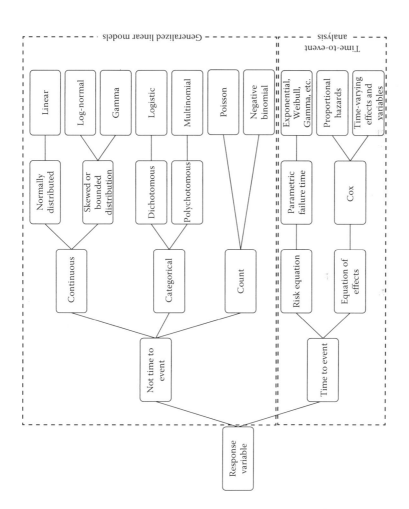

FIGURE 6.7

Classification of regression techniques for different types of response variables. This diagram illustrates the classification of regression types based on the properties of the response variable being analyzed. These can be first separated based on whether they measure a time to event or not; those that are not event time measures can be further classified by whether the variable is continuous, categorical, or counts. Time-to-event variables can be analyzed with two broad types of regressions: those that allow direct estimation of risk (and event times) and those that only measure relative effects.

might also be usable with bounded response variables if the data are reasonably distant from the boundary. In some cases, boundary issues can be mitigated by analyzing changes in values rather than the values themselves, and the natural logarithm transformation yields unbounded values for non-negative response variables. It is preferable, however, to consider other distributions that accommodate skewed or bounded response variables better, and this can be done using the GLM framework.

In GLM, different distributions can be tested and compared based on fit. In addition to the choice of statistical distribution, GLMs require specification of a link function, which imposes a functional form on the relationship between the response variable being regressed and its predictors. For example, the normal distribution regression described above is fitted with an *identity* link, that is, the mean of the distribution is assumed to be a linear function of its predictors. This is expressed in Equation 6.3, which shows that the identity link implies an additive effect for predictors. Alternatively, a normal regression could be fit with a logarithmic link (commonly referred to a *log-normal regression*). This is appropriate for response variables that are not negative and right-skewed. In this case, the equation for the mean is:

$$\ln(\mu_i) = \beta_0 + \beta_1 X_{1i} + \ldots + \beta_k X_{ki} \tag{6.7}$$

or

$$\mu_i = e^{\beta_0} \times e^{\beta_1 X_{1i}} \times \ldots \times e^{\beta_k X_{ki}} \tag{6.8}$$

The log link implies that predictors have an additive effect on the log of the mean, which is the same as a multiplicative effect on the mean. Log-normal equations are more appropriate for costs, for instance, because, in addition to having a skewed distribution, costs tend to change proportionately with predictors. A regression based on a gamma distribution can also be considered in such cases. The gamma distribution also accommodates response variables that are not negative and right-skewed. Other probability distributions for continuous distributions handled in the GLM framework, but less commonly used for DES equations, include the exponential and inverse normal distributions.

6.3.1.2 Categorical Data

Categorical response variables can be analyzed with logistic regression which can handle both dichotomous outcomes (e.g., success or failure), as well as multi-category or polychotomous outcomes (e.g., mild, moderate, and severe). The response variables are assumed to arise from a Bernoulli (or binomial) distribution in the dichotomous case and a multinomial distribution for polychotomous outcomes. In both cases, the distributions are parameterized by event probabilities, which can be related to predictors to capture variation across individuals. The GLM uses a logit link, which corresponds to the log of the odds of the event, that is,

$$\text{logit}(p) = \log\left(\frac{p}{1-p}\right) \tag{6.9}$$

where p is the probability of the event.

In logistic regression, the outcome is coded as one to indicate occurrence of an event or success, and zero otherwise. The equations are

$$Y_i \sim \text{Bernoulli}(p_i) \tag{6.10}$$

$$\text{logit}(p_i) = \beta_0 + \beta_1 X_{1i} + \ldots + \beta_k X_{ki} \tag{6.11}$$

$$\text{Correlation}(Y_i, Y_j) = 0, \text{if } i \neq j \tag{6.12}$$

The coefficients of the equation (i.e., the β) represent the natural logs of the odds-ratios associated with differences in the predictors.

In multinomial regression, the response variable is coded to indicate which of the possible outcomes occurred. If the number of possible outcomes is c, the multinomial distribution is parameterized by a series of c probabilities, with the constraint that these add to one. As with logistic regression, the probabilities can be related to predictors with a logit link (Equation 6.9). One of the possible outcome categories (e.g., the mild severity level) must be selected to serve as the reference for the calculation of the logits; it was not necessary to do this with dichotomous outcomes because one of the levels (the one coded as 0) is inherently implied as the reference. Multinomial regressions can be specified with two different formulations for the relationship of the log-odds with predictors. The first formulation is referred to as an ordinal logistic regression or a cumulative logit regression, which assumes that the effect of predictors is the same for all outcome categories. The equation can be written as follows, taking the last category (c) of the outcome as the reference:

$$Y_i \sim \text{Multinomial}(p_{i1}, \ldots, p_{ic}) \tag{6.13}$$

$$\text{logit}(p_{ij}) = \log\left(\frac{p_{ij}}{p_{ic}}\right) = \beta_{0j} + \beta_1 X_{1i} + \ldots + \beta_k X_{ki} \tag{6.14}$$

$$\text{Correlation}(Y_i, Y_j) = 0, \text{if } i \neq j \tag{6.15}$$

In this representation, j indexes the outcome category, and ranges from one to $c - 1$. The regression allows a different intercept for each category of the outcome, allowing the probabilities to vary for each category, but assumes that the effect of each coefficient is the same when assessing the log odds of category j versus c for any $j = 1, ..., c - 1$. This is sometimes referred to as a proportional odds equation, since the log-odds across different values of a predictor (e.g., age) for different categories of the outcome are parallel on a plot, with gap sizes determined by the differences in the intercepts for each. This formulation is often considered appropriate when there is a natural ordering to the outcomes (e.g., mild, moderate, and severe).

Proportionality may not hold, however, even if outcome categories are naturally ordered. It may be that the difference in log-odds (given by the regression coefficients) between severe and moderate is greater than the difference between moderate and mild, for instance. This implies that a separate coefficient should be allowed for each of the non-reference outcome categories. This can be achieved with a generalized logit regression, which can be expressed as:

$$\text{logit}(p_{ij}) = \log\left(\frac{p_{ij}}{p_{ic}}\right) = \beta_{0j} + \beta_{1j}X_{1i} + ... + \beta_{kj}X_{ki} \tag{6.16}$$

This is effectively the equivalent of fitting a series of $c - 1$ dichotomous logistic regressions with a common reference category. Fitting the equation with a unified multinomial framework ensures that the probabilities will add to one. It also provides a practical approach for coding such analyses. The disadvantage of this generalized approach is that interpretation becomes considerably more complicated due to the large number of coefficients involved ($k \times c + 1$). This becomes particularly problematic when the data used in the analyses are relatively scarce. It is possible to simplify the result by collapsing outcome categories that have similar coefficient values.

6.3.1.3 Count Data

When the response variable reflects counts of an event over some period of time, Poisson regression can be used to relate predictors to the underlying rate of the event. The response variable in these analyses is the observed count of the event for each individual. The expected event count (or mean of the Poisson distribution) is usually parameterized as:

$$\mu_i = \lambda_i \times T_i \tag{6.17}$$

where:
  λ_i is the rate at which the event occurs (per unit of time)
  T_i is the duration of time the individual is observed

Poisson regressions are fitted with a log link:

$$Y_i \sim \text{Poisson}(\mu_i) \tag{6.18}$$

$$\ln(\mu_i) = \ln(\lambda_i \times T_i) = \beta_{0j} + \beta_1 X_{1i} + \ldots + \beta_k X_{ki} \tag{6.19}$$

$$\text{Correlation}(Y_i, Y_j) = 0, \text{if } i \neq j \tag{6.20}$$

The equation is more conveniently expressed in terms of the event rate, as this is independent of the duration of follow-up. This is given by

$$\ln(\lambda_i) = \beta_{0j} + \beta_1 X_{1i} + \ldots + \beta_k X_{ki} - \ln(T_i) \tag{6.21}$$

The natural log of time on the right-hand side of the equation is handled in implementations of these analyses by specifying these as offsets to the regression in statistical procedures. The coefficients of the equation can be interpreted as rate ratios associated with unit differences in the predictor. Poisson regression may also be performed with an identity link, which amounts to modeling the rate directly as a function of predictors, and implies an additive effect for predictors on the rate. In this case, the coefficients of predictors represent rate differences.

There are two fundamental assumptions of a Poisson regression. One is that the event rate is constant over time, and the other is that events occur completely independently and do not cluster together in time. In other words, one occurrence of the event does not trigger occurrence of additional events. When these assumptions fail, the data exhibit greater variability in event counts across individuals than is accounted for by the Poisson distribution. This is referred to as over-dispersion. One approach to deal with this is to base the regression on a negative binomial distribution. The interpretation of the regression components remains similar. The choice between this and a Poisson assumption can be based on the fit of the equation with each distribution.

6.3.2 Handling Longitudinal Data

As noted in the previous sections, GLM equations rely on an assumption of independence between the observations used in the analyses. In other words, the value of any one observation is not associated with the value of other observations. This is the case when the observations do not cluster in any way. The most obvious form of clustering occurs when subjects provide multiple observations for the same outcome over time. For example, if blood pressure is measured repeatedly, then for any one individual, the values would relate to each other—they would tend to be clustered. Clustering may also occur across individuals, if the subjects from the same clinic or study site tend to have more similar outcomes than individuals from different sites.

Analyses must take clustering into account to avoid biasing estimates of the standard errors (which would be lower than appropriate because of the lesser variability in the clustered observations) and to fully explain how patterns vary across individuals. One approach to dealing with this issue is to extend the regression equation to include random effects. Unlike the coefficients for the predictors which have fixed values for all individuals, random effects are individual-specific random quantities assumed to arise from a normal distribution. For a continuous response variable, a normal mixed effects equation is expressed as follows:

$$\mu_i = \beta_0 + \beta_1 X_{1i} + \ldots + \beta_k X_{ki} + \delta_i \tag{6.22}$$

$$\delta_i \sim N\left(0, \tau^2\right) \tag{6.23}$$

$$Y_i \mid \delta_i \sim N\left(\mu_i, \sigma^2\right) \tag{6.24}$$

Here, the random effect δ_i is assumed to arise from a normal distribution with mean zero and variance τ^2. Alternatively, this can be expressed as follows:

$$\mu_i \sim N\left(\beta_0 + \beta_1 X_{1i} + \ldots + \beta_k X_{ki}, \tau^2\right) \tag{6.25}$$

$$Y_i \mid \mu_i \sim N\left(\mu_i, \sigma^2\right) \tag{6.26}$$

This highlights the two levels of variability captured in this model: the mean of the distribution for a given individual is subject to both systematic (captured by the regression) and random variation (captured by the random effect); this is referred to as *between-individual variation*. Conditional on this mean, observations taken from the individual are further subject to within-individual variation.

The approach described above is called a *random-intercept one*, since the random effect term is not linked to any of the predictors; thus, it is seen as modifying the intercept of the equation for each individual. The concept can be similarly extended to allow other coefficients in the regression to vary across individuals. For instance, suppose X_{1i} in Equation 6.25 is time and β_1 represents the slope of change in the response; if the slope of change is expected to vary in a non-systematic way across individuals, a random effect could be added to the slope coefficient to account for this heterogeneity. That is, the equation would be

$$\mu_i \mid \beta_{1i} \sim N\left(\beta_0 + \beta_{1i} X_{1i} + \ldots + \beta_k X_{ki}, \tau^2\right) \tag{6.27}$$

$$Y_i \mid \mu_i \sim N\left(\mu_i, \sigma^2\right) \tag{6.28}$$

$$\beta_{1i} \sim N\left(\beta_1, \tau_1^2\right) \tag{6.29}$$

Thus, in this specification, each individual has a unique slope or rate of change, possibly even with different directions for some. When multiple random effects are included in the equation, these are assumed to be correlated since they apply to the same individual. For example, in an equation with a random intercept and slope of X_1, the random coefficients would be assumed to follow a multivariate (two-dimensional) normal distribution:

$$\left(\beta_{0i}, \beta_{1i}\right) \sim \text{MVN}\left[\left(\beta_0, \beta_1\right), \begin{pmatrix} \tau_0^2 & \tau_{01} \\ \tau_{01} & \tau_1^2 \end{pmatrix}\right] \tag{6.30}$$

where:
   τ_{01} is the covariance between the two random effects
   The diagonal terms in the matrix are the two variances

When there are more than two random effects, some structure must be assumed for the covariance matrix to specify the correlation between pairs of random effects. For instance, a simple structure may assume the correlation is equal and, thus, only require the addition of a single new parameter; alternatively, each pair can be allowed to have a different correlation. This can involve multiple new parameters, which leads to potential complications in being able to estimate the parameters of the equation.

Although this presentation of mixed effects approaches has so far only focused on normal distribution-based equations, the same concepts can be applied to the other types of distributions introduced in the previous section. For instance, in the case of a binomial outcome, a mixed approach would be necessary to model the probability of recurrence of an event of interest. A standard Poisson equation could also be used to deal with recurrent events without the need for introducing random effects. The limitation with this approach would be that the equation would predict the total count of the event and impose a constant hazard between occurrences of the event (which are core properties of the Poisson distribution). A logistic regression with random effects, on the other hand, provides the flexibility to capture variation in the probability of the event over time. A normal distribution is assumed for the random effects in these equations.

An important application of the mixed effect approaches in DES is to capture patterns of longitudinal change in outcomes. In addition to the incorporation of mixed effects, the equations pose another challenge in properly capturing the temporal pattern. A first consideration is whether all subjects follow a similar pattern over time. For instance, some patients may be improving over time, while others may be declining and others may be relative stable. Furthermore, within improvers and decliners, it may be that some improve at a slow constant rate, and others improve exponentially. A first step in such analyses is to examine graphically the scores or responses at the individual, population, and/or subgroup levels to determine if a single equation can capture the various patterns, or whether multiple equations will be required. A second complication is the determination of an appropriate functional form for time in the equation. Simple forms like linear or quadratic patterns of change can be detected from exploratory graphs; more complex patterns may be difficult to identify, however. Some patterns may also be so complicated that simple functional forms do not fit adequately. Flexible techniques like splines or fractional polynomials may help in these situations. The fractional polynomial approach involves selecting one or combinations of polynomial transformations of time among t^{-2}, t^{-1}, $t^{-1/2}$, $\ln(t)$, $t^{1/2}$, t^1, t^2, t^3, and so on, to model change. These are systematically tested to identify the best fitting, favoring smaller models unless there is significant gain in fit with the inclusion of additional polynomial terms. In cases where there are different patterns of change in subgroups of the population, the fractional polynomial approach can be repeated within each of these to allow different forms in the equation for each subgroup.

6.3.3 Time-to-Event Analyses

6.3.3.1 Deriving Equations

Response variables representing times to an event (e.g., death, hospitalization, and stroke) require special analytical approaches because it is unusual to have observed the event at issue in every subject in the dataset. Thus, some, occasionally most, subjects have what are known as censored event times. That is, for those subjects, it is only known that the event time exceeds the duration of observation in the study. This means that the time-to-event distribution will be truncated, requiring some sort of projection to estimate the longer-term occurrence of the event, as is often necessary for DES in HTA.

Projection of time-to-event curves is best handled with parametric failure-time analysis methods that fit a function to the underlying hazard of the event. The hazard can be loosely thought of as the instantaneous rate of the event occurring, among those who have not yet experienced it. In health care applications, the term *survival analysis* is often used for these kinds of analyses because a commonly predicted event is death, and, thus, the time until that event is the subject's survival. The more general term *failure-time* could be used. This term comes from industrial applications where these types of equations usually have to do with the failure of some component like the breakdown of a welding robot in a car factory. Since many events are not failures, however, the term time-to-event will be used here.

This technique formally accounts for censored observations and uses statistical distributions that can account for the typically skewed distributions of time-to-event response variables. This last feature is what distinguishes this technique from more broadly used techniques like Kaplan–Meier analyses and Cox regression. The Kaplan–Meier analysis produces an empirical estimate of the distribution of event times without making any assumption about the underlying shape of the risk function and, hence, is said to be non-parametric. Cox regression, which relates the hazard of the event over time, $h(t)$ to various determinants (X_j), assumes only that the effects of these are multiplicative on the hazard and remain constant over time. It is written as follows:

$$h(t) = h_0(t) e^{\beta_1 X_{1i} + \ldots + \beta_k X_{ki}} \tag{6.31}$$

where:
   $h_0(t)$ is the baseline hazard function in a population where the values of all the determinants X_j in the second part of the equation are set to 0
   The coefficients β_j represent the natural log of the hazard ratio associated with a unit change in the predictor X_j

The popularity of Cox regression derives from the underlying estimation technique which can produce estimates of the coefficients without specifying the baseline hazard function. For this reason, the method is said to be semi-parametric.

By contrast, in parametric failure-time analysis, the shape of the baseline hazard function is fully specified as a distribution for the event times, similar to what is done in the GLM framework. The most commonly used distributions for HTA DES are the exponential, Weibull, Gompertz, log-logistic, and log-normal. The properties of these distributions are described in standard text books (Collett 2003) and have been detailed for DES (Ishak et al. 2013). Key elements, such as the shape of the hazard function associated with each distribution, the form of the cumulative distribution function (see Chapter 2), and the hazard function are summarized in Table 6.3. These distributions involve two parameters: a scale parameter typically denoted by λ (but caution is warranted because many other symbols are used in software packages and other texts) and a shape parameter, typically denoted by γ. The latter is assumed to be common to all individuals in the population, but the scale is allowed to vary as a function of predictors as described in Table 6.3. For example, the survival function of a Weibull distribution is given by:

$$S(t) = e^{-\lambda_i t^\gamma} \tag{6.32}$$

$$\frac{-\log(\lambda_i)}{\gamma} = \beta_1 X_{1i} + \ldots + \beta_k X_{ki} \tag{6.33}$$

The first step in parametric analysis of this type is the selection of the best fitting distribution for the event times. This involves both statistical and

TABLE 6.3

Properties of Commonly Used Distributions in Parametric Time-to-Event Analyses

Distribution	Pattern	1-*cdf*	Hazard Function	Graphical Test	Regression Function βX_i
Exponential	Constant hazard	$e^{-\lambda t}$	λ	$\ln(\hat{S})$ vs. t	$-\ln(\lambda_i)$
Weibull	Monotonically increasing or decreasing	$e^{-\lambda t^\gamma}$	$\lambda\gamma t^{\gamma-1}$	$\ln[-\ln(\hat{S})]$ vs. $\ln(t)$	$-\ln(\lambda_i)/\gamma$
Gompertz	Monotonically increasing or decreasing	$e^{-e^\lambda e^{\gamma t}/\gamma}$	$\lambda\gamma t^{\gamma-1}$	$\ln(\hat{h})$ vs. $t,$ or $\int \log(\hat{h})$ vs. t^2	$-\ln(\lambda_i)/\gamma$
Log-logistic	Monotonic change followed by gradual decreasing	$e^{-e^\lambda e^{\gamma t}-1/\gamma}$	$\lambda\gamma t^{\gamma-1}/(1+\lambda t^\gamma)$	$\ln(\hat{S}/1-\hat{S})$ vs. $\ln(t)$	$-\ln(\lambda_i)/\gamma$
Log-normal	Log of event time is normally distributed	$1-\phi\{[\ln(t)-\lambda]/\gamma^{-1}\}$	*Complex: increasing then decreasing*	$\phi^{-1}(11\hat{S})$ vs. $\ln(t)$	λ_i

ϕ is the cumulative standard normal distribution.

clinical considerations (Ishak et al. 2013) and is carried out in four major steps (Table 6.4). The first step is a preliminary assessment of the possible fit based on plots specific to each of the distributions being considered (e.g., exponential, Weibull, Gompertz, log-logistic, and log-normal). These plots rely on linear relationships that can be derived from the cumulative distribution or hazard functions, possibly after transformation. For example, a plot of log cumulative distribution function against time can be used to assess fit for an exponential distribution, log-negative log cumulative distribution function against log of time can be used for Weibull, and so on (see Table 6.3 for full list). A linear pattern in these graphs indicates that the distribution may be adequate, and, conversely, deviation from linearity indicates a poor fit. Some distributions may be cut out from further consideration, but it is possible that all are carried forward, regardless of fit, in order to make comparisons later on, or for structural uncertainty analyses.

Each of the candidate distributions are then fitted to the data. This is done using the appropriate procedure in statistical software. In SAS®, for example, the LIFEREG procedure is used (but to fit a Gompertz distribution, a specialized macro must be designed as this is not a standard option in SAS). Gompertz-based analyses can be performed more readily in other software, such as STATA® with the *streg* command. At this stage, no predictors are included in the equation (with the exception of intervention group indicators, if applicable). In addition to estimates of the shape and scale parameters,

TABLE 6.4

Steps in a Parametric Time-to-Event Analysis for a DES

Steps
1.	*Graphical assessment of fit:* This step provides a preliminary assessment of the suitability of each distribution for consideration in following steps.
2.	*Statistical assessment of fit:* This step involves fitting specific distributions, deriving their parameters, and examining how good the fit is using statistics like log-likelihood.
3.	*Assessment of accuracy:* In this step, the predicted values are compared with the actual observations to determine how good the fit is throughout.
4.	*Assessment of plausibility:* Despite a good statistical fit, the implications of the predictions must also make sense given clinical and other knowledge.

the log-likelihood statistic is calculated. The latter is used to derive the AIC and BIC, which penalize the fit statistic for complexity in the distribution. These can be used as an objective metric for comparison of fits, with lower values indicating better fit.

The fit of one or more of the distributions can be further examined by comparing the predicted cumulative distribution function to the observed (i.e., empirical estimate) over the study period. This allows closer scrutiny of deviations in fit in particular portions of the data span (e.g., near the end of the study period). These may be missed when considering the AIC or BIC alone, because the statistics reflect overall fit.

Predictions from the equation must provide a good fit over the observed period and beyond. The fit over the observed period is assessed in the second step with the AIC and BIC statistics. The most common assessment of the fit beyond the observed period relies on clinical judgment regarding the plausibility of the predictions or extrapolations. This is assessed by examining the shape of the long-term projection of the time-to-event curve and deriving various measures from the predicted curve. For instance, the plausibility of the estimates of the median or mean event time is assessed. If these contradict existing knowledge or perceptions, then this would be indicative of a possible inadequate fit. Comparison with an external source offering longer-term data is another approach.

In addition to selecting a parametric fit, the analyses must also consider whether the same distribution is appropriate for all individuals. For instance, in analyses of trial data, one must consider whether the event times in the intervention groups follow the same distribution (with similar shape parameters, as these are assumed to be population specific), or whether separate distributions or shape values are more appropriate. Similarly, subgroups of patients may differ in and require separate analyses for optimal fit.

Although parametric fitting is best done with data from individual subjects, this is not absolutely necessary. Empirical survival distributions reported in publications can also be fit to derive the time-to-event equations. This requires two additional steps. The first is to digitize the published curve by reading the coordinates of a large number of points on the curves. Specialized software (such as Engauge®) is available to facilitate this step, which yields tables with values of the cumulative distribution function at corresponding times. These—ideally supplemented with additional information such as the number of individuals still under observation at various points in time—can be used to generate virtual individual data. This is done by estimating, in consecutive intervals of the time axis, the number of people who have the event, drop out or are still event-free at the end of the interval (Hoyle and Henley 2011; Tierney et al. 2007). Once the individual-level virtual data are generated, the fitting process follows the same steps as if actual data were available.

Lack of real data on individuals presents some limitations. Predictors cannot be incorporated in the analyses unless separate cumulative distribution curves are reported for each level of the predictor. This is commonly done for the intervention groups, and occasionally for other factors such as sex, but rarely for all the predictors of interest. Furthermore, since the number and timing of events are approximated, the standard errors and confidence intervals for shape and scale parameters are inaccurate. These can be considered to be, at best, crude approximations in cases where the number of people under observation over time are available and incorporated in the analyses.

With actual or virtual data, the best single fitting does not necessarily imply a good equation for the DES. The best-fitting distribution may still deviate from the observed data in important ways that make use of the equation in DES questionable. This may happen when fluctuations in the hazard function are more complex than what standard distributions can handle—something that becomes more likely as the observation window gets longer. Alternate methods that allow greater flexibility involve partitioning the observation period into consecutive segments. A separate distribution can then be fitted within each of these or a piecemeal linked equation where each segment is a different distribution (for instance, exponential).

6.3.3.2 Incorporating Time-Dependent Predictors

Parametric regression equations can only incorporate predictors whose values are fixed at some point in time (e.g., at baseline). This can be limiting in a DES, where the aim is to capture dynamic relationships and estimate consequences due to changes in predictor values. Ideally, the equations would include the impact of changes in the predictor values. This is better handled in the context of Cox regressions rather than parametric techniques. Therefore, in situations where time-dependent predictors are required, a hybrid approach involving the two techniques can be applied. First, the reference hazard function is estimated using parametric techniques, without inclusion of predictors. This provides the distribution and estimates of its scale and shape parameters—and is tantamount to fully specifying the hazard function $h_0(t)$ in the Cox regression (Equation 6.31). Predictors are then analyzed with a Cox regression that allows values to vary over time. The latter can be thought of as a hazard ratio function (or *effect regression*). This step yields the other portion of the Cox regression, with the exception that some of the predictors are time-dependent. That can be written as:

$$\text{HR}_i(t) = e^{\left[\beta_1 X_{1i} + \dots + \beta_k X_{ki} + \alpha_1 Z_{1i}(t) + \dots + \alpha_m Z_{mi}(t)\right]} \tag{6.34}$$

where time-dependent predictors are denoted by $Z_{mj}(t)$ with coefficients α_j.

The hazard ratio portion can be applied to the reference hazard function derived in the first step to adjust the prediction to the individuals' current status.

The adjustment must consider the profile of the population on which the reference hazard function is based. In other words, the hazard ratio must be one when the values for the reference population are entered in the equation, otherwise the result will no longer correspond to the observed hazard. To ensure that the sum-product of the coefficients and the predictor values is zero when applied to the source population profile requires that all the predictor values be zero. This is done by redefining the predictor as the difference between the source value and the simulated one. Since the simulated one and the source one are the same when the source population is entered in the equation, the difference is zero and the hazard ratio will be one. This is known as centering all predictors at their source mean values. Of note, if the hazard function and the effect function are not derived from the same population, the centering must be based on the profile of the population used to derive the hazard function. Thus, the full equation with both components incorporated is given by:

$$h_i(t) = h_0(t) e^{\left\{\beta_1\left(X_{1i} - \bar{X}_{1i}\right) + \dots + \beta_k\left(X_{ki} - \bar{X}_{ki}\right) + \alpha_1\left[Z_{1i}(t) - \bar{Z}_{1i}(t)\right] + \dots + \alpha_m\left[Z_{mi}(t) - \bar{Z}_{mi}(t)\right]\right\}} \tag{6.35}$$

This ensures that if a prediction is being made for an individual who is 68 years old while the mean age in the source population is 65 years, then the adjustment of the hazard is based on this 3-year difference, rather than generating a prediction by setting age to 68 years in the equation.

6.4 Selection of Predictors

Once a distribution for the response variable is chosen and its parameters are estimated, attention turns to the selection of predictors to be included in the equation. The approach to doing this is the same regardless of the type of response variable analyzed.

6.4.1 Role of Predictors in the Equation

The factors included in an equation are all predictors in the sense that any change in their values leads to a different prediction for the individual modeled in the DES. The first step is to list all potential predictors and ascertain their expected role in the equations in relation to the response variables. A predictor may play several roles, which are not necessarily mutually exclusive.

The most obvious role is as determinants of the response. In an HTA, the most important determinant has to do with the intervention that is under assessment. Other determinants are included, however, because they provide for a more precise estimate of the response for each individual. In some equations, the interventions have no role but other determinants do (e.g., the equation predicting the time of death due to other causes according to patient characteristics).

In developing the equations, the effect of each predictor is sought so that when its value changes in the simulation, the response is accurately predicted. For example, if the effect of sex on the risk of death is estimated correctly, then modeling a male instead of a female should yield the corresponding difference in risk. This may not be accurate, however, if in the source population used to estimate the effect of sex, the males differed from the females in another characteristic that also determines the risk of death. For example, the males might have been older than the females and, thus, part of the apparent effect of being male is actually due to the age differences. Including age as a determinant in the equation helps remove this confounding of the effect of sex. More generally, the relationship between a predictor and the response is said to be confounded if another predictor is a common cause of these (or associated with these) without being in the causal pathway of the predictor and outcome. In the presence of confounding, the true association can be under- or over-estimated or masked altogether. To remove, or at least reduce the bias, the analyses must adjust for confounding by ensuring that these predictors are included in the equation.

Confounding is usually a concern when estimating the relative effects of interventions from analyses of observational studies where groups are not randomized and, thus, are prone to differ in terms of their profiles that determine the outcome. This situation may also arise in analyses of clinical trial data, when subgroups of the populations are selected for analyses, or data from different trials are pooled together. Confounding can become very complex in the context of longitudinal analyses where time-dependent predictors are included, because inclusion of intermediate predictors or common effects of the predictor and outcome can actually induce or exacerbate confounding. Special analytical methods are needed in these situations.

In some cases, a determinant alters the effect of the interventions—it is a modifier of the relation. Effect modification occurs when the effect of one predictor depends on the value of another. For instance, if women respond to a treatment better than men, then it is said the sex modifies the effect of treatment. This is handled analytically when developing the equations by including a product term (i.e., an *interaction* term) in the equation. An equation with such an interaction would be as follows:

$$HR_i = e^{(\beta_{tmt}treatment_i + \beta_{sex}Sex_i + \beta_3treatment_{i0} \times sex)} \tag{6.36}$$

Here, β_3 reflects the difference in the effect of treatment in females. This equation can be thought of as a composite of male- and female-specific equations:

$$Female\,HR_i = e^{(\beta_{tmt}treatment_i + \beta_{sex} + \beta_3)} \tag{6.37}$$

$$Male\,HR_i = e^{(\beta_{tmt}treatment_i)} \tag{6.38}$$

Effect modifiers can be difficult to identify without prior knowledge of where these may exist. It is not recommended to test all possible interactions, as the chances of false-positive findings can be very high. This can lead to equations that over-fit the data in the sense that they reflect random variations in the dataset being analyzed, which do not represent real effects. Instead, interactions should be focused specifically on key predictors in the equation and based on clinical considerations. In longitudinal equations, interactions between predictors and the time parameters should be considered to capture variation in the patterns of change across different types of individuals.

A final role for a predictor is as a linking variable (Table 6.5). These serve to maintain logical consistency across one or more equations. For instance, in the influence diagram, the biomarker is a linking variable between treatment and death in the equation for the latter. It ensures that improvements in the biomarker brought on by treatment lead to corresponding decreases in the risk of death. These types of factors play an important role in the equations.

TABLE 6.5

Potential Roles for a Predictor in a DES

	Roles
1.	*Determinant of response*: The predictor directly affects the response in the sense that a change in its value alters the response.
2.	*Confounder of response*: A predictor that is also related to other predictors and can, thus, bias the estimates of the effect of other predictors if it is not included in the equation.
3.	*Modifier of response*: A factor that interacts with another determinant by altering the latter's effect.
4.	*Linking*: A factor that is used to link one concept with another via inclusion in two or more equations.

The starting point for assessing the role of predictors is the influence diagram which presents *a priori* expectations of how components should relate to each other. Some of these relationships may not be borne out by the data, however. And some elements are not specified in detail in the diagram, such as the individual characteristics that are expected to be predictors (X). The list may include a number of characteristics from which only the relevant ones are to be retained. Automated procedures such as forward, backward, or stepwise selection are available in all standard statistical software but are not a recommended approach, since only statistical significance is taken into consideration. The latter is largely influenced by the sample size of the data and, thus, may leave out clinically meaningful predictors. A manual process is recommended, as it gives the analyst more flexibility in building the equation. If the anticipated relationships are not confirmed in the data, the relationships represented in the diagram should be reconsidered, and the DES may necessitate an alternative formulation. One must also consider the possibility that the relationship between linking variables and responses is confounded, as this can influence the logical consistency of results produced by the simulation.

6.4.2 Building the Equation

The equation building process starts with the listing of each of the factors to be considered for inclusion, as well as of any interactions to be tested. The association of each of these predictors with the response is examined by fitting equations with each predictor by itself. This serves to determine whether any relationship exists with the response and to assess the direction, size, and statistical significance of the relation. Once this is done for all potential predictors, the ones that appear to be predictors are tested together in a single equation. Those that fail to show any sign of an effect in the single-predictor analyses can be omitted. The others can be added to the equation sequentially, assessing at each step, whether the factor remains predictive and what impact its inclusion has on other coefficients in the equation. For instance, a factor that is highly correlated with other predictors in the equation may no longer be important (i.e., its coefficient may be very weak and/or not statistically significant) when included with the others. Also, a change in the coefficients of predictors already in the equation would reveal that the added predictor confounds their association with the outcome and should be retained. The process continues this way until all potential predictors are tested in the single equation. It is worthwhile to reassess factors that were *a priori* expected to be important but ended up being excluded in earlier steps. It is possible that they do have an effect that was masked and is now revealed in the presence of other predictors in the model. Sometimes a factor with very strong *a priori* credibility as a predictor is retained despite lack of statistical significance because of the boost in face validity. Equation building should be seen as being an iterative rather than an algorithmic process, with prior decisions having to be revisited or reconsidered, including the possibility of adjusting the influence diagram to reflect what the data reveal.

6.5 Validation of the Final Equation

Once a final equation has been constructed, its validity must be carefully assessed to ensure that results from the DES can be considered reliable. The following aspects of the validity of the equations should be considered. These are more broadly discussed in the context of the entire DES in Chapter 8.

The minimum requirement is that the equations driving the DES are able to reproduce the data used to create them—that is, that the equations have good dependent validity. This is done by implementing each equation in a spreadsheet or other software, using it to calculate values of the response variable for individuals like those in the source population and checking whether the predicted values from the equation correspond to the observed values in the overall population or important subgroups. The term "dependent" refers to the fact that the equation is tested against data that were used in its construction. If the equation includes linking factors that are themselves predicted by other equations, the observed values of these linking factors are used in generating the predicted values. In the example discussed so far, the equation for death includes time-dependent values of the biomarker as a predictor. Dependent validation of this death equation would use observed values of the biomarker to generate predicted times of death.

Another aspect of validation is face validity—that is, whether the direction and magnitude of the relationships encoded in the coefficients align with prior expectations based on clinical, epidemiologic, and other knowledge in the disease area. For example, if separate from the equation and its source data, it is known that declining values of a biomarker like body weight are associated with improved mortality in morbidly obese people, but the equation predicts the opposite—that weight gain improves mortality—then it would fail face validity. This should prompt careful consideration of the context of the equation, the source data, and the purpose of the DES. A failure of face validity may prompt revision of the equation but this is not necessarily the right thing to do. It is possible that the equation is accurate in its context. For example, weight loss may be associated with increased mortality in underweight people, and if that was the source population, the equation would be correct, but not applicable for an obese population. The other important thing is to be open to the possibility that the equation building has uncovered new relationships.

In the final stage of verification, all equations are tested jointly to determine whether they produce credible predictions when fully linked. Thus, rather than using observed values for linking factors, predictions are generated using values produced by the linking equations. This is effectively a test of the equations as they would be used in the simulation—in fact, this validation can be performed within the DES (see Chapter 8). The primary aim is to assess whether errors in predictions from one equation propagate and amplify errors in other predictions that rely on them. As with face validity and dependent validity assessments, problems noted on this verification should prompt review and refinement of the equations, and possibly also of the influence diagram.

6.6 Combining Inputs and Equations from Different Sources

As noted in Section 6.2, it is not uncommon for the inputs for the DES to be gathered or derived from different data sources. A particular instance of this is when the simulation aims to assess multiple treatments in a therapeutic area, but equations driving the simulation are derived from a data source reflecting a specific treatment or mixture of treatments. In what follows, the treatment captured in the equations of the simulation is denoted by S; this may be a new agent for which clinical trial data were available and used to derive the equations, or may represent a mixture of treatments representing standard of care derived from an observational data source. Treatments that are external to the simulation or, more specifically, to the data sources used to derive the equations of the simulation are denoted by E. The following sections address the considerations involved when integrating data on the effects of treatment E with equations based on treatment S.

There are two main ways in which this kind of combination arises. One scenario is where the effect of the new treatment E is obtained from a study that expresses it as a comparative result relative to the same treatment S modeled in the simulation. That is, comparative data on E versus S are available directly from a particular head-to-head study, or from two or more studies comparing these to a common *reference* treatment (C) (i.e., data are available for E vs. C, and the simulation is constructed using equations that were derived from studies comparing S to C. The second situation arises when there are no direct data informing the comparison of E to S or when available data compare E to a different reference D (while S is compared to C in the simulation equations). Considerations in these two cases differ in various ways and are detailed below.

Another common situation where data from different sources must be integrated arises when a time-to-event distribution is obtained by fitting a published Kaplan–Meier curve reported by a study with a particular follow-up time, but no information is given on what the determinants of the underlying hazard function might be. Thus, the impact of determinants of interest like patient characteristics and, possibly, even the effects of treatments of interest, have to be taken from a different source. That alternate source may be a study that reports a regression analysis that incorporates the determinants of interest, often a Cox proportional hazards regression equation. Frequently, however, the regression equation is derived from a study with a different follow-up time. The following sections address these kinds of situations.

6.6.1 Incorporating External Data on Intervention Effects

A simple illustration of a situation requiring the incorporation of external data on intervention effects is where the effect of a new antihypertensive treatment must be considered in light of what is known about the relation of blood pressure to various patient characteristics over time. The latter may be represented by an equation derived from an observational study carried out in routine clinical practice, chosen for use in the model because the population of patients studied is representative of the target population that is to be simulated. The equation for change in blood pressure in routine care may have the form of the linear mixed model described in Equation 6.22, describing how blood pressure varies over time and as a function of patients' characteristics. This observational study would not contain information about the effects of the new treatments, however. To include the efficacy of any new interventions that are to be considered in the simulation, data would be sought from available publications and could come in various forms.

One possibility is that efficacy results have been published that provide for estimating the comparative effects of interest. In other words, there are published results for each new treatment of interest as observed in their respective trials compared to some older, established treatment or placebo (i.e., the *reference* intervention), and measured in terms of incremental change in blood pressure at a specific point in time (rarely given as function over time). Another possibility is that a network meta-analysis has been reported that produced pairwise comparisons of some or all of the new treatments of interest versus various other reference treatments (which may include placebo) and reported these as incremental change in blood pressure at one or more points in time.

An alternative situation arises when results are available only for a particular treatment in the absolute (i.e., a single cohort without a suitable reference intervention). In other words, these are not comparative results. Instead, for a particular cohort of patients receiving a certain treatment, they provide the changes in blood pressure at one or more time points.

Determining whether the external data can be used in the simulation along with the equations requires consideration of the compatibility of the data sources and of the measures used to quantify the effects. In the noncomparative situation, some additional analytical steps may be required to ensure proper use of the observational information.

6.6.1.1 Considerations with Comparative Data on External Treatments

Comparative data on E can only be used in combination with the equations obtained from other sources if the reference treatment, or treatments (those given to patients in the populations studied in the other sources), are sufficiently similar to the reference intervention used in the clinical trial reporting the effect of the new treatment. The best scenario is where the clinical trial directly compared E to the treatment modeled in the simulation (S). More often, however, the only data one can find provide comparisons of E relative to a common comparator C. For instance, the study used to develop the equations may have reported results where the reference intervention was the *standard of care* and the clinical trial of E also states that it used as a reference the *standard of care* (usually plus placebo to preserve blinding). These two *standards* may not be the same, however. Thus, it is important to assess carefully the similarity of the mixtures of treatments allowed as *standard of care* in the two sources.

A further consideration is whether the estimates from the clinical trial are applicable in the population used to derive the equation. This does not require that the two populations be identical, but they should overlap to a significant degree so that it is reasonable to assume that the effect of E observed in one is applicable in the other. Even if the populations look like they may overlap sufficiently, this may not be enough for a valid application. If treatment E is suspected of having different effects in subgroups of patients, then the distribution of these subgroups must be sufficiently similar and comparisons of the population overall would not suffice. In such cases, the treatment effects should be obtained for each subgroup and applied accordingly. This may be facilitated by analyses of the trial that incorporated predictors of treatment effect.

Another requirement for compatibility of the simulation equations and estimates of the effect of treatment E is that these use common metrics. For instance, if the simulation equations predict absolute change, then the effects of treatment E would have to be reported as incremental absolute change to be usable directly. The effect of treatment E obtained from the clinical trial publication, Δ_{E-S}, as the incremental change of E versus S can then be applied as follows:

$$\mu_{iE}\left(X_i,t\right) = \mu_{iS}\left(X_i,t\right) + \Delta_{E-S} \tag{6.39}$$

where:

$\mu_{iE}(X_i, t)$ represents the desired predicted mean change produced by the treatment of interest E for a patient with characteristics X_i and at time t

$\mu_{iS}(X_i, t)$ is the predicted mean change for the same patient and time t with the treatment S reflected by the simulation equation

Here, it is assumed that the observed effect of treatment E obtained from the clinical trial publication (Δ_{E-S}) is constant over time and common to the entire population. If the effect is shown to vary over time or across subgroups, then values specific for each subgroup at the various times will have to be applied in Equation 6.39. Even if data on the effect of treatment E suggest that it may be constant over time in the clinical trial where it was measured, it may not be reasonable to assume that it remains constant indefinitely in the simulation, where the time horizon may exceed considerably the follow-up duration of the clinical trial. The clinical plausibility of this constancy must be carefully considered. If the effect of E is expected to vary over time, then a suitable adjustment must be applied to taper it off. For example, it may be considered more realistic to adjust the effect of E downward by some constant rate (e.g., 10% lower each year after some period of time in the simulation) to gradually reduce and possibly nullify the impact of treatment E in the long term. This is the usual presumption—that effect wanes over time—but there may be reason to believe that it will increase instead. In this case, the adjustment must produce the desired intensification of the effect of E.

The simulation can also work with other types of treatment effect measure. For instance, if the clinical trial publication reported the effect of treatment E in terms of relative (or percent) change from baseline, a transformation would have to be applied to $\mu_{iS}(X_i, t)$ to convert it to a relative change prior to applying the effect of treatment E. This transformation involves dividing $\mu_{iS}(X_i, t)$ by the patient's baseline level. The relative effect of treatment E reported in the clinical trial can now be applied to obtain the values that are simulated for treatment E.

Sometimes, the clinical trial may report the results in terms of the proportion of patients who achieve a minimum (presumably clinically meaningful) reduction in blood pressure. Often this result is provided as an odds ratio comparing the odds of success with treatment E versus the odds observed with the reference intervention. In this case, the predicted values in the simulation could be transformed according to this response definition to obtain the odds of response in the population, to which can be applied the odds ratio giving the effect of treatment E. Alternatively, the simulation equations could be reworked to predict the odds of response directly using a logistic regression so that the effect of treatment E given in terms of the odds ratio can be applied. Each type of regression used for the simulation is directly compatible with a specific type of measure of the effect of treatment E that corresponds to the way the outcome is defined in the simulation equation. These are summarized in Table 6.6.

Just as with the linear example, alternative measures of the effect of treatment E can be accommodated with some of the other regression types. For instance, if the effect obtained from the clinical trial is a relative or absolute difference in means, then the estimate obtained in the simulation using a log-linear regression should be converted to an absolute change, to enable application of the effect of treatment E. In logistic regression, predictions are expressed as the log odds of the event, making this directly compatible with an odds ratio effect estimate. Odds can readily be converted back to a probability, or risk, so that a relative risk or risk difference can be applied. A similar manipulation with a Poisson regression allows using rate difference

TABLE 6.6

Types of Regression Analysis Used for the Simulation Equations and Compatible Measures of Effect of Treatment E

	Measures of Effect of Treatment E	
Type of Regression	**Directly Compatible**	**Additional Usable with Manipulation**
Linear	Absolute difference in means	Relative (%) difference in means or ratio of means
Log-linear (log-normal, gamma)	Ratio of means, or difference in log of means	Absolute difference in means
Logistic	Odds ratio	Relative risk or risk difference
Poisson	Ratio of rates (per person-time) or hazard ratio	Difference in rates
Time-to-event: Exponential, Weibull, Gompertz	Hazard ratio	Rate ratio
Time-to-event: Log-normal	Ratio of mean time to event or ratio of quantiles of distributions	
Time-to-event: Log-logistic	Survival odds ratio or ratio of quantiles of distributions	

measures. If sufficient data are available, it is also possible to transform or recalculate the effect of treatment E to be directly compatible with the measure used in the simulation equation.

Time-to-event equations align with effect measures that are compatible with the parameterization of the underlying distributions. For instance, if exponential, Weibull, or Gompertz distributions are used in the simulation, then effects expressed using proportional hazards are appropriate since this parameterizes the hazards in different groups of patients to be a constant factor of each other. That is, the proportional hazards method assumes that the ratio of hazards between two groups depends only on the differences in the characteristics of the groups but does not change over time. Thus, hazard ratios calculated from Cox regression analyses of the effect of treatment E can be applied directly. Log-normal regression, on the other hand, does not assume proportional hazards, but rather assumes that the ratio of quintiles (e.g., median) is constant. This is also often referred to as *accelerated failure time* regression (the original Framingham Heart Study equations are a particularly well-known example) (Anderson et al. 1991). While it is still possible to use hazard ratios with this type of distribution (or with the log-logistic) as described in Equation 6.35, the resulting distribution will no longer be of the log-normal (or log-logistic) type. In this situation, it is more appropriate to calculate the quintile ratio using the reported medians of the distributions with treatment E and with the reference treatment. This ratio is then applied in the time-to-event equations in the simulation.

6.6.1.2 Considerations for Non-Comparative Data on Treatment Effects

The previous section addressed the use of compatible comparative data with equations in the simulation. In many situations, however, effects of the treatment of interest (*E*) may only be measured in a single group of patients all of whom receive that treatment (i.e., not in a comparative study). An analogous situation arises when the information on the effects of treatment *E* is taken from the experimental arm of a clinical trial where the reference intervention does not correspond to any of the treatments used in the population from which the simulation equations were derived. Outcomes like changes in biomarkers, rates of response, and time-to-event distributions may still be useful inputs for the simulation, but it is important to consider that these results may be highly dependent on the profile of the population in whom they were measured. Thus, it is not appropriate to apply them directly in the simulation, where the population may differ substantially. This is especially so if the simulation analysis will involve scenarios that deliberately vary the population profiles. Instead, the outcomes reported in the non-comparative study, or in the single arm of the incompatible clinical trial, can be used to derive an estimate of the effects of treatment *E* relative to the treatment *S* incorporated in the population used to derive the simulation equations. That is, the data can be used to estimate (Δ_{E-S}) in Equation 6.39, and this can then be incorporated in the simulation as described in the preceding section.

Two approaches can be used to transform the observed non-comparative results from a clinical trial regarding the effects of treatment *E* to something that can be used in the DES. The first method is known as a *simulated treatment comparison* (Caro and Ishak 2010; Ishak et al. 2013). This approach leverages the equation used in the simulation to predict what the outcome of interest would have been if the simulated treatment (*S* in the notation above) had been used as a reference in the clinical trial for treatment *E*. In other words, a virtual head-to-head comparison is created. This is accomplished by setting the values of the determinants in the simulation equation to their corresponding mean values in the clinical trial reference arm. In effect, this assigns treatment *S* to the subjects randomized to the reference arm, instead of whatever incompatible intervention was actually used in the clinical trial. With this adjustment, a prediction can now be derived of what the results would have been like had *S* been used as the reference in the clinical trial, and this can then be contrasted with the observed results for treatment *E* to derive a suitable measure of the effect of treatment *E* that can be used in the simulation.

Another approach that can produce a similarly adjusted effect estimate is known as the *matching adjusted indirect comparison* (Signorovitch et al. 2010, 2012). This is conceptually similar to simulated treatment comparison but differs in the way the adjusted prediction is generated. Instead of simulating the patients from the reference arm of the incompatible clinical trial as if they had received treatment *S*, an attempt is made to adjust the simulated population so that it matches the clinical trial's reference. This is done by deriving the likelihood that a simulated patient would have been included in the reference arm of the incompatible clinical trial and then using these probabilities as weights to reconstitute the simulated population, which will now match, on average, the clinical trial's reference population. The simulated results with treatment S are obtained for this reweighted population and compared to what was observed for treatment E in the incompatible clinical trial. This estimate can then be used to represent (Δ_{E-S}) and derive the course with treatment E in the simulation.

6.6.2 Joining Time-to-Event Distributions and Predictive Equations

Section 6.3.3 touched on the possibility that the components of the proportional hazards model described by Equation 6.35, specifically $h_0(t)$ and $\exp[\beta_1(X_{1i} - \bar{X}_{1i})] + \ldots + \alpha_m[Z_{mi}(t) - \bar{Z}_{mi}(t)]$ may not be available from the same data source. For instance, it is possible that patient-level data are available from a clinical trial for a treatment that will be incorporated in the simulation. These data can be used to derive the second component of the equation, which addresses the impact of various determinants. These would include the treatment of interest as one of the predictors. The equation may even allow for the values of some determinants to change over time. The clinical trial often observes subjects over a limited follow-up time, however. Thus, the data from such a study may not produce a very reliable estimate of the underlying hazard function $h_0(t)$. Longer-term data have to be sought to obtain a good estimate of the hazard function. These may be available from published sources, either in aggregate form (e.g., from a published Kaplan–Meier curve) or as a dataset containing the actual patient-level information.

Considerations in using inputs derived from two different sources for the time-to-event distributions are similar to those described in the previous section. The populations from which the two components are derived must be consistent. That is, it must be reasonable to assume that the underlying hazard function obtained from one source is applicable to the population of the other source used to estimate the predictors, after adjustment for any differences in the profiles of the two samples. By the same token, the effects of the predictors must be applicable to the population used to estimate the hazard function. If they are not, then the combination of the predictor equation with the hazard function may not produce reliable estimates. Furthermore, correct application of the predictors requires that they be centered to the mean values in the population from which $h_0(t)$ is obtained (as described in Section 6.3.3.2).

Another consideration (as noted earlier) is whether the shape of the hazard function is in accord with how the effects of the predictors have been estimated. For example, the relationship described by Equation 6.35 is based on a proportional hazards assumption. This can be applied to a hazard function that follows an exponential, Weibull, or Gompertz distribution because the resulting distribution retains the specified shape. If the hazard function has a different shape, however, then applying a proportional hazards component will produce an adjusted hazard function that does not necessarily follow the specified form. If patient-level data were available to derive a predictor equation, one would have more flexibility to align with the properties of the hazard function to which it will be applied. For instance, a Cox regression could be used for predictors if the hazard function is exponential, Weibull, or Gompertz. If the hazard function is of a different type, an accelerated failure time equation could be used for predictors. Furthermore, if patient-level data were suitable in terms of events and length of follow-up, one can simply develop a parametric time-to-event equation in which both the hazard and predictors are fitted jointly, avoiding any issues of compatibility.

7

Efficiency and Variance Reduction

As described in Chapter 5, there are many ways to implement a particular design of a discrete event simulation (DES). If the implementation is correctly done, the variations should produce equivalent results in the sense that they convey the same information to the decision maker (they may not be numerically identical because of the stochastic behavior of the simulations). The variants may not be equivalent, however, in terms of their efficiency. Efficiency here refers to both the number of calculations required to obtain a set of results, which determines the time required to run the analyses of the model, and the effort needed to build the model. Although with the vast speed of current computers, this may seem like an unimportant aspect of simulation, it can have a major impact on the quality of the analyses and the usability of the model. If the simulation takes a long time to produce results, then analysts may be reluctant to carry out the necessary range of analyses required to validate the model and to represent structural and parameter uncertainty. In addition, unsound simplifications may be implemented, all in the pursuit of faster execution times. Among any series of implementations that yield equivalent results, the most efficient one should always be preferred.

There are several ways to improve the efficiency of a DES. Some of these relate to how the DES is constructed and, thus, depend in part on the choice of software and the features implemented. Others are more conceptual in nature—they involve the reduction of nuisance variance—and do not depend per se on the software package (although the way they are implemented is still, of course, software-dependent). In this chapter, both approaches to efficiency gains—in the implementation of a DES and in reducing variance—are introduced. As might be expected, this topic is a very large one. For full coverage of variance reduction techniques, interested readers should consult one of the excellent texts on the subject, such as Kelton et al. (2010) and Rubinstein and Kroese (2011). Tips on implementation techniques that improve efficiency are harder to glean, given their dependence on software, but for the major specialized packages there are also comprehensive textbooks (Kelton et al. 2010) and user manuals (AnyLogic 2009; Rockwell Automation 2013; TreeAge Software Inc. 2014; Visual Thinking 1998). For DES constructed in general programming languages or in spreadsheets, guidance on efficient implementation is an unmet need.

7.1 Reducing Unwanted Variance

Individualization of the modeled experience (i.e., the pathways taken through the logic) is one of the major strengths of DES, but it also introduces volatility in the results for the same reason. This volatility masks the information that is meaningful for the health technology assessment (HTA)—it is like the noise around any signal. This unwanted variance should be minimized as much as possible, whenever feasible. By the same token, the more of this variance there is, the larger the number of individuals required in the simulation in order to accurately estimate expected costs and outcomes (i.e., find the correct signal among the noise). Therefore, a major improvement in the efficiency of a DES can be attained by reducing the impact of this unwanted stochastic variability.

There are many techniques (Adewunmi and Aickelin 2012; Avramidis and Wilson 1996; L'Ecuyer 1994) used to reduce variance in DES, including antithetical variates (Fishman and Huang 1983; Hammersley and Morton 1956), control variates (Kwon and Tew 1994), lattice rules (L'Ecuyer and Lemieux 2000), indirect estimation, and conditioning (McLeish and Rollans 1992). Unfortunately, these methods are not suitable for HTA models because they don't work when alternative interventions are being compared—the main problem addressed by DES for HTA. Fortunately, there are several other techniques that can be implemented in a DES to reduce the impact of the unwanted stochastic variance and, thus, minimize the number of entities required in a model replication and the number of replications. This can greatly increase the efficiency of a model run. All of these approaches are adaptations of the same concept, specifically minimizing the degree to which the experience of the groups being compared differs for reasons that have nothing to do with the interventions of interest. These issues are analogous to the use of strict protocols and blinding of participants in randomized controlled trials, to minimize differences in baseline characteristics and the care received by study participants, other than that associated with the use of the interventions being compared, respectively.

7.1.1 Duplicating Entities

A preferred, but seldom feasible, way to run a clinical trial is a *twin study* where identical twins (ideally living identical lives and exposed to the same risk factors at the same time and in the same way) are enrolled and one of the pair is assigned to each of the study arms and then treated according to a strict protocol. Indeed, the ideal would be having n-tuplets available with as many identical clones as needed for each of the interventions of interest (leaving aside issues around the external validity, or generalizability of such a study and any ethical concerns). The reason such a clone study is so attractive is that the only thing that differs between the arms is the use of the intervention of interest. All other determinants of the outcome are identical among the study arms, eliminating the need for randomization. This massively reduces the likelihood that factors other than the intervention are driving any observed differences between the intervention and control groups and, thus, helps establish the causal connection between intervention and effects. This noise reduction, in turn, permits the researcher to obtain accurate estimates of the effects of the interventions with much smaller studies—the *signal* is more easily detected because it is isolated. When studying rodents, for example, very small sample sizes are sufficient because cloned animals are used and they are kept in identical environments.

In the real world, where sufficient identical twins or *n*-tuplets are not available for inclusion in clinical trials, the studies must enroll individuals who can differ considerably in their relevant characteristics, behavior, and environment. Even though randomization helps disperse these differences equally among the trial arms to reduce confounding, the nuisance variance is still there. The trials must then massively increase sample size relative to what a clone study would require, to overcome the noise produced by differences among the study arms in the distributions of determinants other than the intervention of interest. The more nuisance variance there is, or the smaller the effect being sought, the bigger the study must be. This problem is aggravated if there are various subgroups that are also of interest as the sample size requirements now apply to each one.

What does this explanation of a fundamental aspect of clinical trials have to do with DES? Well, in a DES, the enormous advantages of clone studies can be realized because the models are working with virtual people, and neither these simulated entities nor ethics committees object to the entities being cloned as many times as is necessary to run the comparisons of interest! The virtual world provides an unlimited supply of *x*-tuplets. The basic idea is straightforward (Figure 7.1). The entities are assigned the values of any attributes that are not specific to the intervention before the intervention is assigned. That way all the clones of an entity will have the same values for these identifying attributes. Then, each entity is duplicated as many times as necessary, and each clone is separately assigned the attributes corresponding to the specifics of the intervention (e.g., efficacy, side effects, cost, dose, and route of administration). After that the entities can be brought back to a common path to continue their simulation.

The steps for implementing the duplication of entities are also straightforward (Table 7.1). As noted in Chapter 3, this can be done in a spreadsheet by simply copying the table that contains the entities and their baseline attribute values, so that there is one entity table per intervention of interest. In general software, the equivalent task is to copy the array that contains the entities and their baseline attribute values. Specialized software usually contains a module, often called *separate* or *split*, that accomplishes the duplication. These modules can create many duplicates at once, but it may be difficult to separate them into individual copies after the cloning, so it is often simpler

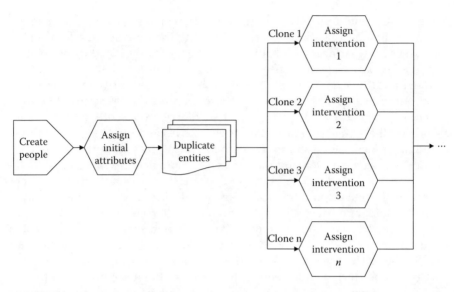

FIGURE 7.1
Duplication of entities to reduce nuisance variance. After creating a set of entities, the initial values for all attributes are assigned to ensure that they are identical in persons receiving each of the interventions. The entities are then duplicated *n* times, and each copy bears the same initial attribute values. The values for those aspects (e.g., efficacy and side-effect profile) that are specific to each intervention are assigned subsequently.

TABLE 7.1

Steps to Duplicating Entities in a DES

Steps
1. *Assign attribute values to original entities*: Assign all the characteristics that are not affected by intervention to each of *n* entities that will be simulated per intervention.
2. *Duplicate entities*: Create copies of each of the *n* entities, one for each of the interventions *I* that will be assessed, creating *n* × *I* entities.
3. *Send each entity to assign intervention*: After duplication, send one copy of each identical set of entities to have an intervention assigned until all interventions are allocated.
4. *Assign intervention-specific attributes*: Assign attributes to each cloned entity that are specific to its intervention.

to chain together a sequence of duplication modules, each one creating one copy (see Figure 3.19). This way, at each duplication, one entity goes to a different intervention, while the other triggers the next duplication.

Deciding which attributes are assigned before duplication and which are assigned after is, for the most part, uncomplicated. Before duplication, the attributes to be assigned are those that are not affected by the interventions (Table 7.2). For most types of attributes, this is easy to discern (e.g., age, sex, ethnicity, treatment history prior to that point, and physiologic values at

TABLE 7.2

Examples of Attribute Values That Should Be Assigned Prior to Duplication and of Those That Should Be Reserved Until After Duplication

Before Duplication	After Duplication
Demographic	Intervention type
Physiologic	Duration of intervention
History (personal, medical, family)	Dosing, frequency, mode, venue
Living situation	Efficacy
Disease specifics	Adverse effects
Caregivers, health care context	Cost of intervention

that time), but it is possible that there may be doubt about some attributes. If the modeler is fairly certain that the characteristic (e.g., time to other-cause death) is not associated with intervention, then it should be assigned prior to duplication. Attributes that are obviously affected by intervention include characteristics such as dose and frequency, which may influence efficacy, costs, occurrence of side effects, adherence, and any monitoring or other clinical activities that are associated with a given intervention.

Although the times to various events are affected by intervention, their assignment should not be left entirely until after duplication. If that were to be the case, then it would be possible to assign by chance a worse time (e.g., sooner if the event is a bad one) to the cloned individual who receives the more effective treatment. To avoid this, and thus effectively reduce the nuisance variance, relevant baseline event times that reflect comparator effects should be assigned initially before duplication. This is tantamount to assuming that without intervention all the clones in a set will experience events at exactly the same time. After the intervention is assigned, relevant event times (i.e., those affected by the specific intervention) are modified according to the assigned intervention's effects. This ensures that all the duplicates are equally *lucky* in terms of these events, and that the intervention's effects are applied relative to that luck (Figure 7.2).

An alternative approach is to assign each individual before cloning a separate probability that represents the point along the cumulative distribution function for each time-to-event parameter (Figure 7.3). These attribute values, which are common to all clones in a set for each time-to-event parameter, are then used to select each entity's time-to-event from the relevant cumulative distribution function, reflecting the intervention assigned to each entity. This approach involves storing the random numbers that determine an individual's luck, a technique more broadly known as the use of common random numbers (CRN) (Bair et al. 2010; Swisher et al. 2003). This is described in more detail in the next section.

The advantage of duplication is that the copied entities will have an identical set of attribute values assigned before receiving the intervention. These

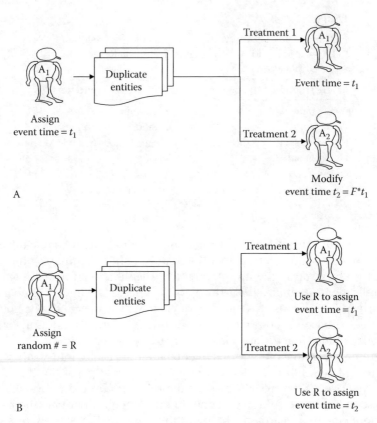

FIGURE 7.2
Two alternatives for ensuring clones are equally *lucky*. In panel A, the original entity is assigned an event time t_1. This event time is duplicated (along with other attributes), and after duplication, it is modified for the duplicate entity by a factor F reflecting the efficacy of the intervention. In panel B, the original entity is assigned a random number R which is duplicated and afterward is used to select the event time from the equations for treatments 1 and 2.

values can be used in predictive equations to ensure times to events, physiologic changes, and costs, and other responses are the same except for any intervention-specific parameters. It is important to note, however, that this resemblance holds only immediately after duplication—the duplicated entities are only identical at that point, but not necessarily throughout the simulation unless something is done to ensure that their experiences unrelated to the intervention are also made to match.

7.1.2 Duplication of Pathways

Another form of duplication relates to events experienced within the model that don't depend on the initial values or on the intervention but are still specific to the individual and should, therefore, remain alike for all duplicates.

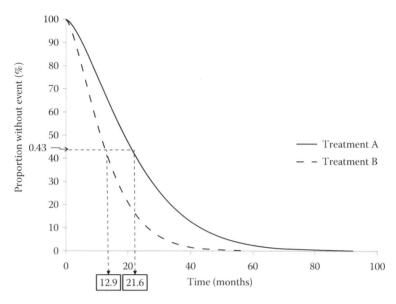

FIGURE 7.3
Use of an identical random number to choose an event time. The same random number (0.43)
is applied to identical entities, one receiving treatment A and the other treatment B. Applying
this random number to the different event-time distributions for the two interventions yields
times that appropriately reflect the impact of each intervention.

For example, if a disease event happens, say a mild stroke, some people may
be hospitalized while others are treated as outpatients. If the probability of
being hospitalized does not depend on which intervention the individual
has received, then in order to reduce nuisance variance further, all duplicates
should follow the same path when they have a mild stroke: if one is hospitalized, then all the others should also be admitted upon the occurrence of a
mild stroke.

Another example arises when a composite endpoint is used. In this situation, the time to something like a major acute cardiovascular event is estimated, and this can be appropriately different for each of the duplicates if it
is influenced by the interventions at issue. Since the health, costs, and other
consequences of such an event differ according to the specific component that
occurs, the composite has to be disaggregated. This is typically achieved by
applying a discrete distribution characterizing the proportion of each type
of event included in the composite. Thus, to reduce variance, it would be
desirable that all duplicates in a model of coronary artery disease prevention
by cholesterol-lowering agents make the same selection from that discrete
distribution, provided that the interventions affect the time of occurrence
but not the type of event.

Forcing experiences to be identical can be handled in two ways (Figure 7.4).
One is to ensure all of them are set before assigning the intervention. For example,

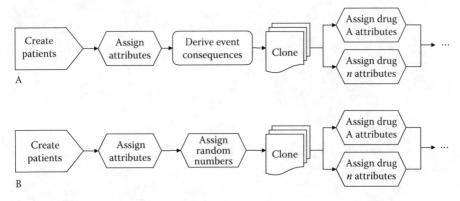

FIGURE 7.4
Two approaches to ensure that duplicate entities continue to have identical experiences except where dictated by the different interventions. In panel A, the consequences (e.g., hospitalization and death) of various events are computed before duplication and assigned to each entity. Duplication then ensures these consequences are carried by all the cloned entities. In panel B, the random numbers to be used to determine consequences are assigned before duplication but the consequences are not derived. When an event happens, the clones use the stored random numbers to ensure the same consequence ensues.

whether a patient will be hospitalized after a mild stroke could be preset in an attribute before duplication. Thus, all duplicates would have this value in their attributes and will follow the same path should they suffer a mild stroke. By the same token, in the cholesterol-lowering model, the type of major acute cardiovascular event, should one occur, could be derived and stored in an attribute for each entity before duplication. That way all the clones will have the same specific type of cardiovascular event.

An equivalent alternative approach is to assign all the random numbers that could be used in predictive equations or branch points prior to duplication. The assigned values are stored as attributes, and called upon as they are needed (e.g., at the time a patient experiences a mild stroke). This avoids having to calculate all the predictive equations beforehand, but the sampled values are predetermined by the stored random number (Chen 2013; Murphy et al. 2013). For example, an entity might be assigned before duplication the value 0.23 as the *hospitalize if stroke random number*. This means that, after duplication, all the clones will have this as the random number to determine if hospitalization occurs when they experience a stroke. If the probability of hospitalization for mild stroke were, say, 17% in the health care system of interest, then none of the duplicates will be hospitalized when they have a mild stroke (because 0.23 exceeds 0.17). Depending on the intervention effects, the times to stroke will be different among the duplicates, some might not ever experience the initial stroke and, therefore, never use the random number—but those who need it have it available in an attribute, and the value is identical for all duplicates.

This method of CRN can be used for as many events as needed in a model. Indeed, the duplication of entities is, itself, an indirect example of CRN. Sampling attribute values before duplicating is equivalent to applying common random numbers to multiple entities because the sampled attribute values are copied to each duplicate. Duplicating the entities is more efficient programming, however, since it limits the impact on memory requirements of added attributes to store random numbers and the number of steps that need to be taken (one sampling per attribute value needed, rather than as many samplings as there are duplicates). The efficiency gained depends on the specific software used.

7.1.3 Synchronizing Entities

Another alternative that avoids storing the random numbers or assigning all extraneous experiences before duplication is to synchronize relevant event pathways at the time one of the copies first experiences an event for which common subsequent pathways should be defined (Figure 7.5). To do this, the first of the duplicates that experiences the event *signals* the other copies that an event of this type has happened and transmits all other relevant information so that an identical event pathway occurs when it is triggered for each duplicate. For example, if the event is other-cause death, then whenever one of the duplicates experiences it (by definition

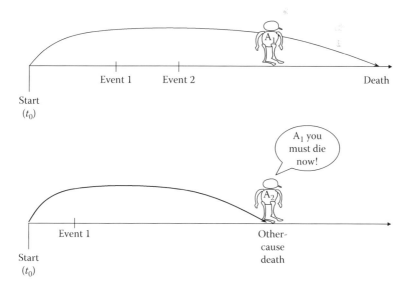

FIGURE 7.5
Use of *signaling* to ensure the experience of cloned entities remains synchronized except where affected by intervention. In this case, entity A_2 is dying due to causes unrelated to intervention 2, so it signals to its clone A_1 that it too must die at this point, despite the fact that death for A_1 was not scheduled to happen for some time yet.

in this case, this will be the first time among the duplicates), the affected entity sends a message *I am dying due to other cause right now* to its clones and the same event *other-cause death* is immediately triggered for those entities as well.

The method to accomplish this differs according to which software is used. In specialized DES software, there may be a built-in module that permits this signaling, but with many programs, it will require going to a deeper level or to the macro language that is built-in for accessing special functionality. Regardless, the common requirement is that the duplicates have stored the *entity IDs* of all their copies so that they all *know* who their clones are. In a spreadsheet, where the entities are defined by rows, this can be the row number of the entity. When an event requiring signaling occurs, the affected entity can then trigger an immediate event that uses each *entity ID* to change the corresponding *time-to-event* attribute for all the copies so that it happens right away, or even to send all the copies directly to the event that must happen. The modeler needs to think carefully through what should happen. For example, if one duplicate entity is in hospital when one of its clones dies due to unrelated causes, should the unrelated death occur immediately in hospital or should the hospitalization finish first?

7.2 Other Efficiency Improvements

Apart from the reduction of variance to increase statistical efficiency, there are several other ways to improve the efficiency of a simulation and reduce its runtime (Table 7.3). The execution time of a model depends on several things.

TABLE 7.3

Various Aspects to Consider When Minimizing Inefficiency in DES

- Reduce nuisance variance
- Increase hardware capabilities
- Diminish the number of calculations
  - Use time-to-event to advance the clock
  - Limit replications
  - Minimize number of updates
  - Keep logical checks to only those necessary
- Use global variables when individual information not needed
- Remove frivolous attributes
- Cut input/output operations
- Eliminate all debugging checks
- Don't use animation
- Avoid interactions with other software

An obvious one is the capabilities of the computer used to do the calculations. Both the speed and number of processors available are important determinants. If the software can utilize multiple cores (e.g., in the computer or in the cloud) and the model can run in parallel across various processors (Fujimoto et al. 2010; Taylor et al. 2011), then the execution time can be cut down substantially.

The next obvious factor is the number of calculations required by the model. The volume of computations in a replication depends on the number of entities simulated, how often their experience is updated, and the number of events and the complexity of the logic that needs to be processed. The more transactions the model has to deal with, the greater the required computer power or the longer the runtime. This is one of the reasons to structure the model based on time-to-event approach rather than periodic checking. How sensitive the processing is to the number of calculations can differ according to the software chosen for implementation and how the required code is written.

Other factors that can have a significant impact are the number of attributes and, to a lesser extent, the number of variables in the model. For a given computer configuration and chosen software, the modeler can determine the absolute and relative effects of these factors by running a simple test. This test involves creating a model with just a few steps and 1,000 entities at the start. Then add 10,000 variables each storing, say, a random number and record the execution time. This can be done quickly by assigning that dimension to a vector. Then, rerun the model with 100,000 and 1,000,000 variables (re-dimension the vector accordingly) and compare the execution times. Then do the same adding 50, 100, 200, and 500 attributes and time the executions again.

Certain model constructs require considerable amounts of computing power because they are evaluated at every time increment. This can occur when a model needs to repeatedly scan all entities for a specific value (e.g., checking to see if a case of an infectious disease like meningitis has occurred in a college dormitory) or a condition must be verified (e.g., *If* no resource is busy *then* close the clinic). Despite their adverse impact on execution times, it may still make sense to use these kinds of constructs because they more accurately reflect the problem.

Interaction with other software, like reading from and writing to spreadsheets or databases, usually increases execution time, as does calling other software to execute pieces of code. Sometimes, even interacting with a built-in macro language slows the model down. The number of external calls should be kept to a minimum with as much as possible being done at each call. For example, if information for attribute values is stored externally, then it is better to call the entire table (and any calculations) at once rather than having each entity doing it individually at various times. The efficiency loss differs depending on the DES software used, the software being called, and the operating system.

7.2.1 Time-to-Event Approach Instead of Periodic Checking

Since the number of calculations is a critical factor in the execution time, it is interesting to revisit the earlier discussion on *time-to-event* versus *periodic checking* (see Chapter 3). The efficiency implications can be seen with simple comparisons. For example, a model concerning a single event (which can occur more than once) and run with 1,000 entities being duplicated to compare 3 interventions will carry out calculations for 3,000 entities. If the time horizon is 10 years and daily periodic checking is implemented, then the occurrence of the event will need to be tested 3,650 times per entity. Thus, periodic checking implies almost 11 million calculations for this very simple model (3,000 entities × 3,650 checks = 10,950,000 calculations). This number of computations is massively reduced by using the time-to-event approach. Initially, an event time will be computed for each of the 3,000 entities. If, say, 50% of the event times fall within the 10 years, then the time for a second event will need to be calculated another 1,500 times. Thus, the number of computations drops by more than 99% to only 4,500 calculations. Even if some entities were to suffer a third event during the time horizon, the computations would be a tiny fraction of those involved in periodic checking. Since most of the conditions modeled for an HTA do not produce events frequently, the majority of calculations yield no events, and, yet, all of them have to be carried out for every time period over which the model is checked (e.g., hours, days, and weeks), regardless of the futility of doing so. This is massively inefficient: in this very simple example with only one event, the ratio is almost 2,500 to 1 for the number of calculations with periodic checking versus time-to-event approach.

7.2.2 Simultaneous Inclusion of Entities

There are two main ways to execute a DES, and they differ in their implications for efficiency, particularly if periodic checking is implemented (Figure 7.6). The basic approach, often taken if conventional programming in general software is used, is to simulate one entity at a time in a sequential manner. Once one entity reaches the end of the time horizon, it is removed and the next entity starts to execute the model steps. At any one time, there is only one entity being simulated. The other approach to executing the model is to allow multiple entities to exist in the simulation at the same time. As many entities as defined by the analyst are created at the appropriate entry points and admitted into the model simultaneously. They experience events according to the combined event calendar for all entities in the simulation and, thus, move through the model much as people do in the real world, each one living his or her life concurrently, without regard to others doing the same thing.

The main advantage of the one-at-a-time approach is that it is straightforward to program and can be implemented easily in any general software. The focus is on the single entity that is being simulated and the logic does

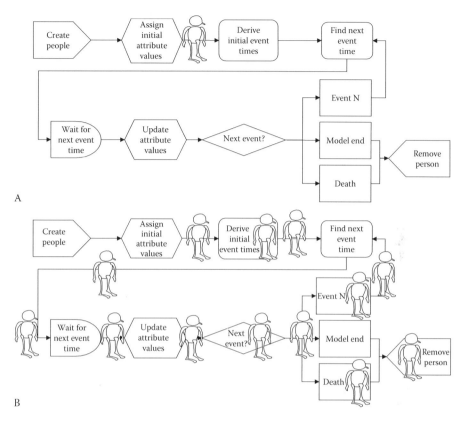

FIGURE 7.6
Two approaches to running the simulation. In panel A, the simulation is run one entity at a time. This may be inefficient, if the software has to allow time for the entity to pass from one module to the next or if processing of a single entity could take care of many entities at once. In panel B, multiple entities are in the simulation at once, so the software can process instructions according to whatever needs doing next and, if possible, can handle the processing of many entities at once (e.g., at the time of creation).

not need to consider what else might be happening to another entity. Since there is only one entity in the model at any time point, there is no need to distinguish between attributes and global variables. Entity-specific data and more general information all relate to the current entity in the model. The event calendar is singular and easy to maintain and searches to find out what the next event is for the entity.

Despite the simplicity, there are important downsides to executing one entity at a time. From an efficiency point of view, running the model repeatedly for a single entity is detrimental. All of the code that initializes the program, reads data from sources, processes and outputs the outcomes, and so on, has to be executed for every entity instead of only once for all entities. Although the exact loss of efficiency is software-dependent, one empirical test revealed that the penalty for running 1,000 entities one at a time was more than a five-fold increase in execution time compared to simulating the 1,000 entities simultaneously. This efficiency loss is magnified if the model is implemented via periodic checking for events. In this situation, the simulation is checking every day (or whatever the period is set to) to see if an event has happened to the single entity in the model at that time and must advance the clock at each step. If all entities are simulated at once, then this clock advance only has to happen once per period and not separately for each entity.

Beyond the efficiency problems, simulating entities only one at a time precludes any competition for resources and, thus, makes it impossible to consider capacity constraints, queues, and so on. This makes periodic checking inappropriate for problems like modeling transplant waiting lists or any health care situation where the interaction among entities is important (e.g., when patients' caregivers are also simulated).

In DES software, it is always possible to simulate as many entities as appropriate at the same time. Indeed, this is the default mode for DES software, although not absolutely required—the *create* event can be set to generate only one entity at a time and be triggered by the *model end* event so that the termination of one entity's experience allows the next entity into the simulation. Though possible, this is a rather clumsy way to execute the model. In general software, simultaneous simulation of many entities requires an appropriate *event calendar* to keep track of all event times for the entities in the model. It also presumes that there is separation between global model data and entity-specific *attributes*. In spreadsheets, the natural way to implement a DES is to model entities simultaneously because of the way calculations are processed. For example, if the experience of each entity is represented by entries in a row of a worksheet, and there are as many rows as there are entities, then the spreadsheet runs all the calculations simultaneously (at least from the modeler's point of view). Despite this, some spreadsheet implementations take the one-at-a-time approach by structuring the logic so that its sequence is represented across the rows of a worksheet and then using a macro language to run the calculations for one entity, store the outcomes, and loop back to run the next entity.

7.2.3 Using Super Entities

If a model implementation implies simultaneous simulation of a vast number of entities, the execution time can be prohibitive. For example, to assess the effects of implementing a public health policy (e.g., reducing salt in the diet) or a case finding approach (e.g., sending a mobile laboratory to test all people of a certain age for diabetes) would entail creating an entity to represent each individual at issue—everyone in the country if the intervention is considered at a national level. Modeling tens or hundreds of millions of entities taxes even the most powerful computers and can lead to impractically long execution times. Much efficiency can be gained in this sort of situation by grouping the simulated people by some relevant characteristics such as age and gender, risk level, or other appropriate features. Each group is then incorporated in the model as a single entity—a *super entity* that represents the entire category of people. The *super entity* can then experience events on behalf of everyone in that group, provided that all members are homogeneous with respect to the relevant prediction equations. When this is no longer the case (e.g., because one of the individuals in the group has experienced an event that alters his or her risks), either the *super entity* can release that individual from the group to be modeled as an ordinary entity from there onward or the entire *super entity* can be dissolved back into single-person entities, which are subsequently able to individually experience events (Figure 7.7).

If, at any one time, a large number of individual entities meet the homogeneity criteria, they can be re-grouped back into the same or another *super entity*. The *super entity* must carry in its attributes its defining characteristics,

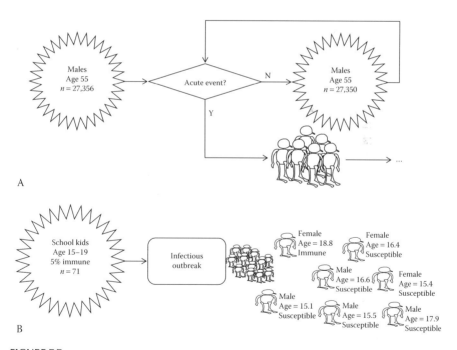

FIGURE 7.7

Two uses of a super entity. In panel A, a super entity is used to group individuals by demographic characteristics (the super entity for 55-year-old men is shown). Periodically, some members have an event (6 in this case), and they are *ejected* to begin individual pathways through the simulation, and the count in the super entity is reduced accordingly. In panel B, the super entities (the one for high school kids is shown) are dissolved by an infectious outbreak and the individuals are assigned appropriate attribute values. After the outbreak, survivors can be grouped back into super entities until the next outbreak.

counts of the number of people it contains, and any other information that is required to properly process its experience and assess whether it can remain as a group or should be dissolved. If one or more entities are separated from the *super entity*, they should carry in their own attributes information about which group they belonged to, and any information accumulated in the super entity's attributes that is relevant to their subsequent experience (e.g., time spent in the group).

For this approach to be worth the additional processing involved in grouping and splitting, the time spent as a *super entity* should represent a substantial proportion of the model time. An example of this is a model simulating an epidemic infectious disease over a century (Table 7.4), where there was sufficient time between the outbreaks for it to be worth creating super entities to hold the population while it waited for the next epidemic.

TABLE 7.4

Outline of a Model Using Super Entities to Simulate a Large Population Exposed to an Epidemic Disease

---

At the Start of the Model

Action 1:   Create super entities representing each risk location (pre-school, elementary school, high-school, college dormitory, college non-dormitory, adult not in educational facility, child not in educational facility)

Action 2:   Assign each super entity appropriate attributes, including its type and the number of individuals in the group

During the Simulation, While No Outbreak

Action 1:   Model transitions between super entities (e.g., graduation from high school and enrollment in college with residence in dormitory) by increasing or decreasing the value in the corresponding *group size* attributes

Action 2:   Wait for an outbreak to be triggered

At the Start of the Outbreak

Action 1:   Identify the super entity where the index case occurs

Action 2:   Split the affected super entity into its individual entities and assign common attributes (e.g., college dormitory dweller) and individual attributes (e.g., age, gender, family size, predisposing conditions) as needed

During the Outbreak

Action 1:   Expose other super entities to appropriate risks of also developing cases and split if cases arise

Action 2:   Model the infection in the affected individual entities as in any other model of that disease

Action 3:   Expose other individual entities to the risk of infection

Action 4:   Monitor the population for the end of the outbreak

At the End of the Outbreak

Action: Re-group the surviving entities back into their super entities (may have changed due to outbreak)

The gain in efficiency is obvious. For example, if the population is stratified by age and gender because these are the only two factors on which incidence of the condition depends, then 202 super entities would be enough (2 genders × 101 age groups) if age is classified by year and everyone from birth to 100 years of age is to be modeled. This replaces a population of any size—the entire planet of several billion if necessary. Even adding more factors and, thus, stratifying into more categories would still be considerably more efficient than modeling every individual.

7.3 Settings Affecting Model Execution

In DES software, many aspects of how the simulation is executed can be controlled via various settings that can be changed before a run. There is often the possibility to run the model in different *modes* that display varying amounts of information (with correspondingly lower execution speeds for more details). The details of these options differ among software packages, but one thing they all have in common is that the choice of settings will affect the execution time. For example, if a run is to activate various features for use in debugging, then the execution must be slowed down so that the information can be displayed and the analyst can examine it. This execution speed is often much slower than the maximum possible for the software.

7.3.1 Animation Settings

Animation of the entities and their experience can be useful for model debugging and walking model users through the model. DES software that offers animation as a built-in feature also provides various modes for controlling the degree of animation that is implemented during a replication, with the precise specifications of each mode differing according to the software package. The mode that is chosen will affect the execution time because the number of calculations required in the background to keep the animation updated will differ. In a mode that provides full animation, the status and position of all objects in the animation (entities, number displays, graphs, etc.) are updated repeatedly (the frequency is often controllable), and each update requires many calculations and refreshing of the display. At the other extreme is a mode, often referred to as a batch analysis, where the model disregards any animation instructions and performs none of the related calculations or updates. This mode provides for the fastest execution but makes it impossible to turn on animation during the run because the code that is executing has no links to any of the animated features. The batch option is most often used when a validated model is in final form and the objective of the replications is to collect the outcomes that will inform the HTA. Some software has an intermediate mode where the animation code is active but the screen display is temporarily disabled while calculations of the positions are being done. This reduces the execution time somewhat while still providing some animation.

7.3.2 Speed-Destroying Leftovers

In addition to using animation when building and validating a model, modelers usually add bits of logic, such as modules, processes, temporary variables, and calculation, and turn on special functionality, such as display of intermediate values and step-by-step execution. This is done to facilitate the process of construction and to help ensure that the model is behaving as designed. This can be an important component of *verification* (see Chapter 8). For example, various temporary *write* modules may be incorporated to output every entity's attribute values so the modeler can check if the values make sense. *Break* points that stop execution if certain event combinations occur may also be added to enable checks of the logic. Some software offer *debug traces* that display on the screen everything that is happening to one or more entities.

While extremely useful during construction and validation, these components are speed destroyers because they can involve the execution of many additional instructions and interaction with the rest of the model logic, and with the display screen for many of the things happening in the model. Thus, it is very important to keep track of all of these extraneous components and to ensure they are deleted or, at least, turned off when they are no longer needed for debugging or validation. One efficient way to do this is to condition as many of them as possible on a *debug-mode flag* that can be set on or off. This way, the components can be easily turned off when no longer needed, but turned back on if, for example, a revalidation is required after the model has been modified.

8

Validation

To properly inform decision makers, a health technology assessment (HTA) must provide information that is not only relevant to the problem they face but also credible (Mandelblatt et al. 2012). For a model used to inform the HTA, this credibility has to do with: how well the structure represents the disease, its course and management; the correctness of the programming and calculations; and the degree of accuracy with which the model outputs reflect the real world that the model is trying to simulate. If the decision maker is uncertain or, worse, suspicious of the model's credibility, its results are unlikely to be given much stock in making decisions (Watkins 2012). Thus, it is important that a modeler evaluates the performance of a simulation against suitable criteria, and that the results of this evaluation be made available to the decision makers or, for that matter, to any stakeholder who requests them (Afzali et al. 2013). Without this documented validation, it is difficult to assess whether the simulation results are worth paying attention to (Caro 2015; Caro and Möller 2014).

Validation of a discrete event simulation (DES) is, conceptually, no different to that for any other type of model. It involves three major steps. The first step is to confirm that all aspects of the design and structure of the model accord with what is known about the problem (commonly known as *face validity*, the extent to which the model appears to be appropriately representing the disease and the decision problem). The second step requires careful testing of all logic, programming, and calculations to make sure they are doing what they are supposed to do (commonly known as *verification* or *technical* or *internal validity*). Finally, the third step is to assess how well the results of the model reflect what happens in the real world (commonly known as *external validity*). Sometimes, an additional activity involving comparison with other models (commonly known as *cross-validation*) is implemented as well (Guo et al. 2014; Hoogendoorn et al. 2014). The process of validation should start from the moment it is decided to construct a model and continue so long as the simulation is still in use (Figure 8.1).

Guidelines for validation of models used in HTA have been issued by a task force that brought together many expert modelers (Eddy et al. 2012), and these are frequently cited by those publishing model-based analyses. In this chapter, we detail each type of validation with special attention to how the various aspects apply to DES. The first section addresses face validity, including the choice of modeling technique. This is followed by

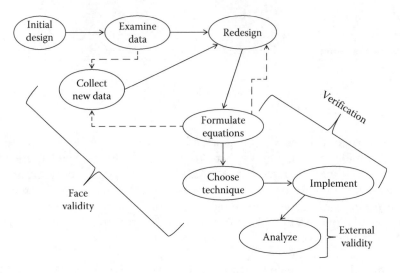

FIGURE 8.1
Process of modeling and validation phases. The ovals represent the major steps in the process of modeling, from the initial design, through the examination of data and consequent modifications of the design, to formulation of the equations and their implementation using a chosen technique. The process culminates with the analyses that inform the HTA. The brackets encompass parts of the process to show the various types of validation and their overlaps.

consideration of the main aspects of verification, although this sphere of validation can be subject to very personal styles and tactics. The very challenging topic of external validation is dealt with next and the chapter concludes with a brief look at other related areas, namely calibration and cross-validation.

8.1 Face Validity

The design of a model involves making many choices about what to include, how to include it, which inputs to parameterize, what data to use to inform these and how to analyze them, which scenarios to consider, what outputs to produce, and which analyses to carry out. Each of these choices can obey to multiple criteria, some of which may be conflicting. The modeler wants to reflect the problem as accurately as possible yet without unduly burdening the development and analysis of the simulation. While one may strive to use the best data possible to inform the inputs and equations, the feasibility, time, and expense of acquiring those data also need to be considered. One wants to be flexible to accommodate many stakeholders' points of view and possible requests for outcomes, but one must also remain mindful of the cognitive burden imposed by a massive set of scenarios and corresponding output.

TABLE 8.1

Roles and Responsibilities in Validation of a DES

Role	Responsibilities
Clinical expert	• Is the HTA problem definition consistent with clinical knowledge? • Are the objectives consistent with the HTA problem? • Does the influence diagram capture the necessary concepts and relationships? • Do the flow diagrams represent clinical practices adequately? • Is the event list sufficient? • Are all relevant event consequences considered? • Is the population appropriately defined? • Are the data sources the best available? • Do the equations reflect clinical understanding? • Are attribute value updates frequent enough and in the right context? • Do the results look reasonable?
Biostatistician	• Have effects (intended and unintended) of the interventions been estimated correctly? • Are the choices of distributions appropriate? • Have the equations been correctly developed, tested, and incorporated? • Is uncertainty correctly specified?
Modeling expert	• Is DES appropriate for the problem? • Is the time advance mechanism proper? • Do the flow diagrams reflect the problem and model conceptualization? • Is the implementation correct and sufficiently verified? • Has discounting been implemented properly? • Is variance reduction sufficient? • Are the number of entities and replications suitable? • Is the software adequate for the task?
All	• Does the model design meet the needs of the HTA? • Do the analyses address the problem fully? • Has uncertainty been relevantly and sufficiently assessed? • Is the model validation sufficient?

Although it is laudable to want to comprehensively address uncertainty—particularly that resulting from structural variations—planning to conduct a vast number of analyses can substantially impair the eventual completion of the simulation project.

All of these aspects require assessment by expert parties other than the modeler (Table 8.1). Their role is to ensure that the proposed model design, data inputs, and choice of modeling technique are face valid. This should be done before the model is implemented because once the programming has been completed, the modelers tend to become resistant to making changes that imply significant reprogramming. Needless to say, the assessment of face validity is very subjective, and a design choice that is considered acceptable by one expert may be turned down by another. Thus, it is highly desirable to have a number of people with the appropriate expertise and representing a spectrum of perspectives assess a model for face validity. This process may inform amendments to the model structure or may lead to a decision to reject the advice to alter the representation of the problem. In either case, the process that was undertaken to address face validity and who was involved should be reported, and justification of the final model structure should be provided in light of the expert advice.

For a DES, face validation should include an assessment of the characteristics of the model that led to the decision to use DES. While clinical experts cannot be expected to judge the technical details, they should consider whether the decision to use DES makes sense. This face validity assessment should cover the specification of the problem and its translation into the model logic. The list of events to be considered and their proposed consequences should be reviewed. The choice of approach to advancing the simulation clock should be addressed. Which population is to be modeled, how the population is specified, and the proposed sources of data for doing so; the set of attributes that will be incorporated; the proposed sources for initial values and how those values are to be updated are all important to consider. The clinical experts should also look at what information will be recorded during the simulation and when. A biostatistician should assess how the interventions' effects are to be estimated and applied and the development of equations and their inclusion in the model as well as the specification of all distributions and their incorporation into the simulation. Modeling experts need to examine how discounting will be implemented; the variance reduction techniques that are proposed; how many entities will be simulated in a replication and how many replications will be carried out in a run; and the choice of software and any peculiarities it may present, for example, in terms of its random number generator. All should consider which analyses are proposed, what aspects of uncertainty are to be considered, and how. Many of these aspects have been presented in previous chapters in terms of their implementation. In this section, they are addressed from the point of view of face validation. Some of these elements should also be re-assessed during verification to ensure that the implementation was accurate.

8.1.1 Problem Specification and Model Conceptualization

As the initial stage in the model development process, it is important to ensure that the problem being addressed has been appropriately conceptualized. This aspect of face validity requires confirmation that factors such as the disease or condition, eligible population, interventions to be evaluated, outcomes, and evaluation perspective have been specified appropriately. This will require consultation with clinical and policy stakeholders, in particular.

Once the specification of the problem has been validated, the next step is to ensure that the model is correctly conceptualized. Many assumptions and choices are made in the conceptualization of the HTA model and how this will be translated into the model logic. As well as assessing the validity of individual assumptions and choices, the assessor must ask whether groups of assumptions fit together well and, eventually, consider broader portions of the model, until the entire proposed simulation logic is reviewed. It is worth emphasizing that these assessments rarely yield a clear pass or fail. Instead, the assessor is evaluating how reasonable the model logic is and how far the assumptions may deviate from accepted ideas. Of course, there may not be a consensus in the field or the evidence may be conflicting. It behooves the modeler to justify the choices made and to consider building in a means of considering alternative structures in uncertainty analyses.

In constructing the model logic, it is important for the modeler to seek advice from experts in the relevant disciplines. Clinicians with the appropriate specialty are, of course, essential consultants. Epidemiologists and biostatisticians can also be helpful. As HTA has to do with resource use and costs, experts in the workings of the health care system of interest should also review the proposed handling of these elements.

To facilitate review of the model logic by the experts, it is very important to present the model structure clearly and as devoid of modeling jargon as possible. Early on in the process, the modeler should check whether the concepts selected for inclusion sufficiently encompass the problem and whether they are correctly interconnected. In addition, redundant or extraneous elements should be identified. This conceptual model should be presented to the experts using a device like an influence diagram, which clearly represents the proposed elements and how they link up. Later on in the process, flow diagrams provide the blueprints for the model and, with suitable symbols like those introduced in Chapter 1 and used throughout this book, offer an excellent means of conveying the simulation's structure. Some specialized DES software supply graphical displays of the model logic which can be leveraged in presentations to experts. Showing spreadsheets or programming code is rarely useful.

8.1.2 Is the Choice to Use DES Justified?

When a researcher identifies the need to construct a model to inform an HTA, one of the choices to be made concerns the technique to be used. While it might be tempting to stick to whatever method is familiar, it behooves the modeler to justify the choice in light of the specifications and characteristics of the problem (von Scheele et al. 2014). One strong justification may be that a DES is a good choice when the problem involves events because these are naturally incorporated in this type of simulation. Many diseases, perhaps most, entail events. There is the onset of disease, its diagnosis, testing, start of treatment, possible occurrence of side effects, disease sequelae, death, and so on. All of these are best thought of as events. It is awkward to try to conceptualize them in terms of states, even though one can try by defining states before and after the event. For example, the diagnosis event could be represented as the transition between a pre-diagnosis state and a diagnosed disease state but this is less straightforward than having a *diagnosis* event. Moreover, the latter facilitates incorporating a cost of diagnosis, a decision point, perhaps some implications for other events, and so on.

In some cases, a disease may be readily described by a series of states. These may represent, for example, stages of a cancer or severity levels. Does that argue against the use of DES? Usually no because, even in these situations, there is typically a mixture of events and states, not a pure set of states. DES allows the modeling of states as values in corresponding attributes such as *cancer stage* or *disease severity*. These can be combined with events that reflect the transitions between states, as well as with other events that are not easily viewed as states. The converse is not true. The state-transition (*Markov*) technique has no place for events.

Another justification for using DES that may be provided is that there is known to be heterogeneity in the population regarding relevant determinants (Davis et al. 2014). For most diseases today, something is known about the things that help determine what happens. It is important to consider these *risk factors* because their values modify a person's path through the simulation. If there are a small number of such factors and each one can only be one of a few values, then it may be possible to run the model by simulating a set of cohorts which cover all the relevant combinations. In the extreme, if no risk factors are known, then a single set suffices. If there are more than two or three factors, however, or if some have many possible values (e.g., a continuous range), then this cohort approach is not viable. (The common solution of modeling a cohort defined by the mean values of each factor is invalid.) The only proper approach in this situation is to simulate individuals, each one bearing a particular set of values. DES facilitates this via the entities and their attributes which can contain the factor values.

Needless to say, if the problem involves the explicit modeling of resources with limited capacities and ensuing queues, then DES is the best choice and easily justified. Indeed, this was one of the main reasons for developing the technique.

So, what grounds may those who are assessing face validity consider for coming to a determination that DES should not be used? One of the features of a DES is that the entities are unaware of each other or anything else in their environment. Each entity passes through the simulation blindly, guided purely by the logic, the various equations, random numbers, and so on. A very limited amount of interaction can be incorporated by using special attributes. For example, the cloned entities in a set can know about each other by storing the identifying numbers in attributes. Contagion can be modeled by including attributes such as *exposure* and *number of contacts*. A much better way to do this, however, is to use a different technique which extends DES to give entities, now called *agents*, additional features that specify their reactions to other entities and the environment. This *agent-based simulation* method is described briefly in Chapter 9.

Another concern that may argue against use of a DES is a paucity of data to inform the simulation. In order to represent the population of interest as individual entities, it is necessary to have information on the identifying attribute values. Although this is best accomplished by obtaining data on real individuals, this is not essential. Knowing, or being able to surmise, the relevant distributions can be enough. If even this cannot be achieved, then a DES may be overkill, unless its main purpose is to model constrained resources or an event conceptualization is paramount.

8.1.3 Is the Specification of Events Adequate?

At the core of a DES for HTA are the events that the entities can experience. In assessing face validity, then, it is important to ascertain that the proposed events are consistent with the problem at issue and what is known about the disease, the interventions to be compared, and whatever other aspects are to be represented. For each event, the modeler should indicate how the occurrence of that event will be determined (e.g., will a hazard function be used to determine a time of occurrence, or will a conditional probability for a particular time period be implemented, or will the event be a consequence of a logical test). Face validity addresses whether the proposed approach makes sense for that type of event.

The modeler also needs to enumerate the consequences of that event. These can be on accruing values (e.g., costs and quality-adjusted life years [QALYs]), on the value of the entity's attributes (e.g., disease severity and quality of life), on what happens to the entity (e.g., if the event is a logical gate), on future instances of that same event (e.g., how likely a recurrence is or how severe), or on the occurrences of other events (e.g., the risk of death). An event may even impinge on other entities (e.g., the probability of infection of contacts) if the model has been set up that way. For each consequence on the list, the appraiser of face validity must assess how reasonable it is and to what extent it is consistent with available data and extant knowledge. The proposed data sources for estimating input parameter values corresponding to events in the model and methods of analysis should be considered (Goldhaber-Fiebert 2012). In defining the timing of a particular event, for example, will the most relevant data source be used, will all important patient characteristics be represented, and is the approach to estimating the hazard function appropriate?

If there will be a specialized *updater* event (see Chapter 3), then the rationale for including it and which values will be updated should be reviewed. It is also necessary to gauge whether the proposed periodicity of this updating makes sense given how the affected parameters are assumed to change over time. Will it be frequent enough or, perhaps, too frequent? Is the way in which each updated value will be determined appropriate?

8.1.4 How Will Time Be Advanced?

Two methods for advancing the simulation clock were detailed in Chapter 3. The time-to-event approach jumps from one event occurring now to the time of the next event on the event calendar, while the periodic step technique advances the clock in fixed increments of time. If the latter has been selected, then it is necessary to examine when the events that occur during an interval will be assumed to occur and how accruing quantities will be determined. The appraiser should consider whether this approach creates too much inaccuracy, and this is largely a function of the period length and how rapidly events are occurring. More frequent (i.e., higher risk) events necessitate shorter intervals. The modeler may also be thinking of implementing an adjustment to minimize the inaccuracy; for example, by considering that all events happen at the mid-point of the interval. The appropriateness of this, and its potential impact on discounting costs and health outcomes, should be addressed.

Although this happens rarely in HTA, sometimes the modeler implements a delayed time zero to ensure that there are sufficient entities in their correct places at the start and attributes and variables have the appropriate values. If this will be done, then the appraiser should ensure that the rationale is sensible and the proposed starting configuration at the delayed time zero is appropriate for the problem being modeled.

8.1.5 What Entities Will Be Simulated?

During the design of the simulation, decisions will have been made regarding what types of entities are to be modeled. Almost certainly for an HTA, the main type of entity will be people who are subject to experiencing the events in the model. Often these entities will represent patients, but, in some cases, many of them may not yet have a disease (e.g., in an assessment of a screening strategy, the entities are candidates for screening). If illness occurs in these healthy people during the simulation, it is not necessary to change the entity type as the disease status can be managed via an attribute, provided that only a single type of *people* entity was implemented rather than separate ones for *candidate* and *patient*.

Other types of human entities may be incorporated but the face validity of the decision to do so should be appraised. A good reason for including other types of entities is that they will follow separate logic and have their own attribute sets. Making them a distinct entity type facilitates implementation of these differences. A common additional type of entity is people who provide care to the patient entity. They may be family members or paid caregivers. Rather than create many additional types to represent this mix, a single caregiver type should be included, and the relation to the patient should be addressed via an attribute. This parsimonious allocation of entity types should prevail. For example, instead of creating separate male and female entities, sex should be an attribute of people entities.

Another type of human entity that may be conceptualized is clinical personnel (e.g., nurse and doctor). In this case, the question is whether they are best represented as an entity type or as a resource. If the clinician needs to experience events and follow logic, then employing an entity type is reasonable, whereas if their only purpose is to account for usage and costs, considering them as a resource should be sufficient.

Other non-human types of entities may be included in a model (e.g., a laparoscope, an ambulance, and a bacterium). In these cases, the appraiser should assess the rationale for including them, using the same criteria: Do they need to experience separate logic and events? Do they need to have their own sets of attributes? For example, a *laparoscope* entity type may have been conceptualized because these instruments go through various sterilization procedures, can move from one surgical suite to another, and may bear attributes concerning their features, time they have been in use, longevity, and so on.

Occasionally, a simulation may include entity types that serve only modeling purposes. For example, if the model is to read out interim results at particular times, the modeler may opt to use a *report* entity type. Its sole function would be to trigger the *read-out* event and then wait until the next reporting time before triggering it again. Another example is an entity that sets off epidemics. This entity could experience a *start epidemic* event at appropriate times and cause immunity in the population of human entities to diminish. The face validity of including such types of entities in a model is not usually a concern—their purpose is transparency and ease of implementation.

Entities of each type will need to be created at appropriate times. This aspect is not specific to DES but can be more obvious because the technique usually makes the creation explicit (in spreadsheets it may be implicit in the table that lists the entities and their attributes). The appraiser needs to check whether the proposed times of creation are appropriate to the specification of the problem. For example, if incidence of a condition that identifies the entity type is to be considered, then are the creation events generating those entities scheduled at the appropriate times and in the correct numbers?

8.1.6 What Attributes Will Be Modeled?

Although other types of individual simulation also assign characteristics to individuals, a DES makes this very obvious because the attributes are conceptualized explicitly. The appraiser should check that the list of characteristics that will be incorporated as attributes includes all the necessary features but is not overly inclusive. Attributes whose values are not expected to affect the course of the simulation, or be modified by it, and which do not alter the accumulation of outcomes, should not be included. This aspect overlaps with the validation of event specification, for example, the following attributes related to cancer staging may influence the experience of recurrence events in a cancer model: *tumor size*, *node involvement*, *metastases*, and *location*. Sometimes an attribute may be built-in to enable future consideration of an element that is not yet active. For example, if it is expected that a biomarker test will become available and its value will alter the patients' course, an attribute *biomarker value* may be included but left unconnected to the course until the evidence is available.

Initial values may be assigned to attributes by sampling corresponding distributions or by using the values manifested by real people. If the former approach will be used, then the main face validity concern has to do with whether the resulting profiles will reflect people that actually exist in the population and do so in the right numbers. If values from real people will be used, then the issue is whether the resulting simulated population will be representative of the one required by the problem at hand. Neither of these concerns is specific to DES, and methods for checking the face validity of the proposed sampling are addressed in general modeling texts. Similarly, the sources of data used that will be used to set the initial values need to be relevant and credible, and this too has been addressed elsewhere (Caro et al. 2014).

Some of the attribute values may be updated during the course of the simulation, and it is a component of face validity to assess whether the updates are indispensable, and, if so, do they happen at appropriate intervals (not too far apart but not unnecessarily frequent either). For example, is it necessary to update a biomarker value such as glycosylated hemoglobin (HbA_{1c}) in diabetes at a constant interval (e.g., every day) or is it sufficient to do so when it is actually measured at a doctor's visit? The reason for updating the value also needs to be considered. Is it only needed for informing a treatment decision or does the value affect the course of disease? If the latter, then another aspect to consider is how the value is used in the risk equations. Is it appropriate to use an updated value or does the equation subsume value changes over time and, thus, recalculating a time-to-event is an overuse of the equation?

8.2 Verification

Even a modestly complicated simulation involves many moving parts, with dozens or hundreds of calculations, logic checks, Monte Carlo samplings, value modifications, output recordings, and so on. Depending on the software used, code has to be written, modules executed, or functions called. At each of these steps, there is potential for bugs to creep in (Tappenden and Chilcott 2014). In addition, the integration of all the steps into a simulation may generate additional errors even if the individual components are correctly implemented. Thus, it is essential that the entire process be comprehensively tested formally. This is best done as the pieces are built but must be repeated as interacting blocks are incorporated. It is best if the testing is done by someone different than the person constructing the model. Developers of software have elaborated detailed approaches to completing this kind of verification, and there is an extensive literature on the subject (Beyer 2014). The following sections address common aspects of DES for HTA that should be checked as part of the verification process.

The whole verification procedure should be carefully documented, including what was tested, by whom, how it was tested, when, and what were the results of the testing. In practical terms, an external assessor addressing this aspect, after the model is completed, can only rely on this documentation as repeating the verification is very laborious and not usually feasible.

8.2.1 Model Logic and Implementation

As noted in Chapter 3, whereas clinical episodes usually have a duration, simulation events are instantaneous. The way in which the modeler implemented clinical episodes should be reviewed, particularly to confirm that the occurrence of the corresponding events will not inadvertently shelter the entity from exposure to other risks. For example, if an exacerbation of disease can lead to a hospitalization, will the *admission to hospital* event still allow other risks, such as that for *infection*, to apply or is the entity protected from them because the admission logic isolates the entity?

The manner in which the simulation is terminated needs to be assessed. Does this happen after a fixed period of time? Is it triggered by the occurrence of a specialized event? Is the simulation allowed to run until all entities have left the model? It is important to verify that when the simulation ends all accrued outcomes will have been accounted for and reported. In particular, if the termination can occur while there are still entities in the simulation, provision needs to have been made to record their accumulated survival and costs to that point. This may have been done by forcing them to pass through an *end* event with the appropriate consequences.

If the time-to-event method for advancing the clock has been implemented, then the computation of the initial event times should be assessed, but the main concern is with any recalculation of times implemented upon the occurrence of events. The appraiser should verify that the times that are updated are only those for events whose risk is affected. For example, if an entity experiences a *cancer diagnosis* event, then his or her *death* time should be recalculated but not that for *onset of diabetes*. For each of the affected events, the new time should be determined so that it correctly reflects the appropriate direction of change in the risk. For example, if *cancer diagnosis* is experienced, then the new *death* time should presumably be sooner than the previous one. If the modeler has inadvertently allowed the entire distribution of death times to be resampled, then the new *death* time could be inappropriately much later than the previous one.

As HTA usually compares one intervention to others, the entities are typically duplicated so that identical clones can be assigned to each strategy. For example, if the problem involves a comparison of five antihypertensive treatments, then each entity is duplicated to produce four clones, and each of the five entities in a set is assigned one of the treatments to be compared. When this cloning has been implemented in a model, the appraiser needs to check that it was placed at the correct point in the logic. Before the cloning occurs, the values of all the attributes that should be identical should be assigned; values for those attributes that differ should be assigned after the cloning; and additional modifications may have been implemented to accord preexisting values with the new intervention. For example, if patients with newly diagnosed hypertension are to be modeled and the risk of stroke is altered by the effectiveness of the five antihypertensive treatments, then the *time to stroke* should be computed for each entity prior to cloning based on the risk in untreated hypertension. Then after cloning, this time would have to be modified for each entity in a cloned set, given the effect of each one's antihypertensive treatment.

There are many other aspects that should be assessed to fully verify a model, but these are not specific to DES. A model typically uses a number of distributions, and the assessors should check their specifications to ensure they properly reflect the parameters they are representing. Then, how they were incorporated in the simulation should be reviewed. This depends on the software selected for implementation of the model. In a similar vein, the inclusion of any equations in the simulation should be examined along with the methods used to develop them and the sources of data. Each intervention compared in the HTA will have various effects that need to be applied. How these were estimated from the source data, and how they are applied in the simulation should be assessed. In most HTAs, the values that accrue over time are discounted to account for time preferences. This discounting is often applied discretely in models that use a fixed step, but in DES, where continuous time is operating, a more accurate technique can be implemented for discounting. Either way, the approach used should be considered. Each replication of a model records information during the simulation. The appraiser should determine whether the type of output and when it is recorded appropriate to the problem at hand.

When it comes to the analyses, the appraiser should check that they provide the information required by the HTA, and that the various types of uncertainty have been considered and how. It is also important to assess if sufficient entities have been simulated in each replication and how many replications were carried out in a run.

Another thing to check is the efficiency of the implementation. Although there is no loss of validity either way, it is better if the more efficient approaches have been implemented. For example, if the individual values for a parameter are not necessary, then it is inefficient to carry it as an attribute. A simple check, then, is whether any of the attributes would have been sufficiently represented as global variables. Although not expressly an issue of validity, it is helpful if the modelers have implemented variance reduction techniques. If so, these should be inspected to ensure they are operating properly.

Finally, the choice of software should be addressed in terms of any peculiarities it may present. One that can be problematic for a DES is use of random number generator that does not provide a sufficiently long series of values before repeating.

8.2.2 Equations

All equations used in the model should be tested separately before being entered into the model. This is commonly done in a regular spreadsheet with the purpose of ensuring that they are correctly expressed and give the right results for entered values. The equations should also be calculated with extreme values on the borders of the model scope to make sure that there are no problems in those regions.

8.2.3 Code

To test all aspects of the model when it is completed is a huge task, due to the complexities and interactions (Kimmel et al. 2015). The sensible approach is that the modeler keeps testing the model as it evolves, when it is easier to check that each component is implemented correctly. This is facilitated by constructing the simulation in modular fashion. The module testing consists of checking that the module's output is correct for a given set of inputs. As the modules are added on, repeated testing should take place, thus ensuring correctness of the behavior of the increasingly complex model. A useful approach is to simulate single, well-defined patients step by step to make sure the paths and events he or she experiences are the expected ones.

Another test the model should withstand is an extreme value test. The rationale is two-fold: it will evaluate the effects toward the extremes of the ranges of the equations, indicating if their combined effect is reasonable at the edges, and it will also give some immediate feedback. For example, if mortality is turned off, does everybody survive? This kind of test should be set up in advance and the modeler should specify what the anticipated outcomes are for the different values. Any differences between what the simulation produces and the expected results need to be investigated (Table 8.2).

It is also important to re-verify the model once any errors have been corrected, since the interaction between model parameters may be more complicated than it seems to be. Correcting some of the error may reveal other errors in the model. Similarly, verification has to be repeated after any changes are made to an already tested model.

TABLE 8.2

Situations That Should Be Considered Red Flags during Verification

- QALYs that are higher than the life years lived
- Observing negative life years or QALYs
- Average life years or QALYs that exceed the time horizon
- Negative costs (not only overall but any cost component)
- More events occurring than is possible (for instance, observing 1,003 deaths in 1,000 patients)
- Discounted values that are greater than undiscounted values
- Untreated patients who have treatment costs
- The sum of component costs not equaling total costs

8.3 External Validation

Assessing the degree of face validity is a necessary exercise but is not enough to establish the validity of the model. Verification of the simulation implementation is also a key task but it too is insufficient. A model may be judged as reasonably conceptualized and accurately implemented, but it may still not yield results that are accurate enough for the purpose at hand. Thus, validation must extend beyond face validity and verification, and various additional checks need to be completed to demonstrate that the model outputs reflect what would be observed in the eligible population. There are three broad forms of this external model validation: dependent, independent, and predictive validation. None of these are peculiar to DES, but setting a DES to undertake empirical validation can require significant modifications to the model to ensure that relevant outputs are captured.

8.3.1 Dependent Validation

Dependent validation involves the comparison of model outputs to the results observed in a dataset that was used to inform one or more of the model's input parameter values or equations. As an example, for a model of osteoporosis treatment, a dataset obtained from a registry of women with osteoporosis may record fractures occurring over the follow-up time. If that dataset was used to derive an equation for the distribution of fracture event times, then a necessary dependent validation is to compare the predictions made by the model with what actually happened in the registry. For this to be a reasonable test, the DES must be set so that it is modeling a population that is as similar as possible to that in the registry. Any treatments present in the registry would need to be incorporated as well, and any other aspects that may modify the risk of fractures should be set to replicate the registry as closely as feasible. The model is then run over the same span of time as the registry follow-up, collecting information on the fractures that occur during the simulation.

If published outcomes will be used for validation, then model outputs must be processed so that they can be compared to the published results. For example, if a Kaplan–Meier curve showing the cumulative proportion of patients not yet experiencing a fracture is used, then the model output should be used to re-create the same kind of curve. Once this is done, the model-derived output can be compared to what was actually observed in the registry. If primary registry data are available, it should be possible to generate observed data for a wider range of model outputs. Repeated model replications using populations of the same size and characteristics to that observed in the registry can be undertaken to obtain stable estimates of the mean value and the variance around a model's predictions of observed parameter values.

External validation can use statistical techniques developed for other comparative analyses, but with the strong caution that variance of simulation data can be artificially low because the modeler can increase sample size at will. Thus, statistical significance of any differences is an unsatisfactory measure. By the same token, statistical significance tests the differences over the entire curves while the modeler may view some differences (e.g., early on) as more important than others.

Dependent validation can be undertaken using a run of the base case model to generate estimates of the relevant model outputs. It can also be carried out by incorporating uncertainty in a series of model runs, perhaps using probabilistic sensitivity analysis. Including uncertainty affords the opportunity to compare the spread of the predicted outputs to the observed spread, though the comparison becomes more complicated!

Typically, any given output from a simulation draws from several components of the model. For example, the occurrence of fractures in the preceding example clearly depends on the time-to-event equation formulated based on the registry data. It also depends, however, on any treatment effects that are applied, perhaps on how other aspects evolve (e.g., smoking), and so on. These other aspects may be informed from data sources other than that registry. In this case, comparing the simulated fracture occurrence to the observed one tests more than the time-to-event equation, and this additional testing covers items that are not dependent on the registry. This extended coverage by one external validation test is called *partially dependent* validation. Another example is examination of a different outcome. The fracture equation indirectly informs estimates of survival because the occurrence of a fracture alters mortality. Hence, the survival results are partially dependent on the time-to-fracture data taken from the registry for use as model inputs. Since survival itself was not estimated from the registry, comparing simulated and observed survival over the defined time period is also a partially dependent validation.

Whether dependent or partially dependent validation is undertaken, and whether a single run is used or uncertainty is considered, a decision about how much deviation is acceptable is generally required (it is rare for a model to predict observed outputs with 100% accuracy, even for dependent validation). If a single data source is used to populate input parameters over the time period for which the dependent validation targets are available, and the simulation can be set up to otherwise faithfully reproduce that data source, then the model should predict the validation targets with a high degree of accuracy. If multiple data sources contribute to the predicted model outputs, as in partially dependent validation, some deviation from the observed values may be acceptable but it is not possible to categorically state how much. Any departures from the observed outputs should be investigated, and the rationale for such variation reported.

Dependent validation demonstrates that the model yields outputs that are consistent with the data sources used to populate the model, but it does not necessarily establish that the model outputs are representative of the expected outcomes in the population for whom a technology is being assessed.

8.3.2 Independent Validation

A stronger form of validation than the dependent type is one that compares model outputs to what is observed in a dataset that was not used in any way for the construction of the model. This *independent validation* tests the ability of the model to predict the outcomes in a particular setting that was not used by the developers when deriving the model's structure, inputs, and so on. To do this, the modeler must find a dataset that is unrelated to those used in building the model. That setting of the validation dataset must be specified in sufficient detail so that relevant model parameters can be set to ensure that the simulated population replicates as closely as possible the validation population. For example, if the model considers osteoporosis in terms of the score on a bone mineral density test assigned at baseline, then the validation dataset should describe the distribution of this score in the observed population so that the model can be set to replicate that setting reasonably closely. Needless to say, at least one outcome must be reported in the independent dataset that is also simulated in the model, and it must be possible to reflect the specifics of the real outcome in the simulation. For example, if fractures in the validation dataset are detected by symptoms and asymptomatic events are not considered, then the model must be able to do the same things. It is preferable if the validation targets cover as many components of the model as possible, including intermediate and final endpoints. If these conditions are reasonably met, then the independent validation can proceed following the same steps as for the dependent one.

A major problem with conducting independent validation is finding a data source that meets the conditions and is reasonably representative of the population of relevance to the decision problem addressed by the technology assessment. If such a source has been identified, then the vast majority of modelers will succumb to the temptation to exploit it to inform the model construction, rather than reserving it for independent validation. Splitting of the dataset may be a viable option if there are sufficient data. This way, a good dataset can be used to help build the model but can also be leveraged in the validation later on.

In most cases, independent validation is conducted for the model's outputs pertaining to the technologies that are used in practice and against which the newer technology is compared, because the historical data from which validation targets are drawn will reflect the experience with those older technologies. If validation data are available over a very extended time horizon, it may be necessary to account for variation in clinical practice over that time period or restrict the validation data to a period that is most reflective of contemporary clinical practice.

For a variety of reasons—regulatory pressure, marketing demands, and so on—health care technologies are often investigated in observational studies that register all users of the technology, or all candidates for use. The registries can provide a good source of data for validation because they reflect outcomes in real clinical practice and often provide data over a long time horizon. They can be analyzed for the full population of interest to the decision problem and are typically large enough to enable splitting for use in model construction and validation.

It is very difficult to identify relevant data sources that can be used to validate model outputs relating to costs and outcomes like QALYs. These kinds of data are rarely collected, even in registries. Increasing use of electronic health records, and the routine collection of patient-reported outcome measures (PROMs), as being implemented in the United Kingdom, may provide valuable data for improving the scope of HTA model validation.

The acceptability of the fit of the model outputs to the validation targets will depend on the relevance and quality of the independent data sources, including the extent to which the model can replicate the clinical characteristics and management pathways of the individuals represented in the validation data. Any divergence between the model outputs and the validation targets should be assessed, and a rationale should be reported for the acceptance of the model as being sufficiently valid for the purpose at hand.

If divergence from the targets is judged to be significant, and not sufficiently explained by the lack of relevance and quality of the independent data sources, it may be necessary to revise the model. Alterations to the model structure may be considered, but another option is to calibrate the model's input parameters so that the outputs accord better with the independent validation targets. If this step is taken, it is important to avoid the calibration of irrelevant parameter values. Thus, inputs should be sampled from probability distributions describing the uncertainty around the true value of each input parameter (as required for probabilistic sensitivity analyses). An advantage of model calibration is that it can be integrated with the uncertainty analysis of the model, by assigning probability weights to convergent sets of input parameter values (Haji Ali Afzali et al. 2013; Jackson et al. 2015; Vanni et al. 2011).

Independent validation is not commonly reported for HTA models of any description, including DES, but it is perhaps the most important form of validation because it involves comparison of model outputs to observed outputs derived from data sources that were not used to populate the model.

8.3.3 Predictive Validation

Predictive validation is a special form of external validation that can provide the strongest assurance of validity (Figure 8.2). It involves setting up the model so that it matches as closely as possible an independent clinical trial, registry, or other data collection effort that has not yet reported its results (ideally it has not even completed obtaining the data). The simulation outputs resulting from this setup are then reported and stored until the independent study concludes and reports its results. The simulation outputs are then compared to the observed data. Since the former were produced by a model that could not possibly be influenced by the independent study, which represents events that occur subsequent to the completion of the model, the validation is immune from any conscious or subconscious manipulations to bring them closer together.

Validation against events not yet observed can provide post-hoc support for decisions that were informed by the model, but it is most useful for

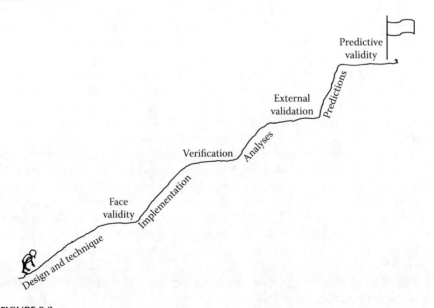

FIGURE 8.2
Attaining the validation summit. In many ways, the process of validating a model is like an arduous climb up a mountain. The effort is considerable but the higher you get, the better it is. Unlike a pleasure trek, however, getting as close as possible to, or ideally attaining, the summit is essential.

models that are expected to be used repeatedly to inform many decisions. Predictive validation provides evidence of the credibility of the model in supporting ongoing or new decisions over time. In the context of HTA models, it is particularly useful to validate model outputs relating to new technologies, for which limited independent validation data may have been available to inform initial validation exercises.

The predictive validity of a model can use routine data sources such as registries that continue collecting relevant data beyond the initial model development period. New research external to the HTA process may be very relevant, but it may be difficult to obtain, all the information required to set up the model to replicate them. It is also possible for a research study to be designed specifically to support the predictive validation of a model.

The concept of coverage with evidence development is gaining in popularity across regulatory authorities such as the Centers for Medicare and Medicaid in the United States, National Institute for Health and Care Excellence in the United Kingdom, and Pharmaceutical Benefits Advisory Committee in Australia. Where there is significant uncertainty around the cost-effectiveness of a new technology, funding for that technology may be provided on a contingency basis. The manufacturer is asked to collect further evidence specifically to better inform the assessment. The funding decision is scheduled for re-assessment at a later date, and there may be additional consequences if the observed efficiency differs from the estimated one used to justify coverage. While the aim of such studies is to reduce uncertainty around the value of the technology by providing more precise input parameters, this kind of research is also a tremendous opportunity to engage in predictive validation of the model.

9

Special Topics

Previous chapters have addressed the central concepts pertaining to a discrete event simulation (DES) developed to inform a health technology assessment (HTA). In this chapter, several additional topics that are not specific to DES are covered. The chapter begins with a delineation of what needs to be documented during the process of designing, constructing, validating, and analyzing a model. This is followed by a brief explanation of a tool not much used for HTA DES, the animation of a simulation. Software, both specialized and general, is described next. The chapter concludes with extensions of DES; first a short section on agent-based models (ABMs) and then the various hybrids with other techniques are considered.

9.1 Documentation

As in all project work, especially programming, documentation of what was done is of great importance. This documentation should cover the conceptualization, data sources, implementation, validation, and analyses of the model—something which applies to simulation of any type not just DES (Figure 9.1). Without proper documentation, it will be difficult to communicate to others what was done and why, and this will impair credibility of the results of the simulation. Moreover, good documentation will facilitate the task for anyone (including the original project participants) who needs to update, modify, or adapt the model in the future; or who needs to respond to questions from authorities, peer-reviewers; and even for other modelers seeking to explain why their results differ. Although at the time things are done, the details may seem very obvious and scarcely requiring note taking, this will diminish rapidly with time and other intervening projects. Thorough written documentation meticulously maintained while the particulars are clear is the only safeguard.

This requirement for careful documentation applies regardless of the duration of the project or the intended life cycle of the model. Even quick and dirty models designed for fast estimation of the potential for a given technology may be subject to review and questioning, and they may evolve into the model to be used for the full HTA process. Therefore, there is little excuse for not maintaining proper documentation.

Project Management:
☐ Meeting minutes
☐ Change requests

Design:
☐ Influence diagram
☐ Flow diagram
☐ Rationale for decisions
☐ Model specifications

Implementation:
☐ Programming notes
☐ Log documenting build actions, changes

Analyses:
☐ Input for each analysis, including settings
☐ Corresponding results

Input Data Sources:
☐ For each input
   ☐ Source
   ☐ Justification for using it
   ☐ Processing of data
☐ Articles listed as sources, ideally with the specific data highlighted
☐ Statistical analyses, resulting equations, and rationale for choices

Validation:
☐ Steps taken for
   ☐ Face validity
   ☐ Verification
   ☐ External validity
☐ Studies used for comparisons
   ☐ Rationale for selection
   ☐ Results of the comparisons
   ☐ Actions taken in response

FIGURE 9.1
A partial checklist of items that should be included in model documentation. The items on this list provide some idea of the extent of documentation. How these are grouped into specific documents for reporting is up to each modeler.

Documentation of a DES follows the major steps of the modeling process (Figure 9.2). The degree of formality imposed will depend on the organization, funding requirements, and other demands. At its most formal—for regulatory work for example—every item must carry the name of the record keeper and the date of the notation, and this must also be done for any modifications. For most DES projects, this level of formality is unnecessary, and it suffices to note who the overall authors of a document are and the date of the report. One exception may be changes to the model programming, as noted below.

9.1.1 Design Stage

During the design stage, the documentation is largely about the decisions that are taken in reducing a large complex reality to a manageable but sufficiently detailed representation in a model. It is important to record not only what was decided but why, and it is helpful to know what alternatives were considered and why, they were rejected. These records should include minutes of the kickoff and other meetings and the successive versions of the detailed specifications as they evolve (Table 9.1). The specifications should cover the problem description, the objectives of the modeling project, the scope of the model (including populations covered, interventions analyzed,

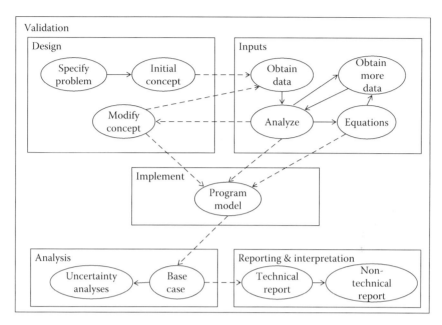

FIGURE 9.2
Depiction of the main elements of the modeling process. Each of these aspects requires detailed documentation, which is best kept as the process unfolds.

TABLE 9.1

Design Stage Records

• Minutes of kickoff	• Scope of the model, including
• Minutes of other relevant meetings	• Populations covered
• Versions of detailed specifications as they evolve	• Interventions analyzed
• Problem description	• Time horizon
• Objectives of the modeling project	• Extent of the disease considered
	• Economic perspective
• Influence diagram	• Analyses to be done
• Explanation and rationale for included items	• Choice of modeling technique
• Rationale for excluded items	• Justification
	• Software to be used
	• Time frame for project

time horizon, extent of the disease considered, economic perspective taken, etc.), the analyses to be done, the choice of modeling technique and its justification, the software to be used for implementation and analyses, and the time frame for the project. The influence diagram should be included as part of the specifications. All of the items included in the influence diagram should be explained and the rationale for inclusion given. It is helpful if the various iterations are included, or at least there is a discussion of what was left out and why. As the design stage proceeds, the flow diagrams for the model will be developed, and these should become part of the specifications. Often, the very process of documenting what is being done uncovers latent disagreements among the modelers and experts and can foster discussion leading to a better understanding of the concepts involved.

The documentation kept during this stage becomes a core element of the reports that are prepared at the end of the project. For studies that are undertaken on behalf of others who are paying for the work, this documentation can be extremely helpful in addressing any contentions that may arise regarding the degree to which the finished model meets the specified requirements. Without these records, the modelers may be faced with undertaking expensive rebuilds as they cannot confirm that the model corresponds to the specifications.

9.1.2 Input Data Stage

The search for data with which to populate the model begins almost at the same time as the design stage, though it should be guided by the design. Necessarily, there will be interaction between these two stages, with the design launching the quest for the required data and the results of these searches and of the analyses of the data obtained leading to modifications of the design. Documentation of the data side of this process (Table 9.2) involves the details of what was sought; where it is searched for, how, and why; what was obtained (and what was not and why); and how the data were processed for use in the model. The specifics of obtaining data also need to be documented, but this epidemiologic work is beyond the scope of this book, as is documentation of the statistical processes followed in analyzing the data.

TABLE 9.2

Data Records

• Data item description	• Data processing
• Search process	• Standard parametric distributions
• Where searched	• Decision on which to fit
• How?	• Decision on which to select
• Why there?	• Rationale for selection
• What was obtained?	• Individual or aggregated data
• What was not found?	• Was meta-analysis of studies done?
• Why?	• Methods used
	• Studies selected
	• Rationale for selection

Decisions about which standard parametric distributions to fit, which ones were selected, and why should be carefully documented. Details should include whether the fits were undertaken using individual data or whether aggregate information was used (e.g., a published Kaplan–Meier curve). If information was obtained via meta-analysis of various studies, the methods used need to be documented. Other types of data collection (e.g., Delphi panels, surveys) will have their own specifics that need to be recorded. Often for HTA, administrative data (i.e., information collected for organizational purposes other than the objectives of the HTA) are used, and the processes followed to obtain access narrow down the data to the ones required for the HTA and analyze them should be documented.

Much of the documentation produced during this stage will be highly technical in nature and meant for experts in the specific methods being described. They will, thus, form part of the core of the technical report produced at the end of the project. Indeed, they may be so technical and extensive that they will be consigned to separate reports which will either be summarized in the main one or incorporated as appendices. Nevertheless, the sources of data, basic decisions about them, and their justifications should be included in the non-technical report as well.

9.1.3 Implementation Stage

Constructing a DES resembles in many ways the development of software, and the techniques that have emerged in that field for documenting the code are very helpful in the DES area as well. As much as possible details should be kept within the model code itself. In this way, the documentation follows the model and cannot be accidentally forgotten or deleted. Somewhat depending on the software used, this recordkeeping can be easy to maintain as modifications of the model code are made. The notes are, thus, always available with the model, rendering them accessible to anyone looking at the model code (Figure 9.3).

General programming languages typically offer a commenting option that distinguishes some text from the actual programming statements. This facility can be used to maintain documentation of the code, though very extensive explanations may distract from the programming components. In specialized DES software, there is usually a way to add notes to the modules, and it may be possible to insert pure text boxes into the graphical depiction of the model. Other elements such as arrows, color, and the use of suitably named submodels can help describe the model logic. As the specialized software takes care of the actual programming code, the modeler does not need to include comments there, but it may be possible to export the actual code to a text file, where additional notations can be incorporated. In spreadsheets, the opportunity for documentation exists via the commenting feature tied to specific cells. In addition, the modeler has the option of using portions of the worksheets as spaces where notations can be made. Indeed, some modelers

```
Option Explicit              'Requires every variable used to be declared
Dim nbrPatients As Long      'Number of Patients in the Model Population
Dim TimeHorizon As Double    'Model time horizon (in years)
Dim WeibullShape1 As Double  'Weibull dist Shape parameter(for PFS)
Dim WeibullScale1 As Double  'Weibull dist Scale parameter(for PFS)
Dim i As Integer             'Used as loop variable in For-Next loops
Dim Treatment(3) As String   'Contains the names of the Tx

Sub InitiateModel()
' ~~~~~~~~~~~~~~~~~~~~~~~~~~~~~~~~~~~~~~~~~~~~~~~~~~~~~~~~~~~~~~~~~~~~~~~~~~~~~~~~~~~~~~
'  Purpose: InitiateModel will initiate all values and set up patients
```

A

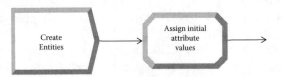

B

```
aSex is 0 for Male and 1 for Female
Age-distribution is in 10 groups with agemax(iAgeDistMx(aGender,1-10))
```

	A	B	C	D	E	F	G
1	Entity ID	Sex	Age (yrs)	LDH (Units/l)	Pro ind		
2	1	Male	55	229	11.5		
3	2	Female	58	285	9.52	0.8242	6.4
4	3	Male	68	195	15.82	0.3939	12.1

jorgen.moller: This is a comment that can be attached to any cell in Microsoft Excel®, and work as a description on the cell content

◄◄ ► ►◄   Patients ⟨ Sources | Assumptions ⟩ Model

C

FIGURE 9.3
Documenting a DES in the programming code. Panel A shows the documentation in a general programming language, using comments within the code denoted by an apostrophe. This example is from VBA and the comments would show up as green in the programming editor. Panel B shows one way to add documentation in a bespoke DES programming platform (here Arena®) by writing free text notes in the model design area. Tool-tip comments that pop up with mouse-over of modules and for definitions in tables could also be added. Panel C shows documentation in a spreadsheet program (here Microsoft Excel®) using a comment that can be added to any cell. Two dedicated sheets with documentation (Sources and Assumptions) are circled. VBA, Visual Basic for Applications.

reserve one or more worksheets purely for documentation, treating these more as text documents than as calculation tables. If a macro language (e.g., VBA) is used in a spreadsheet to implement some of the required components of the DES, then the same commenting particulars apply as for general programming languages.

Aside from comments, the use of descriptive names for all the components of the model will make it much more *readable*. The temptation to use intricate abbreviations should be avoided as they will have meaning only to the originator (and even then, it often is so only for the short time the modeler remembers the abbreviation scheme). If abbreviations are employed, these should be used consistently and documented clearly.

The use of descriptive names is somewhat subject to the conventions of the specific software. Many programs reserve certain words for their own functions or disallow the use of some symbols or spacing between terms. Whether capitalization is distinguished also varies. Within the particular bounds of the chosen software, the modeler can improve documentation of the model by specifically naming components so that it is clear to any reader what each one is. In spreadsheets, the naming function can be used to do this, and then in formulas the name is used instead of the cell references. It is much easier, for example, to understand quickly a formula for determining which coefficient to use in an equation if it reads IF (*male, BetaMale, BetaFemale*) than one that says IF (B2 = 1, Inputs!c21, Inputs!c22). Chapter 4 provides many examples of the use of this kind of naming to simplify the formulas required.

Specialized DES software typically provides default names for all the components, often numbered sequentially. It is not helpful, however, to see a series of modules labeled *Create 1, Assign 1, Assign 2, Separate 1, Assign 3, Assign 4, Delay 1*, and so on (Figure 9.4). Much more useful would be *Create patients, Assign initial attribute values, Select initial event times, Duplicate entities, Assign comparator treatment, Assign new intervention, Delay until next event*, and so on. The same applies to variables, attributes, counters, expressions, sets, and any other components that are incorporated in a simulation.

In general programming software, the provision of default names for the components is much less prevalent and the modeler is, thus, forced to name many of them. This applies to variables, subroutines, functions, and so on.

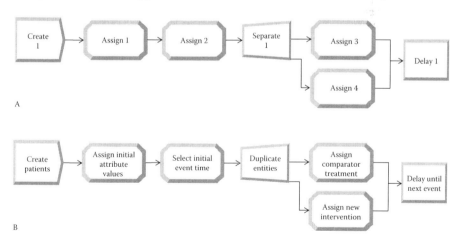

FIGURE 9.4
Two ways of naming model components. In panel A, the modeler has accepted the default names provided by the software (Arena® in this case), while in panel B these have been renamed to describe the purpose of each module. While the two structures may yield the same results, it is much easier to inspect the one in panel B and quickly get a sense of what is being done.

Although some of the naming requirements may be restrictive (e.g., limiting the number and type of characters), the modeler should try to make these as descriptive as possible, and liberally document what the names refer to when forced to use abbreviations.

One handy way of creating names for components that are short but highly descriptive is to use the format: *<type><descriptor>*. The *type* is a standardized abbreviation that indicates the kind of component to which the name applies, and the *descriptor* portion of the name describes what the component is about. For example, the name *aSex* immediately conveys that this is an attribute that contains information about the entity's sex. This way of naming, often called CamelCase and used in various programming standards (Binkley et al. 2009), can be very powerful. By convention, the type indicator is abbreviated using lower case letters and the descriptor portion begins with a capital letter. Space is avoided as is the underscore character ('_'). The descriptor should be brief. Many examples of this kind of naming were used in Chapter 4, and a listing of common ones is provided in Table 9.3. The specific convention used is not material—what is important is consistency. Whatever standard is adopted, it should be used in all models to enhance understanding of the model by both reviewers and the modeling team.

When building the model it is important to keep a record of all variables, attributes, and other names. This record should specify what each component does and details of any special features (e.g., how indexes are used). If a submodel is used, its documentation should describe what the submodel represents (e.g., selection of next event), what is expected of the entities entering, and what will have changed when they leave (Figure 9.5). All build steps and any changes should be documented in a log with the date, brief explanation of what was done, and the model version where that build or that change is saved.

TABLE 9.3

Commonly Used Abbreviated Naming Conventions

Type	Used for
aName	Attributes
iName	Invariant global information, typically an input
vName	Variable global information (changes during the simulation), often used for accumulating outputs
rName	Resources
qName	Queues
fName	Flags (e.g., to indicate which path to take in the structure)
eName	Expressions or equations
cName	Counters
tName	Tallies
sName	Sets

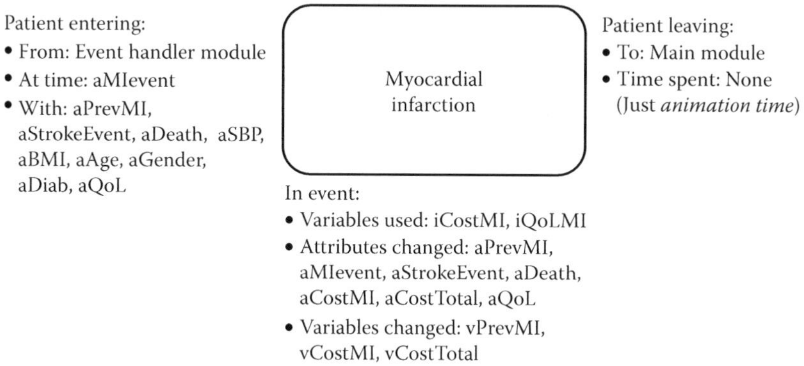

Patient entering:
- From: Event handler module
- At time: aMIevent
- With: aPrevMI,
  aStrokeEvent, aDeath, aSBP,
  aBMI, aAge, aGender,
  aDiab, aQoL

Myocardial
infarction

Patient leaving:
- To: Main module
- Time spent: None
  (Just *animation time*)

In event:
- Variables used: iCostMI, iQoLMI
- Attributes changed: aPrevMI,
  aMIevent, aStrokeEvent, aDeath,
  aCostMI, aCostTotal, aQoL
- Variables changed: vPrevMI,
  vCostMI, vCostTotal

FIGURE 9.5
Documentation of a submodel that processes a myocardial infarction. This submodel handles all the code that processes a myocardial infarction, including its consequences for costs and quality of life. The documentation specifies who enters the submodel, where the departing entity goes, and the main items that are changed in the submodel.

9.1.4 Analysis Stage

Once the model is constructed (and suitably validated), the analyses required to inform the HTA can proceed. During this stage, documentation mainly involves saving the set of inputs used for each run of the model, along with any settings not stored in the inputs (e.g., the number of replications in the run) and storing the corresponding set of results. It is helpful, in case the model is revised subsequently, to also record the version of the model that was used for the analyses. With spreadsheets, a convenient way to do this is to save the entire file, changing its name appropriately for each analysis. If the file is very large, the input and output worksheets can be exported, excluding the worksheets that contain the model calculations. When using specialized DES software, the output file is often saved automatically, but it is also usually possible to export whatever the modeler wants to either a spreadsheet or other format. General programming software offers many ways of saving results and exporting them to other formats.

9.1.5 Validation

Validation of a model (see Chapter 8) should take place throughout the process (Eddy et al. 2012). Documentation of the face validation involves recording how this was carried out, including: who was consulted about the reasonableness of the design decisions; what their detailed assessment was; and how these comments were addressed by the modelers, including why they may have been rejected. Verification of the model should follow established procedures for software (see Chapter 8) and corresponding documentation, including: who undertook verification; of what version of the model; what analyses were done and when; and what the results were. External validation involves comparing model predictions to the results observed in other studies, both those used in informing the model (i.e., dependent validation) and those that were not considered in the model build (i.e., independent validation). Documentation of these steps involves recording how the studies were selected and why, how comparisons were set up and who did this, what the comparisons showed, and the interpretation of these.

9.1.6 Reports

Models constructed for informing an HTA should be accompanied by at least two types of report. One report provides all the technical documentation for the model at a level of detail that is sufficient to ensure that a reviewer can query and understand any aspect of the simulation. The other main type of report is written for a non-technical audience and is intended to provide transparency at a level that can be understood by anyone interested in the model and its results.

9.1.6.1 Non-Technical Documentation

The non-technical report documents the model in an accessible way for reviewers who do not have the expertise to delve into the detailed methods. The use of jargon should be avoided as much as possible and the descriptions should be more about what is done than how it is done. This report should include explanations the model's purpose and how it approaches this. For example, for the model developed in Chapter 4, the description might begin with "this model compares the use of MetaMin with standard of care in patients with cancer XX. The health and cost consequences are estimated using a simulation of what is likely to happen to individual patients who may receive MetaMin." The document should explain what applications the model is intended for (e.g., This model of MetaMin is intended to inform a HTA in country X.), including the scope covered. The report should state what the sources of funding were and make a statement about the funders' role in the design, input data, implementation, analyses, and interpretation. Then, the structure of the model should be described, ideally with a graphical aid. This depiction need not be fully detailed, as long as it covers the main components. Flow diagrams developed for the model can be very useful for this purpose. What data were obtained, from where and why should be stated along with a description of what the analyses sought and what the results were. Statistical details are unnecessary but any equations used in the model should be provided. The main results should be given along with their interpretations, and an overview of how uncertainty was handled and its implications should be included. The steps undertaken to validate the model should be listed and the overall assessment of the different types of validity should be provided. Handy, non-technical tools exist for this purpose (Caro et al. 2014).

9.1.6.2 Technical Report

The non-technical report provides an overview of the model but is not meant to document in detail everything that was done. The technical report explains all aspects of the model at a level sufficient for someone with the required expertise to understand the approach and be able to reproduce it, at least in principle (in practice that can be a very laborious undertaking for a complex model). Since modelers may wish to retain the intellectual rights to a simulation, the sharing of such a document may require that the reviewer sign an agreement that protects intellectual property. This is sanctioned by current guidelines (Caro et al. 2012).

The content of a technical report would cover all the items of design, data inputs, implementation, analyses, and validation. Apart from all the aspects covered in the non-technical report, each item is described here in full technical detail, providing a description of what was done, how, and why. These accounts amount to reports appropriate to the discipline involved, be it epidemiology, biostatistics, meta-analysis, outcomes research, economic analysis, etc. All analyses are fully detailed, including data sources, assumptions made, statistical specifics, and results. It can be very useful to include the analytic code as an appendix. It may also be helpful to include a copy of the model itself, along with a user's guide describing the software and hardware required to run the model, how inputs are modified, where these inputs can be found in the model, and how the model is run and results extracted. Sample inputs and results that a reviewer can use for testing should be provided as well.

9.2 Animation

One of the distinguishing features of specialized DES software is that most offer the possibility of implementing different levels of animation in the simulation. These range from simple graphs that update periodically as the simulation runs to fully-featured three dimensional moving pictures similar to a video or computer game. This is not surprising since quite a few games are based on the DES concept and share the technology and even some software components.

9.2.1 Purpose of Animation

Animation of a model is often regarded as an unnecessary luxury, but that is a misconception. The incremental effort required to incorporate a basic animation is small when it is built throughout the model implementation and the benefits can be substantial. During construction of the model, it is very helpful to see what is happening in the simulation. This often provides rapid and important feedback regarding what the latest build step or change actually accomplished. For example, a change made to the implementation of an equation that specifies the time until a myocardial infarction (MI) event may be shown on animation to produce a sudden flow of entities to the *MI event* near the start of the simulation, something which was unexpected and not supported by the actual equation that was being incorporated. This surprising result of the modification might not be caught for some time if the outcome only reports the count and costs of an MI event across all the patients receiving a given treatment. Those might still look reasonable since they comprise everyone's results. Based on the visual feedback provided by the animation in this example, a revisit of the equation modification would be prompted and the modeler would find that a mistake had been introduced by the change. These kinds of illogical flows are better detected via animation and corrected before proceeding further with construction of the model.

For presentation of the model to others, both within the modeling team but particularly to external reviewers and decision makers, animation is a great facilitator. It is much easier to explain the structure of a model while both you and the audience simultaneously see the entities moving on the screen, experiencing the various events, accruing consequences, and exiting the simulation. For those unfamiliar with DES, the animation can provide a user-friendly introduction to the modeling technique and help further their understanding of how it works and what the implications might be for the specifics of the HTA. Animation also fosters questioning of the model structure and behavior of the simulation by demystifying what might otherwise seem highly technical and unapproachable. Some of the purposes of animation are listed in Table 9.4.

It is difficult in a printed text like this to illustrate what animation can do for a model, but some sense can be gained by comparing explanations in the text in previous chapters with their graphical representations in figures. An even more obvious illustration can be seen in Figure 9.6 where snapshots of an animated graph display how it would build during the simulation, clearly revealing where the effect of the new treatment begins to appear and how the gap widens.

TABLE 9.4

Purposes of Animating a Model

- Debugging (verification)
- Face validity
- Illustrating the movement of entities during simulation time
- Live view of counts and results over time
- Model understanding
- Increase confidence in results

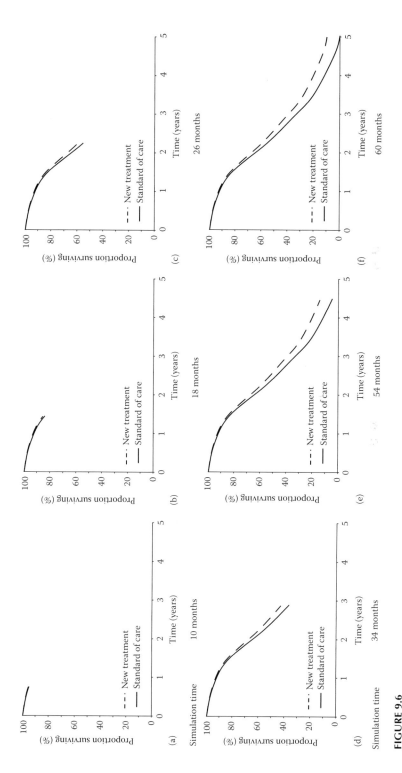

FIGURE 9.6

Successive snapshots of an animated graph. These six panels (a–f) show snapshots of a survival graph as it builds during a simulation (where it would be displayed continuously). In the first snapshot taken at 10 months of simulation time, there is no separation in the survival curves and this remains the same at 18 months. By 26 months, the animated graph begins to show some advantage for the new treatment and the gap widens progressively. By 5 years, all the patients treated with standard of care have died, while more than 10% of those receiving the new treatment remain alive.

9.2.2 Difficulties with Animation

The most common use of DES in other fields is modeling of physical environments where entities reflect production, distribution, queuing for services, and other such flows. In models built for these purposes it is easy to conceptualize what a DES animation should represent given that a real factory, waiting room, or other physical location actually exists, and these can be represented as virtual places that the simulated entities pass through, along with the physical processes, that can happen. In a DES for HTA, the problem has to do with disease processes but these are not typically modeled directly. For example, although the structure and mineralization of bones may be deteriorating in osteoporosis, the HTA DES conceptualizes this disease in terms of the fracture events, pain, and perhaps a score reflecting bone density, but neither the bones nor much less the osteoclasts and osteoblasts and their pathophysiological process are considered. Thus, none of the elements in such a DES have a clear visual counterpart that can be animated. Even if patients in the simulation go to the hospital for treatment of a fracture, and to a clinic or a doctor's office for monitoring, these locations are not typically explicitly incorporated in a DES for HTA. It is difficult, therefore, to animate the flows, events, and processes in these DES.

This does not mean, however, that animation should be abandoned altogether. As shown in Figure 9.6, it is possible to leverage animation to display results of interest as they accrue during the simulation. Moreover, some idea of what is happening can be gained by animating the entity flows between spots that represent events (Figure 9.7). Symbols can reflect the type of event and counters can keep track of how many entities have experienced each animated event. Entities can be colored or otherwise imprinted with changing indicators of their status (e.g., a t-shirt that changes color to reflect disease severity). Even though this is a very artificial animation that does not correspond to physical reality per se, it can still reflect the main components of the simulation and address many of the objectives in Table 9.4, particularly enhancing the understanding of the model.

9.2.3 Components

Most specialized DES software provide several components for animation (Table 9.5). The background is the canvas on which other components are placed. It can be a simple white box or it can contain a static drawing that can be anything from a few graphics to a detailed photograph. In an HTA DES, the pictures placed on this background can represent major clinical events (e.g., fracture, death), locations of care (e.g., home, hospital, nursing facility), or even states (e.g., in pain). The pictures themselves are not typically animated but are chosen so they reflect the element they are representing (Figure 9.8).

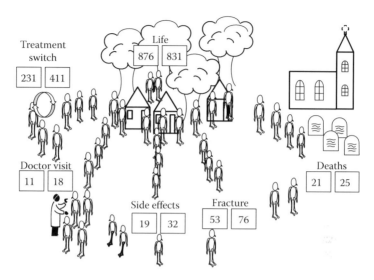

FIGURE 9.7

Simple animation of an osteoporosis model. This display of the animation for an osteoporosis model shows six *locations* for the entities. They spend most of their time in Life where there are at the time of this snapshot, 876 patients taking treatment 1 (first box in scoreboard) and 831 taking (second box). From Life, entities can experience a fracture, after which they may die or return to Life. Entities may also have a side effect, after which many of them will go see a doctor and perhaps go through a treatment switch. Regular visits to the doctor with neither a side effect nor a fracture are also possible.

TABLE 9.5

Animation Components

Static backgrounds
Dynamic entities moving in the model
Pathways
Resources
Graphs
Numerical scoreboards
Other specialized components

On top of the background are the dynamic objects that will be animated during the simulation. The most common of these are the entities. These are represented by a picture which has been assigned as a special attribute to each entity (i.e., *entity picture*). An entity's picture can change during the course of the simulation to reflect what is happening to that entity. If there are many entities in the model at once, the animation may become very crowded, so it may be necessary to limit the number of entities pictured. Animated counters can be used to show how many entities are in each location at any point in time.

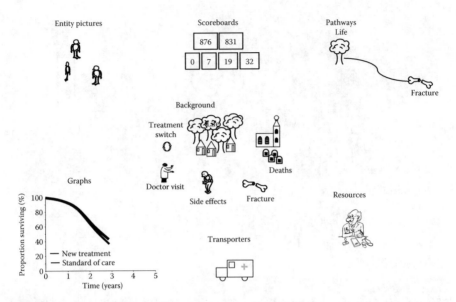

FIGURE 9.8
Components of animation of a DES. The main components used for animation of a DES for
HTA are illustrated. The Background is a set of static pictures or labels that depict where the
entities may go during the simulation. Entities are represented by an entity picture, which
may reflect their state(s). The entities move along pathways that connect the static pictures.
Usually this movement is *walking* but they may be carried by transporters. At some sites, they
may engage with a resource, whose picture may also change to reflect the state. Outcomes are
displayed in scoreboards and live graphs.

Entities move between the static pictures representing events, locations of
care, states, and so on. These movements occur along pathways that *guide*
the entities from one location to another (since they are unaware of their
environment and, thus, incapable of directing themselves—see the discussion of agents in Section 9.4). These pathways are typically concealed during
the simulation so the animation appears cleanly, without many crisscrossing
path lines.

In an animation, it is also possible to include scoreboards that display the
ongoing values of global variables and statistical accumulators. These can
be shown as the numbers themselves (Figure 9.7) or displayed as dynamic
graphs of different types (Figure 9.6) or both. These displayed values or
graphs are updated in simulation time. In some cases, the images can also be
exported after the replication.

Another type of dynamic object is a resource. If these have been explicitly modeled—an uncommon occurrence in DES for HTA—their state can
be animated with different graphics to show whether they are busy, free,
unavailable, broken, or whatever is appropriate. The entity using a resource
is displayed in the neighborhood of the resource during the processing time.

DES software intended for use in other fields also offer many specialized animation objects, like conveyer belts, transporters, cranes, different types of vehicles, switches, and so on, which are not usually used in DES for HTA.

9.2.4 How to Animate

Each software package provides its own tools for building the animation. Some software offer very sophisticated graphic symbols bundled with the logic. For example, there may be a welding robot which when placed in the model brings up a dialog with the characteristics for the robot as well as a suggested picture for it. Some software even allow detailing of the possible pre-defined movements. As of the time of writing, there were no packages of this type specifically designed for HTA. Other software provide the animation as a component that is separate from the model logic. With these types of packages, the connections between the animation and the model logic need to be defined by the modeler explicitly when designing the entity flow and other dynamic components. Animating using this type of software will require adding components to the logic that are purely for animation purposes. Animations for HTA DES are typically constructed using this type of package.

The first step in building an animation is to assign the entities the picture that will be used to represent them (Figure 9.9). This picture can reflect the value of one or more attributes. For example, a picture of a man can be used for entities with *aSex* set to male and a picture of a woman for those with this attribute set to female. The specific treatment a person is receiving can be represented by the color of the entity's clothes or a number stamped on their chest, and their disease status by a change in picture from standing to using a cane, to wheel chair and then to bedridden. These pictures can be designed in a graphics package (sometimes included in the software) or taken from another source, such as an online repository.

In a DES constructed using the type of software that specifies the animation separately, the logic needs to allow for places between which the entities will move (Figure 9.10). These *stations* can represent physical locations where the entities are living, or they can reflect conceptual locations such as model events or states. For example, a woman at the start of an osteoporosis DES can be thought as residing in a station called *home*, or she may simply be in a state *life*. When she suffers a fracture, the animation may involve creation of a station called *fracture*, to which the woman will move. Alternatively, the station may represent where the fracture is managed: *hospital, clinic, emergency department,* and so on. Both types of station may be implemented so that the woman first goes to the *fracture* station, reflecting the onset of the event, and then moves to a station reflecting the specific location where care of the fracture will be provided. Of course, movement between the stations representing locations of care could also take place. To implement these stations, the modeler must define them in the logic and then create their counterpart in the animation area as well.

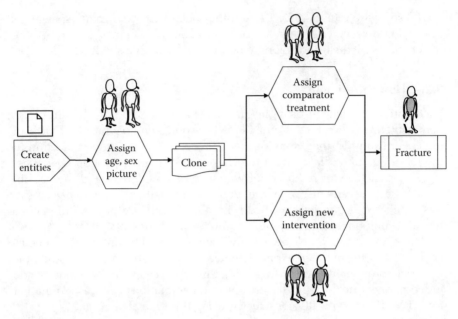

FIGURE 9.9
Changing entity pictures. In this sketch of a DES, the initial entities have a default picture which is changed to a male or female stick figure, as appropriate, once sex is assigned. After cloning, the duplicates who get the new intervention are given a new shaded t-shirt and upon one of them suffering a fracture, the picture changes to indicate this.

The next step is to make sure that the movement of the entities in the animation is properly directed. In the logic, this is specified either by linking the modules as has been shown in previous chapters, or by implementing a *wireless* transmission (indicated by the *Send* module in Figure 9.10) that specifies which station the entity is to be sent to next. In the animation, the stations are linked with pathways along which entities will travel, regardless of the method implemented in the logic. Since it takes some simulation time for an entity to move along the pathway when it leaves a station, it is necessary to specify a *travel time* between stations. This time must be greater than zero but can be very short relative to the timescale on which events are happening so that the event timings are not materially affected.

If resources are explicitly simulated, then their pictures must also be assigned, taking into account that these should be able to reflect the usage of the resource (e.g., an unoccupied doctor might be represented by a picture of a person reading the newspaper, while the occupied one is shown performing surgery). It is also essential to define how long it takes to use the resource so that the entity pauses during the animation and visually interacts with the resource. This duration can correspond to the time it might actually take in reality. The major steps in creating an animation are shown in Table 9.6.

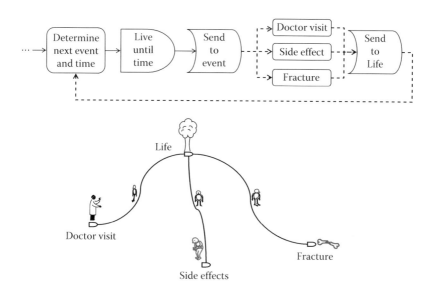

FIGURE 9.10
Partial logic and corresponding animation for a DES of osteoporosis. In the upper panel,
a snippet of the DES logic is shown. In the initial submodel, the next event and time are determined, the entity then lives in a delay module until the time of the next event. At that time, the
entity is sent (indicated by the new symbol) to the corresponding event submodel where there
is a receiving station. The logic for that event is processed and then the entity is sent back to the
next event selection submodel (where there is also a receiving station). In the animation, shown
in the lower panel, the entities flow along pathways between the Life station and the stations for
each event. When the logic at each station is finished, they can retrace their steps back to Life.
Their travel along the pathways takes some time, which in the logic has been specified in the
Send module.

TABLE 9.6

Steps in Creating an Animation

- *Background and structure*
  Design the background and static pictures representing stations
  Connect the pictures with pathways that represent allowed movements
  Create locations in the logic where the stations should be
- *Entities*
  Design entity pictures to reflect attributes to be shown in animation
  Assign entities an appropriate picture and change it as needed
  Make sure entity movements are allowed to take some time
- *Scoreboards and graphs*
  Select global variables and counters for display
  Design appearance of the displays
- *Resources*
  Design resource pictures to reflect state of the resource
  Allocate resource use a duration

9.3 Software

As of the time this is written, there is no specialized DES software specifically designed for use in HTA. Therefore, this book has presented the options for implementation based on what is commonly used by modelers in the HTA field: spreadsheets, general programming languages, and specialized DES software packages. In this section, the focus is on the latter. Many other publications address simulation with spreadsheets (Greasley 1998; Moore and Weatherford 2001) and with general programming languages (Garrido 2013; Nutaro 2011; Thulasidasan et al. 2012).

9.3.1 DES Software History

First tested in 1954, and commercially released in 1957, Fortran marked the beginning of computer software (Figure 9.11). Before it, most computer programming was done in machine language—a cumbersome, complex set of instructions. Fortran gave computer users the first accessible *high-level* language and enabled computers to carry out operations many times more efficiently. Within a decade, Fortran became the first national computing standard (in 1966) and was used in most major data centers in the United States and parts of Europe (IBM Corporation 2011). Given its speed and relative ease-of-use, it was adopted by the early simulation community (Schriber 1977).

Almost simultaneously with the development of Fortran, languages specifically designed for simulation started to appear. In 1958, Keith Douglas Tocher, working at United Steel Companies, developed the first DES programming language, called the General Simulation Program (GSP). It was

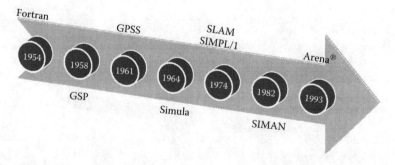

FIGURE 9.11

Some significant developments in software for discrete event simulation. The use of DES started within general software but early on specialized software started to create a separate development path within software. Interestingly enough, there has been constant cross-breeding between general software and DES software and the object orientated paradigm of computer programming has its origination in Simula, a DES software from 1964.

the first of many tools for building models of industrial production. GSP dealt with the tasks pertinent to most simulations: initialization of the model, stepping through time, changing states of the system, consumption of resources, and logging and generating result reports. It enabled queues and had machines that could be in one of various states: busy, idle, unavailable, or failed.

Shortly afterward, in 1961, GSP was followed by the General Purpose Simulation System (GPSS), created by Geoffrey Gordon (Schriber 1974) (see Figure 9.12 for an example). There are still versions of GPSS being sold and used for education and analyses. GPSS has been the foundation for many of the modern simulation tools (Goldsman et al. 2010) like Simul8® and Arena® (Kelton et al. 2010). The next remarkable contribution was Simula, created in 1964 by Ole-Johan Dahl and Kristen Nygaard, working at the Norwegian Computing Center (Dahl et al. 1966). They based their software on an existing programming language, Algol 60 (ALGOrithmic Language). With their next version in 1967, they invented the concept of object-oriented programming, where the designer creates a type of code—an object—which can then be reused throughout the program, customizing only its specific details as needed. This greatly facilitates the coding

```
* The GPSS model consists of one file
SIMULATE                    ;Define Triage model

* Part 1 Create patients

        GENERATE 3,2        ; Create patients Inter-arrival time 3 min SD 2
        QUEUE   TriageQ     ; Queue for triage nurse
        SEIZE   rNurse      ; Seize the resource Nurse
        DEPART  TriageQ     ; Leave the triage queue
        ADVANCE 2.9,1       ; Be triaged by the nurse for 2.9 min SD 1
        RELEASE rNurse      ; Release the resource Nurse
        TERMINATE           ; Remove the patient from the model

* Part 2 Create the entity triggering the End Event
        GENERATE 480        ; EndEntity created at 480 min
        TERMINATE 1         ; Shut off the run

* Model administration

        START   1   ; Start one run
        END         ; End model
```

(Inspired by Thomas J. Schriber, Simulation using GPSS, 1974)

FIGURE 9.12
GPSS code for a very simple model. GPSS was born in 1961 but is still being developed and available. The example here of a nurse triaging patients in an Emergency Room shows the compactness of GPSS. The disadvantage of any text-editor-based programming language isn't evident until the first misplaced comma can completely change the way the model works (if it still compiles).

of simulations. Other languages from that period are SimScript and GASP, followed a decade later by SIMPL/I and SLAM which were much improved languages that could harness the growing power, speed, and memory of the computers of the time.

Those languages were developed and used in the era of mainframe computers, where the simulation program was uploaded to the computer, followed by a period of intense suspense waiting a night or even a few days for the outputs, and hoping that the program didn't have any major bugs, actually worked and the results were usable. This was a challenging process, not least since the mainframe computers ran at a price of $600–$1000 per hour (Kelton et al. 2007). During the 1980s, personal computers emerged and made simulation much more accessible. Since then, the number of available software has exploded, both in the academic and commercial worlds. The ease of use has improved, with the programming paradigms varying from pure text code to full graphical user interfaces.

Animation was the next big breakthrough as it allowed the modeler to see what took place during the execution, rather than having to wade through inch-thick stacks of paper normally printed out after each run of the model. The developments started toward the end of the 1960s with display of graphical results on the not-so-graphic monitors of the time (Donovan and Jones 1968). In 1971, an animation of the actual process flow was presented (Reitman 1971).

Soon, domain-specific software started to emerge. These programs contained elements that were particular to the automobile industry, to telecommunications, to computer networks, to air traffic, and so forth. Simulation in the health care field started with models of physical environments, most often emergency rooms (Jun et al. 1999). Some companies producing general simulation software released versions that addressed this rather complex area, with detailed scheduling of multiple-competence resources, multiple entry points, random arrival patterns, diagnoses requiring different pathways, and physical constraints. Examples are MedModel® from ProModel® and a hospital template from Arena® (since discontinued and replaced with Hospital Navigator®). These simulation tools are focused on the physical flow, queues, and resource constraints that are a critical and unavoidable part of modeling physical health care settings.

9.3.2 DES Software Future

More statistical analysis capabilities have been added to simulation software, both for pre-analyses of the inputs and for analysis during and after the model execution. This is supported by the capability to design *dashboards* that present the results during the runs. Another development is the possibility to communicate with other programs before, during, and after the run. This has led to embedding of DES into other systems like Enterprise Resource Planning software, and scheduling and delivery planning software. The purpose in this case is to create plans that are less sensitive to disturbances and to be able to reschedule and re-plan as the availability or capacity of critical resources change due to failures or delayed deliveries of critical components. There are applications where the model is connected to the actual manufacturing machines and the model is fed with their current status in real time, so when a new urgent order comes in, it is possible to simulate in great detail what will happen if the order is accepted.

Now, much interest is focused on using the *cloud*, or online storage and execution of models. This reverses the trend that moved models from multi-user mainframes into local personal computers—only the mainframes have been replaced by large servers interconnected online. With this development, a complex model can be designed to run in the cloud so that it can be executed on hundreds or thousands of cores (i.e., computer processors) simultaneously. This radically cuts down the time to obtain results, even though the actual execution time of the model is quite long.

9.3.3 Specialized for DES

There are many examples of DES software available, both commercial and academic. Wikipedia (http://en.wikipedia.org/wiki/List_of_discrete_event_simulation_software) provides a regularly updated list, but none of these have been developed to meet the particular requirements of HTA models. Generally, they are focused on the simulations created to address manufacturing, distribution, and supply chain questions. Some packages are highly specialized for particular industries, but many are flexible enough that they can be used to create HTA-oriented DES.

All modern DES software have graphical user interfaces to make the modeling easier and not force modelers to remember the exact syntax, where each comma and semicolon should be or the precise wording of a command. Typically, they supply dropdown menus to help remember the names of functions or the options available at a particular point; many will also store the labels created by the modeler so that there is no need to remember the exact spelling used. This substantially cuts down on errors. For example, in Figure 9.12 is the GPSS code for a very simple model. In this example, the entities are individuals who arrive at a health care facility and need to see a nurse. They arrive, on average (mean), every 3 minutes (standard deviation 2 minutes) and must wait in a queue until the nurse (a resource) is available and can interview the individual. The interview (i.e., when the nurse is busy) lasts, on average, for 2.9 minutes, and afterward the individual leaves the model. The modeler writing this code has to remember that the resource was called *rNurse* not *TriageNurse*; that the queue was called *TriageQ* and so on; to write the command creating the queue before the resource is seized; plus all the command names. In a brief snippet of code like this, that is not so difficult, but with thousands of lines of code and labels, it is very easy to make mistakes.

By comparison, the same model built in a graphical user interface environment (here Arena®) uses only three modules (Figure 9.13). The first one creates the individual and allows specification of the frequency of arrival. The second one addresses use of the resource, including duration, and the last module takes the entities out of the simple model. For all three, a form is available that guides the modeler to fill in the information required for that module and in all cases where it makes sense, the options are provided via a dropdown list. There is no need to remember labels or syntax.

Of note, these three modules generate SIMAN® code, as this is the language underlying Arena®. As can be seen in Figure 9.14, even for this very simple model, there are several additional lines of code added by the software to manage the simulation—none of which the modeler has to worry about.

The inner workings of DES software are rather complex. Considerable effort is spent on getting the random number generation to be as truly random as possible. Efficient and flexible handling of multiple entities and resources modeled simultaneously are required to keep the execution time reasonably low—a critical requirement the larger and more complex the model becomes. Another important component is the handling of the *event calendar* (see Chapter 3) to ensure the order and timing of all events is tracked correctly; updated as needed at every event happening, with any resulting reordering if it affects the current, or any other entity's events on the list. On top of this, the software must also keep track of any statistics required to be able to write out the results at the end of a replication (Schriber et al. 2013). Table 9.7 lists some of the available DES software and links to where more information can be found.

9.3.4 General Programming Languages

The greatest flexibility to code a model exactly the way the project team wants is attained by using a general programming language like Fortran, Java, Python, or Visual Basic. There are many such languages available, each one with its advantages and disadvantages. If the flexibility is their biggest advantage, then the flip-side of the coin is that they are meant to be general so they do not natively offer the basic DES components like entities, attributes, events, the event calendar, etc. When programming a DES in general software, the modeler has to code all of these processes, unless a library of relevant subroutines is available in that language. For some languages, there are libraries available that have some elements of a DES; and for others it is possible to find partial solutions like a random number generator, or a distribution implementation and those can be used as a basis for development. The price of the general programming language is a fraction of the price of a full-blown DES software but the model programming requires many more hours, at least initially.

With the flexibility comes also execution speed. It is possible to minimize unnecessary routines and overhead and optimize the code to run the particular model

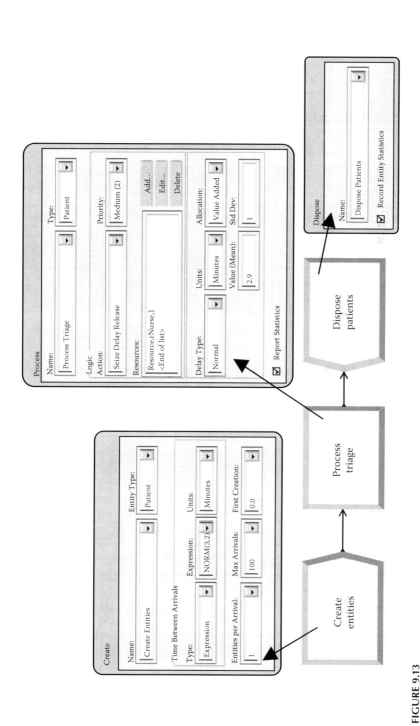

FIGURE 9.13
Arena® modules for specifying simple model. Arena® by Rockwell Software has a history reaching back to 1982 when Denis Pegden developed SIMAN. With it, a DES can be fully programmed with graphical modules. The simple example of the nurse in an emergency room requires three modules, and the required specifications are typed into the dialog boxes in a very guided way with error checking during the process.

```
;Model statements for module:  BasicProcess.Create 1 (Create Patients)
2$  CREATE,  1,MinutesToBaseTime(0.0),Man:MinutesToBaseTime(NORM(3,2 )),100:NEXT(3$);
3$  ASSIGN: Create Patients.NumberOut=Create Patients.NumberOut + 1:NEXT(0$);
;Model statements for module:  BasicProcess.Process 1 (Process Triage)
0$  ASSIGN: Process Triage.NumberIn=Process Triage.NumberIn + 1:
9$  QUEUE,  Process Triage.Queue;
8$  SEIZE,  2,VA:
        rNurse,1:NEXT(7$);
7$  DELAY:  Normal(2.9,1),,VA;
6$  RELEASE: rNurse,1;
54$ ASSIGN: Process Triage.NumberOut=Process Triage.NumberOut + 1:
        Process Triage.WIP=Process Triage.WIP-1:NEXT(1$);
;Model statements for module:  BasicProcess.Dispose 1 (Dispose Patients)
1$  ASSIGN: Dispose Patients.NumberOut=Dispose Patients.NumberOut + 1;
57$ DISPOSE:     Yes;
This is the Triage.exp  -the "experimental" file
PROJECT,     "Triage","Jorgen Moller",,,Yes,Yes,Yes,Yes,No,No,No,No,No,No;
VARIABLES:  Process Triage.NumberIn,CLEAR(Statistics),CATEGORY("Exclude"):
        Dispose Patients.NumberOut,CLEAR(Statistics),CATEGORY("Exclude"):
        Create Patients.NumberOut,CLEAR(Statistics),CATEGORY("Exclude"):
        Process Triage.WIP,CLEAR(System),CATEGORY("Exclude-Exclude"),DATATYPE(Real):
        Process Triage.NumberOut,CLEAR(Statistics),CATEGORY("Exclude");
QUEUES:      Process Triage.Queue,FIFO,,AUTOSTATS(Yes,,);
RESOURCES: rNurse,Capacity(1),,,COST(0.0,0.0,0.0),CATEGORY(Resources),,AUTOSTATS(Yes,,);
REPLICATE,   1,,,Yes,Yes,,,,24,Minutes,No,No,,,Yes,No;
ENTITIES:    Patient,Picture.Person,0.0,0.0,0.0,0.0,0.0,0.0,0.0,AUTOSTATS(Yes,,);
```

FIGURE 9.14

SIMAN code for simple Arena® model. The SIMAN language still underlies Arena® and is an intermediate step when the graphic modules are being compiled to the executable model. As seen, each graphic module represents several lines of SIMAN code. It is still possible to code models in SIMAN using a text editor and compile them.

TABLE 9.7

Some of the Many Specialized DES Software

Name	Website	Model Types
Arena®	www.arenasimulation.com	DES
Simul8	www.simul8.com	DES
Anylogic	www.anylogic.com	DES, ABM, SD
Simio	www.simio.com	DES, ABM
ProModel	www.promodel.com	DES
FlexSim	www.flexsim.com	DES
SAS Simulation Studio®	http://www.sas.com/en_us/software/ analytics/simulation-studio.html	DES

DES, discrete event simulation; ABM, agent-based simulation.

as efficiently as possible. It also opens up the possibility of using multiple-core processing, either locally on a large server or in the cloud, where the number of cores that can be used is limited only by the hourly cost per unit.

One major disadvantage of the general programming languages is the lack of transparency. A model coded in a general programming language is very hard to digest for a non-programmer, and even for experienced programmers, it can be difficult to sort out what is happening unless the code has been documented in great detail. Before using one of these languages for an HTA, it is advisable to check with the relevant HTA agency if such a model will be accepted.

To get the best of both worlds, large commercial models are sometimes built by first constructing a *prototype* in a specialized DES software package and then, when that is securely working, porting the model to a general programming language. If the anticipated long-term usage of the model warrants it, this can provide substantial gains in efficiency and customized user interaction that may make it worthwhile. Once the model is reprogrammed, the code can be compiled (increasing the efficiency of execution) and distributed without the limitations of license restrictions typically imposed by the existing specialized DES software packages.

9.3.5 Spreadsheets

In Chapter 3, some of the advantages and disadvantages of using a spreadsheet program like Microsoft Excel® to implement a DES for HTA were mentioned. The most prominent advantages for Microsoft Excel® are its widespread *availability* and the *familiarity* of most HTA stakeholders with this software. As with a model programmed in a general language and then compiled, a DES constructed in Microsoft Excel® will not obligate most users to incur license fees for the software platform, since in our field, Microsoft Excel® is usually installed already on computers. Although the level of expertise varies considerably, every researcher, modeler, reviewer, and analyst in our field has had at least some exposure to Microsoft Excel® and many are quite adept at using it for modeling.

The main limitation of using a spreadsheet for the construction of a DES for HTA is the way the spreadsheet implements calculations. All cells in a spreadsheet are calculated virtually at the same time (guided only by the sequence of references from one cell to another). The modeler has little control of this process, and, thus, it is difficult to implement the sequential nature of event processing inherent in a DES. With a small number of events, as shown in Chapter 4, this limitation does not pose a major problem, but as the number of possible events increases and the sequences become dynamic (i.e., they can change during the simulation), it becomes increasingly difficult to build the DES using only the spreadsheets. These limitations can be overcome by using a macro language or appealing to external utilities that are designed with the purpose of assisting the spreadsheet program.

9.3.6 Spreadsheets Plus

9.3.6.1 *Visual Basic for Applications*

In 1964, John George Kemeny and Thomas Eugene Kurtz at Dartmouth College in New Hampshire created the programming language BASIC (an acronym for Beginner's All-purpose Symbolic Instruction Code). This has (against all odds and predictions) survived to this day. It has been modified and added to in order to allow for the new kind of graphical interface access that has become common. There are many flavors of BASIC languages, but the most commonly used is Visual Basic, which, in turn has various incarnations.

Microsoft Visual Basic® 1.0 was introduced in 1991 as a platform for developing applications. Version 6 released in 1998 was the end of that path, though it continued to be available. In 2002, it was replaced by Visual Basic.NET, which made major changes and, thus, posed some problems in terms of backward compatibility. VBA was introduced as the macro language in Microsoft Excel® in 1993 to replace the native macro language in the original spreadsheet package. VBA is a subset of Visual Basic. Arena® was the first specialized simulation software to be licensed by Microsoft to use VBA as its macro language. Thus, there are VBA *dialects* in order to be relevant to the different parent software. In Microsoft Excel®, for example, VBA needs to be able to handle sheets and cells, while in Arena® it has to handle entities. Each VBA dialect therefore has an *object model*, which states what objects are available to manipulate in that application, and what specific Events, Properties, Methods, and Functions they can have. These are typically documented in the Help menu.

VBA has most of the abilities of a general programming language (but it executes at a much slower speed). Thus, it permits the modeler to overcome most of the limitations of using Microsoft Excel® to construct a DES for HTA. For example, a standard loop in VBA code can sequentially execute instructions and solve the problem of the non-sequential calculations in the spreadsheet. The modeler can then implement the required processes for the DES, while appearing to keep the benefits of working in a familiar environment. The model can be built with VBA code doing much of the DES processing and interacting with the spreadsheets, which are used to store values and for basic calculations and output processing. Such a model can be more transparent than a pure spreadsheet model if the code is well structured and documented.

9.3.6.2 *Monte-Carlo-Simulation Extensions*

Crystal Ball® and @Risk® are examples of a software category called Monte-Carlo-simulation extensions. They are utilities that are added on to spreadsheet programs, most commonly used with Microsoft Excel®. Their main purpose is to implement distributions and handle the selection of values. They also can store the results for replications carried out with different sets of random numbers. Thus, they make it easier to implement stochastic processes in spreadsheet models and incorporate uncertainty in the analyses.

These utilities permit the modeler to specify one or more cells as containing inputs whose values are to be selected from a distribution. A standard probability distribution (e.g., Weibull, normal) can be selected for each cell, and its parameters specified or an empirical distribution, either discrete or continuous, can be described. The utility then selects a value from the specified distribution and provides it to the spreadsheet for use in the calculations. Other cells can be specified as containing results that will change from one replication to the next and the utility will keep track of the outcomes and display them after the run is completed.

These utilities strengthen Microsoft Excel® in areas where it is weak, without forcing the modeler to write extensive VBA code.

9.3.7 TreeAge

TreeAge is a software (TreeAge Pro 2015, Revised version 2.0, TreeAge Software, Williamstown, MA, http://www.treeage.com) that is widely used in the HTA community. This is a program that was originally developed with the goal of assisting its users with modeling decision analyses based on decision trees. A capability to include Markov components in those models was added early on. As part of the TreeAge update released in January 2014, a DES element was introduced. This new component makes it possible to incorporate time-to-event sequences as a *node* in a decision-tree-based model. While the addition of a DES node does not enable a full-blown DES given that there is no ability to handle competition for resources or the resulting queues, it does allow construction of the type of DES that is common in HTA.

A DES constructed using TreeAge is executed at the individual level, running one person at a time through the model and accumulating the data. In Figure 9.15, a decision tree that includes a DES node is displayed for the example detailed in Chapter 4.

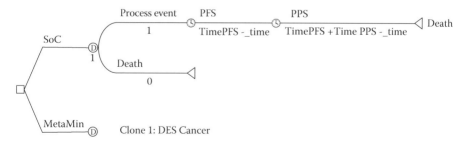

FIGURE 9.15
Inclusion of a DES component in a model constructed using TreeAge. This diagram of a decision tree shows how the cancer model detailed in Chapter 4 would be structured using TreeAge. The DES nodes are indicated by the encircled D. They are followed by time nodes depicted with a clock icon. Branches end with a value node indicated by a triangle. The label *Clone 1: DES Cancer* on the lower branch indicates that the structure of the upper branch is repeated here. The variable *_time* stores the current simulation time.

As there are no recurrent events, there is no re-sampling of the times of PFS or PPS in the example displayed in Figure 9.15. This is possible within TreeAge, however. It may be needed, for example, to implement additional recurrences in a model with multiple lines of antineoplastic therapy. The software makes it simple to mix decision tree and Markov functionality with the strength of some DES elements.

9.3.8 R

The open-source programming language R is based on a statistical language developed by AT&T called S. Open source means it is free for use, under the *GNU General Public License,* and this allows the language to be constantly modified by the R community. It also means that there are plenty of forums where questions can be posted and from which much can be learned. In its original form, R was a pure text language, but several graphical user interfaces have now been made available.

In Figure 9.16 an excerpt is displayed from an R program for the example DES presented in Chapter 4. A more extensive example of a DES can be found

```
# Generate initial patient set.
d <- patients[sample(1:nrow(patients), size=n, replace=T),] # with
replacement?
d$LY <- 0
d$QALY <- 0
d$cost <- 0

# Clone patients and assign treatment
d <- rbind(
  cbind(patient=1:n,      d, treatment=1),
  cbind(patient=n+(1:n), d, treatment=2))
# Determine time to first event.
d$time.to.progression < 1/12*(-log(runif(n)) /
  (0.000916 * (d$sex*(-0.458) + d$age*0.032 + d$ldh*0.003)))^(1/1.67) *
  ifelse(d$treatment==1,1.681,1.00)
# The DES model.
# Simple disease course (pathway through system) enables sequential
evaluation.
d$ldh <- d$ldh * ldh.change(d$sex)
d$age <- d$age + d$time.to.progression

d$LY   <- d$LY   + d$time.to.progression
d$QALY <- d$QALY + d$time.to.progression*utility1
d$cost <- d$cost +
d$time.to.progression*ifelse(d$treatment==1,cost1,cost2)

d$time.to.death < 1/12*(-log(runif(n)) /
  (ifelse(d$treatment==1,0.73,1.00) * 0.00019 *
  (d$sex*(-0.188) + d$age*0.086 + d$ldh*0.004)))^(1/1.24)
```

FIGURE 9.16
R code for part of the model in Chapter 4. The R language is an open-source statistical programming language which is coded using a text editor. Graphical user interfaces are available from the R community.

elsewhere (Matloff 2011). One major advantage of building a DES in R is the extensive statistical toolbox that is available for direct access from the model, without having to call any external routines. This is especially helpful when building models which are dependent on statistical analyses for decisions during the run (e.g., when simulating a randomized clinical trial with interim data analyses).

R is widely accepted by HTA agencies and is extensively used and taught in universities. It is an object-oriented software and flexible enough to use for DES, although it is not meant for this application per se. It is designed to handle vectors and matrices as easy as it handles scalars, with the same code. One limitation is that it needs to keep the full datasets in memory, which requires some planning when working with large models that create and use large data tables.

9.4 Agent-Based Models

Most health technologies can be evaluated without regard to the impact of interactions between individuals, but certain problems may be more accurately modeled by allowing for such interactions. The most obvious examples are evaluations of interventions to prevent or manage infectious diseases, where the prevention or cure of disease in one individual reduces the rate of transmission within the modeled population and proximity to an infectious individual is relevant. Another example might be the evaluation of a health promotion technology, where the representation of the spread of information from individuals with the information to those without the information could be modeled.

Such problems can be handled using differential-equation models, but that approach presents limitations with respect to representing variation or heterogeneity among individuals within the modeled population. A simulation-based alternative is an ABM, where the *agent* is equivalent to an entity in a DES but with additional features (Chhatwal and He 2015). ABMs are an extension of DES, in which interactions between individuals (i.e., *agents*) are represented. As with the entities in a DES, agents have attributes that determine their experience of events within the model, but they may also determine their interactions with other agents in the model and how they behave in the simulated environment (Figure 9.17). Interactions may be between agents of the same type (e.g., university students living in a particular dormitory) or they may be across types (e.g., parents and newborn babies). The form and frequency of the interactions between individuals (e.g., between those infected and individuals who are susceptible to infection) are described. Alternative types of agents may also be defined. For example, if modeling

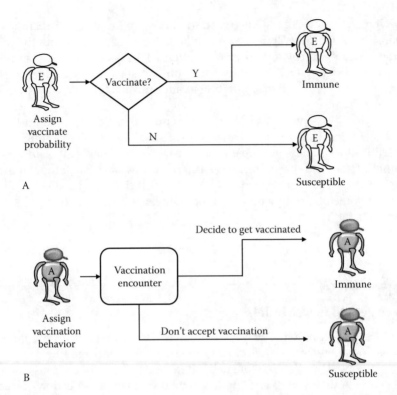

FIGURE 9.17
Agents versus entities. In panel A, an entity *E* in a DES is assigned a probability of being vaccinated and this gets tested at a branch point with the result that the entity becomes immune or remains susceptible. In panel B, the agent *A* is assigned a set of rules that determine based on the attributes and conditions whether the agent will decide to get vaccinated or not when such an encounter takes place.

the transmission of disease within a hospital, alternative agent types may be specified for patients, nurses, doctors, and administrative staff.

Unlike the entities in a DES, agents are not controlled by the simulation logic. Instead, they are given rules and heuristics that are used to determine what pathways to take and what to do when encountering other agents or model elements. Agents can also learn from other agents or from the environment. Given the autonomous behavior of agents, the simulation may yield effects that were not specifically included as inputs into the model. For example, herd immunity may emerge as more individuals become immune through vaccination or an outbreak can occur because of agents electing not to be vaccinated (Epstein 2009). These emergent phenomena are a major difference between ABM and DES.

An important aspect of an ABM is the environment in which the agents find themselves. This typically represents the real physical space within which the agents are interacting. The environment may be broadly defined,

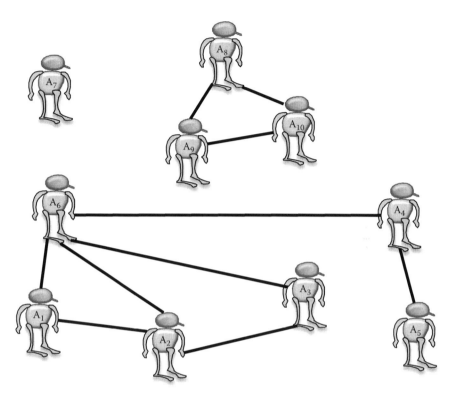

FIGURE 9.18
Networks of agents in ABM. The sketch shows how two networks (agents 1–6 and 8–10) form leaving one isolated agent (7). Lines between agents indicate the permitted interactions.

for example, as an entire nation state; or more specifically, such as a particular hospital environment or the university dormitory. The breadth of the environment will influence another key component of an ABM—the network structure (Figure 9.18).

The network structure describes which agents can interact with other agents (i.e., what the potential connections are among the agents). It is necessary to specify the likelihood or timing of interactions between the agents, as well as the probability of infection (if that is what is at issue) when agents interact. These parameters may reflect agent attributes, such as age, gender, education, whether a doctor has washed his or her hands, and so on. The network structure may also represent the movement of agents between locations within the model environment, for example, agents may be in hospital, or out of hospital.

As in DES, time in an ABM can either move forward using a time-to-event approach, or the clock can be advanced using periodic time increments (typically of constant duration). Unlike DES, however, it has been noted that the periodic checking approach may be preferred because the addition of stochastic behavior and times to interaction of the agents may greatly complicate the representation of times to events in model implementation (Railsback and Grimm 2011).

While in a DES for HTA, the entities can enter the model en masse at the start, or sequentially during the simulation, in ABM the eligible population is defined at the start of the simulation. This eligible population can be defined as either an open or closed population. A closed population includes only those agents specified at the start of the simulation, while an open population allows for the entry of new agents as the simulation time advances. Some problems require the specification of an open population. For example, in an evaluation of a newborn vaccination program, it is important to reflect the entry of new babies over the duration of the model's time horizon. For other problems, the additional value of keeping the population open will vary according to the length of the model's time horizon, and the impact of the condition being modeled in the segments of the population who could enter or leave the model (e.g., newly retired people or immigrants and emigrants to the model environment).

In summary, ABM is a modeling technique that provides a useful extension to DES for the representation of interactions between model agents. As an alternative to differential equation modeling, ABMs are particularly useful when interactions between agents are complex, for example, when the likelihood and frequency of interaction changes over the model's time horizon. Although ABMs have the potential for greater complexity, the development, population, and analysis of ABMs for HTA follow the same general principles as described for DES models in the preceding chapters.

9.5 Hybrid Models

Despite its great flexibility and breadth of applications, DES is occasionally not able to properly, or most efficiently, handle a particular HTA problem. In these situations, it is often possible to incorporate elements from another modeling technique in a hybrid model. Apart from the DES components, these hybrids may contain parts from a *continuous* simulation, from a *state-transition* (i.e., Markov) model, from a *system-dynamics* model, from a *dynamic transmission* model, and even from a *decision tree*. The ABM described in the preceding section are, naturally, hybrids because it is not necessary for all the simulated object to be *agents*. An ABM can also include *entities*, and they are simulated just as they would be in a DES without agents.

9.5.1 With Continuous Simulation

In the engineering literature, the term *hybrid model* is usually applied to a mixture of discrete and continuous simulation (Barton and Lee 2002; Cellier et al. 1993). This combination developed because many physical systems have both continuous (e.g., the filling of an oil tank on a petroleum transport ship) and discrete (e.g., the arrival of the ship, its docking and eventual departure back to sea) components. The simulation of such physical systems has been extensively developed and even applied to medical problems (Korsunsky et al. 2014). The continuous element of these models is represented by differential equations that specify how the levels of the continuous variables change and events are implemented discretely as is normally the case in DES. These kinds of models can be implemented either from the continuous side or from the discrete event side, with the latter being easier to do. Some degree of continuous simulation is available in most of the specialized DES software packages.

For HTA problems, continuous aspects are very common. For example, most physiologic parameters are of a continuous nature: blood pressure changes without interruption, increases and decreases throughout the day, kidneys incessantly filter the blood and make urine which accumulates in the bladder in a constant fashion, and so on. Aging is, of course, a continuous process as are many of its disease correlates (e.g., atherosclerosis), and numerous aspects of disease are also in endless flux (e.g., tumor growth in cancer, demyelination in multiple sclerosis, pain levels in a variety of illnesses). Thus, in most cases, a DES should properly be a hybrid with some elements of continuous simulation.

In current implementations of DES for HTA, the continuous elements are usually either fully discretized or handled via discrete updating of values. For example, tumor growth can be represented in terms of discrete events like recurrence and metastasis (see example in Chapter 10). Many physiologic parameters are discretized by defining thresholds for what is considered an abnormal value (e.g., blood pressure level that is viewed as hypertension, hyperglycemia that is labeled diabetes). Some degree of continuity can still be implemented by delineating an ordinal scale (e.g., low, normal, moderate, high, very high). The other approach is to periodically update the continuous value. This is often the case with age, for example. At convenient points during the simulation, the elapsed time is added to the starting age to obtain the updated value. Indeed, in Chapter 3 the use of an *updater* event for just this purpose was described. Other continuous quantities can also be updated this way. In the example detailed in Chapter 4, the value of LDH was updated at the time of cancer recurrence. Although either of these discrete approaches to a continuous level diminishes the accuracy of the model with respect to reality, the loss of fidelity is usually assumed to be acceptable.

9.5.2 With State-Transition Models

In a state-transition (Markov) model, a problem is represented using discrete states and the transitions among them. These types of models can be implemented by applying transition probabilities either to an entire group of people (i.e., a cohort) or to individuals one at a time (i.e., so called microsimulation). Either of these can be combined with a DES but the individual approach is not usually of interest because a DES already subsumes it. In a DES the *states* of an individual Markov model can be represented via one or more attributes (a single one if they are to be true, mutually exclusive, Markov states). Each individual can have only one value for that Markov state, and transitions to other states are implemented as events. In the example provided in Chapter 4, the Markov formulation would have involved three states: *recurrence-free*, *progressive disease*, and *death*. These states could have been easily represented in an attribute *aCancerStatus*. It was not necessary to do that since the relevant sojourn times were stored for each individual in *aPFS* and *aPPS* and the consequences of being in those states were handled at the time of the transition event by accruing the relevant costs and quality adjusted survival. Thus, there was no need to explicitly implement an individual Markov model, and the DES, despite its simplicity, provided additional flexibility (e.g., for updating LDH level) and did not impose limitations like requiring half-cycle corrections.

It may be useful, however, to incorporate elements of a cohort Markov model in a DES. The essence of such a model is that there is no need to consider individuals because the entire group can be modeled as a whole given that there is no heterogeneity in the determinant profiles. For example, in a model considering vaccination to prevent bacterial meningitis, most of the interest may be in what happens when there is an outbreak. At that time, it is important to simulate individuals as their age, living situation, vaccination history, and other characteristics are significant determinants of whether they become infected and what happens to them subsequently. Costs, survival, sequelae, and quality of life, all require individual application. Between outbreaks, however, the individuals may be grouped back into suitable cohorts that are in a *non-outbreak* state because there is no need to differentiate the experience of these individuals. This makes the model very efficient as it considerably reduces the number of calculations. In Chapter 7, these cohorts were referred to as super entities.

Another example of a hybrid of a DES and a cohort Markov model is in the assessment of various treatments for lung cancer (indeed this conceptualization could apply in many cancers). Increasingly, biomarkers are being discovered (Korpanty et al. 2014) that are the strongest, or even sole, determinant of the risk of developing a malignancy, of the response to treatment, or of the associated mortality. Individuals manifesting a given biomarker are homogeneous with respect to the risk that the biomarker identifies. Thus, they can be efficiently modeled using a cohort Markov approach in terms of transitioning to that state (e.g., to *early lung cancer*). Once they are in that state, however, their subsequent course may differ according to other determinants (e.g., age, comorbidity, smoking status) and, of course, according to how they are treated. Thus, from that point onward, they are better modeled as individuals in a DES.

9.5.3 With System Dynamic Models

System dynamic models, as the name implies, examine the behavior of an entire system as it changes over time. In such a model, there are elements called *stocks* that represent accumulations of something of interest (e.g., a stock of people infected with a particular virus and another stock of uninfected people) and flows from one stock to another (e.g., as given by the rate of infection). There can be various determinants of the flow (external to the stocks which are homogeneous with respect to the flows) and feedback loops may exist where, for example, the level of a stock may change the rate of flow. These models are implemented at the aggregate population level using systems of interrelated differential equations that can capture the flows and complex feedback loops.

System dynamic elements can be integrated with a DES (Brailsford et al. 2014), and these hybrid models can provide an efficient way to consider continuous aspects of a problem that do not require individualization. Such hybrids have been used to evaluate infection control techniques (Viana et al. 2014) and have been proposed as a general approach to HTA (Djanatliev et al. 2014).

9.5.4 With Decision Trees

Decision trees are a modeling technique closely associated with decision analysis and are designed for selecting an *optimal* pathway starting from a *decision node*. This is the point in the tree (usually at the start) where the path branches out and a decision regarding which branch to choose has to be made. The branches contain *chance nodes* where additional branching may occur according to the probability assigned each alternative. All branches ultimately culminate in a *value node* where the value of the entire branch is computed based on valuations of the components along the path. In traditional decision analysis, the decision is based on which branch offers the most value, but in an HTA that partitions value into cost and health effects, the optimization criteria are no longer straightforward.

The use of decision trees has a long history in medicine because they conveniently represent the choices that clinicians and patients need to make in the process of diagnosing illness and managing it. A major limitation of decision trees is that time has no explicit place. Although the full decision tree is, therefore, executed in zero time, all the many elements of an HTA that have a time dimension cannot be properly represented. By the same token, nothing has a duration, except that intervals can be included as descriptors of a condition in a branch (e.g., the patient has a rash for 10 days). Another major issue is that the branches of a decision tree are mutually exclusive, so representing a combined condition (e.g., in remission but with a side effect) requires additional branches. In a DES, any number of conditions can co-exist simultaneously.

In essence, any model designed to inform an HTA—regardless if it uses a Markov technique, a DES, or something else—has components of a decision tree embedded in it. There are many places where *decisions* have to be made (either explicitly as in which treatment to select next or implicitly given the logic as in *if patient suffers side effect, stop treatment*). In a DES, these aspects of a decision tree are easily handled with branching nodes (indeed, in Arena®, these are called *Decision* modules). The cloning function of a DES is a direct representation of the initial *decision node*. The values that a decision tree uses to inform a decision are also readily carried in a DES. Thus, there is nothing gained by adding a decision tree to a DES.

10

Case Study: Breast Cancer Surveillance

In order to illustrate the implementation of a discrete event simulation (DES) for a real health technology assessment (HTA), a case study is presented. This case study describes the development and analysis of a DES of alternative options for the use of mammography to monitor women following the diagnosis and treatment of early breast cancer (Bessen and Karnon 2014). The aim of this monitoring or surveillance is to detect recurrent localized cancer at an early stage and, thus, enable treatment of the recurrent cancer with the aim of preventing further disease progression. Currently, it is recommended that most women undergo annual mammography for the detection of recurrent disease. The rationale for the modeling study was that women with early breast cancer have a wide range of prognoses (risk of recurrence), and that the *one-size-fits-all* surveillance strategy may be over-screening women at low risk of recurrence. The aim of this study was to define cost-effective surveillance strategies for women with early breast cancer, at various levels of recurrence risk.

In this chapter, this model of cancer surveillance is developed step by step from its design through its analyses and interpretation. The implementation of each of the main parts of the applied DES is described in some detail. The chapter begins by giving some background regarding the problem this HTA addressed. It then provides the reasons for selecting to use a DES for this model. The next section details the design of the DES, including the influence diagram and model assumptions that guided model development and the flow diagram for the structure. The data sources for the inputs are described next, along with a brief explanation of how they were processed and the calibration that was required (Karnon and Vanni 2011). The implementation of the model is then detailed and the final section addresses the analyses that were done and discusses the results.

10.1 Background

The duration of survival for women diagnosed and treated for early breast cancer is increasing due to earlier detection and the improved effectiveness of treatment. For example, in Australian women (the subjects of this DES for HTA), the 5-year survival from breast cancer increased from 72% to 89% over

20 years, from the mid-1980s to the middle of the first decade of this century (Australian Institute of Health and Welfare [AIHW] 2012). These gains were even more impressive in women between the ages of 50 and 69 years at the time of diagnosis: from 70% to 91% for women in their fifties and from 72% to 93% for women in their sixties. As more women enter these age groups, there will be an increasing need to provide them with follow-up to detect any cancer recurrences that may occur (Karnon and Vanni 2011). This follow-up should include periodic screening with mammography, which is highly effective at detecting recurrence in the same or contralateral breast (National Institute for Health and Care Excellence [NICE] 2014).

In the context of HTA, screening usually refers to the process of identifying individuals who have a sign (e.g., a nodule on an X-ray and a high level of a biomarker) that indicates that they may have an undiagnosed primary disease (Karnon et al. 2007a). Following a positive screening result, further diagnostic intervention is offered, followed by treatment if the suspected diagnosis is confirmed. Surveillance of patients with a diagnosed condition for the recurrence or progression of that condition is also a form of screening (Tomin and Donegan 1987).

The direct evaluation of the effectiveness and cost-effectiveness of screening programs is hampered by the numerous practical difficulties in the application of robust study designs (Feig 2014; Saquib et al. 2015). The primary difficulty concerns the very large sample size required and the accompanying extended duration of follow-up that is often necessary to observe significant differences in meaningful outcomes between groups. For example, new results from breast cancer screening trials are still being reported after 25 years of follow-up (Miller et al. 2014). There may also be limitations with respect to the range of comparators that may be evaluated.

The following factors are all components of a fully specified screening process: eligible population (e.g., by age and baseline risk), frequency and method of screening, diagnostic criteria, and intervention and surveillance options for treated patients. In addition to difficulties in defining the most relevant comparators when setting up such studies, the study timelines introduce a high probability that clinical practice will have altered significantly by the time the results are published, thus reducing the relevance of the study findings.

In response to these difficulties with direct evaluation of screening via randomized controlled trials, model-based studies of the effectiveness and cost-effectiveness of screening programs have become widespread. Possibly the best known set of model-based analyses in this area are the seven cancer intervention and surveillance modeling network (CISNET) models, funded by the US National Cancer Institute to estimate the effect of population screening for breast cancer under a variety of policies (Mandelblatt et al. 2011).

Current clinical practice guidelines recommend annual mammographic surveillance following diagnosis and treatment of early breast cancer, noting the low level of evidence upon which this recommendation is based (Karnon et al. 2007a). If some women can be adequately managed with fewer surveillance episodes, health system resources can be re-allocated to areas of greater need, and patients may experience less anxiety and inconvenience (alternatively, patients may lose some benefits of reassurance following a negative mammographic result).

In the absence of robust clinical studies to inform a move away from this one-size-fits-all approach to mammographic surveillance, a DES for HTA was developed to estimate the costs and quality-adjusted life years (QALYs) associated with alternative surveillance strategies for women with early breast cancer at varying levels of risk of recurrence. The analysis was undertaken from an Australian health system perspective.

10.2 Why DES?

Models of screening programs generally need to represent fairly complex structures to cover the many components of a screening process, including the promotion required to bring candidates in for screening, the screen itself, the follow-up and further diagnostic steps taken in the case of positive tests, and the consequences of the decisions made over the many years that it may take for these to develop. To be able to handle all these components, DES is often applied in screening studies because the technique offers the flexibility to integrate the many aspects required. Indeed, four of the seven CISNET models used simulation (Clarke et al. 2006).

There were three main reasons why DES was chosen as the most appropriate modeling technique for this case study of the cost-effectiveness of surveillance options following diagnosis and treatment of early breast cancer (as summarized in Table 10.1). First, the modeled process represents the experience of clinical events (different forms of recurrence and mortality) and ongoing intervention events (mammographic surveillance). Such combinations of events are more intuitively represented in a time-to-event format, whereby the sequencing of clinical and intervention events is naturally determined by the sampled time to each possible next event for each patient.

TABLE 10.1

Reasons to Use DES for This HTA of Screening for Breast Cancer Recurrence

- Combination of clinical and surveillance events more intuitively described using DES
- Use of individual level data to sample surveillance pathways and calibrate unobserved input parameters
- Given relative rarity of events, time-to-event model progression more efficient than cycle-based (periodic time increments) state transition approach

Second, the input data for the model included data for individual patients describing their relevant prognostic indicators and the timing of mammographic surveillance. The model population comprised a randomly sampled (with replacement) set of the observed patients, for whom observed surveillance pathways were applied. This facilitated the calibration of unobserved input parameters (such as the incidence of impalpable local recurrence) to rates of recurrence and mortality in the observed set of patients.

Finally, the probability of recurrence was low, especially in patient groups with good prognoses. If using a cycle-based approach to advancing model time (i.e., applying periodic increments in time), no event would occur in the vast majority of model cycles. Model running times are significantly reduced through the use of a time-to-event approach in which the model moves forward in time only when the next event occurs.

10.3 Design

To address the needs of the HTA, the model needed to represent the disease pathways related to the possible progression of early breast cancer, and the influence of patients' age, prognosis, and adherence with surveillance on those pathways. The influence diagram (Figure 10.1) represents the

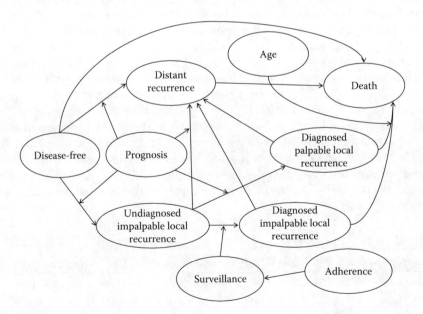

FIGURE 10.1
Influence diagram for DES HTA case study. This diagram shows the key concepts that were required in the DES for an HTA of breast cancer screening.

pathways from the initial disease-free event (following primary treatment of the diagnosed early breast cancer) to local and distant recurrence, or death without recurrence. Local recurrence is initially impalpable, but eventually becomes palpable if patients do not die or experience a distant recurrence in the interim period. Patients' prognosis at the time of diagnosis influences the likelihood and timing of recurrence, as well as progression post-recurrence. Age influences the timing of death in the absence of distant recurrence, while adherence influences the frequency of surveillance, which in turn influences the likelihood of the diagnosis of impalpable local recurrence.

10.3.1 Population

The DES for this HTA considers postmenopausal women who are free of disease following primary treatment for early breast cancer. This aggregate population was split into four alternative prognostic groups: poor, moderate, good, and excellent. This prognostic classification was based on the Nottingham prognostic index (NPI) (Galea et al. 1992), which estimates risk scores based on patients' tumor size, lymph node stage, and histologic grade.

Separate versions of the model were populated for each of the four prognostic groups. Relevant data from 1,100 eligible patients were extracted from routine health systems databases in South Australia, including information on age, tumor size, lymph node stage, histologic grade, timing and outcome of mammographic surveillance, timing and type of recurrence, and date of death (if applicable). The patients were assigned to their respective prognostic groups on the basis of their NPI score.

Model populations of 2,000 patients were created for each of the four prognostic groups by sampling with replacement from the relevant subsets of the 1,100 observed patients. As an example, the set of 407 patients at moderate risk was sampled 2,000 times, with replacement of the sampled patient. Within the DES, patients' age and prognostic group were stored as attributes. The timing of each surveillance episode for each patient was stored in a spreadsheet to inform the appropriate selection of the next surveillance episode in the simulation. The number of surveillance episodes experienced was stored as an attribute.

Data describing the type and timing of recurrence and date of death were used as calibration targets, against which the model outputs could be compared to ensure model accuracy.

10.3.2 Structure

At the start of the model, all women are free of breast cancer but are at risk of developing a recurrence (Figure 10.2). This recurrence may be distant or local. Distant recurrence refers to the spread of cancer from the breast to other parts of the body (e.g., bone, brain, or internal organs). Mammography

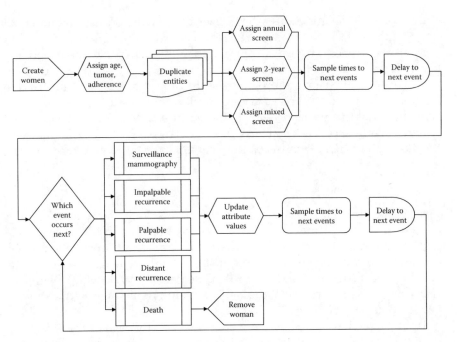

FIGURE 10.2

Flow diagram for early breast cancer surveillance model. After creating women who are disease-free after initial diagnosis and treatment of localized breast cancer, they are assigned baseline characteristics and then duplicated so that one screening strategy can be assigned to each copy. The DES then checks for the next event and processes it.

is not able to detect cancer that has spread beyond the breast, and so the impact of surveillance mammography is via the early detection and treatment of local recurrence. These lesions can be treated, which may reduce the progression of local recurrence to distant metastases.

Three health events related to localized recurrence were defined. One was *impalpable local recurrence*. This is a recurrence within the treated, or the opposite, breast that is only detectable by follow-up mammography because it cannot be felt during manual breast examination. A second event was *removed impalpable local recurrence*. This is a local recurrence detected by follow-up mammography that is surgically removed. The third event was *palpable local recurrence*, a recurrence within the treated, or opposite breast, that is clinically detectable by a doctor via manual examination of the breast or by the patient via breast self-examination. Detection of these palpable recurrences happens prior to their surgical removal.

In the absence of progression to *distant recurrence* or *death*, or surveillance, patients move sequentially from the *disease-free* condition to *impalpable local recurrence* and then to *palpable local recurrence*. Women may develop *distant recurrence* at any time. Women may also die at any time. On the basis of expert clinical advice, it was assumed that women could only die of causes related to breast cancer following the development of metastases. Thus, prior to developing *distant recurrence*, women are subject to the same likelihood of death as the age-matched general population (minus the probability of death due to breast cancer). After experiencing *distant recurrence*, women are at a significantly increased risk of death.

Mammographic follow-up is represented in this DES as an instantaneous event. Women may experience this screening event when they are still *disease-free* or after experiencing an *impalpable local recurrence* (but before the local recurrence becomes palpable or progresses to *distant recurrence*). For women who are *disease-free*, the mammography may be a true negative or a false positive test. In both cases, women return to the *disease-free* condition, though additional costs are incurred following a false positive test to account for the further diagnostic tests that would be undertaken. For women with an *impalpable local recurrence*, surveillance may provide a true positive or a false negative result. Following a true positive result, women move to the *removed impalpable local recurrence* condition. This event represents the surgery to remove the recurrence and results in a reduced likelihood of experiencing a *distant recurrence*. Following a false negative result, women return to the *impalpable local recurrence* condition, with no modification in their course.

10.3.3 Interventions

For each prognostic group defined by the NPI score, three alternative mammographic surveillance strategies were evaluated: ongoing annual surveillance (annual), ongoing two-yearly surveillance (two-yearly), and annual surveillance for 5 years, followed by ongoing two-yearly surveillance (mixed).

10.4 Data Sources

Often in screening models, including in the CISNET simulations, the values for inputs are obtained from results reported overall for entire populations, without access to data for individual patients. This case study, however, illustrates the benefits of using individual-level data to populate a DES. To demonstrate the remainder of the modeling process (i.e., the data, implementation, and analysis), the case study focuses on one of the four prognostic groups: the women at moderate risk. Similar processes were applied to estimate costs and effects of the alternative surveillance strategies in the other three prognostic groups.

De-identified data were obtained retrospectively from the 407 women with a moderate risk of recurrence (based on their NPI). In addition to informing patients' NPI, these data included date of birth; the timing of mammographic follow-up; the experience and timing of local recurrence or distant metastases; and the experience and timing of death due to breast cancer or other causes.

Calibration is a common component of many screening models in order to estimate unobserved input parameters (Vanni et al. 2011). In the case study model, it is not possible to observe the point at which impalpable local recurrence first becomes detectable via surveillance mammography (i.e., the time from becoming *disease-free* to experiencing an *impalpable local recurrence*). Hence, it is also not possible to observe the time between *impalpable* and *palpable local recurrence*, and between *impalpable* and *removed impalpable local recurrence*. Given the inability to directly estimate these input parameters, they were fitted to observed model outputs via calibration.

Model calibration involves sequentially testing values for these parameters and comparing simulation output values with the results observed in a population of relevance to the study population. The process is continued until one or more sets of input parameter values are found that yield an acceptable match (converge on the observed calibration targets).

A calibration version of the model was developed to fit values for the model input parameters describing disease progression and mammography test performance (sensitivity and specificity, which also determine the timing of the detection of local recurrence). Literature searches were undertaken to identify parameter ranges for times from *disease-free* to *removed local recurrence*, from *removed local recurrence* to *distant recurrence*, and from *distant recurrence* to *breast cancer death*, as well as the test performance parameters. The observed disease progression parameter values were presented to clinical experts, and they were asked to define broad ranges of possible values for the model's disease progression parameters (Table 10.2). The calibration model is described further in the next section.

The timing of the death event for women without distant metastases (i.e., death due to causes other than breast cancer) was informed by rates obtained from general population life tables for Australia (Australian Bureau of Statistics (ABS) 2007) minus the age-specific proportion of deaths attributed to breast cancer (AIHW 2015). Cost (Karnon et al. 2007b) and utility (quality of life) weights (de Koning et al. 1991; Peasgood et al. 2010; Tengs and Wallace 2000) following each of the events in the model were also derived from the existing literature (Table 10.3).

A key remaining input parameter describes patient adherence with the intended surveillance strategy. No data regarding women's adherence with surveillance mammograms were identified, but adherence with population-based mammography screening (i.e., primary screening) was estimated to be 55%. Assuming women with experience of breast cancer are more likely

TABLE 10.2

Initial Ranges Developed for Parameters to Be Derived via Calibration

Parameter	Initial Range
Time from disease-free to impalpable local recurrence	
Weibull distribution, alpha parameter	0.6–0.85
Weibull distribution, beta parameter	10,000–20,000
Time from disease-free to distant recurrence	
Weibull distribution, alpha parameter	0.74–0.99
Weibull distribution, beta parameter	5,000–6,000
Time from impalpable to palpable local recurrence (weeks)	57–233
Time from impalpable to distant recurrence (weeks)	318–1,577
Time from palpable to distant recurrence (weeks)	278–1,400
Time from removed palpable local to distant recurrence (weeks)	375–1,700
Time from distant recurrence to breast cancer death (weeks)	57–102
Mammography sensitivity	0.6–0.9
Mammography specificity	0.8–0.95

TABLE 10.3

Cost and Utility Input Parameter Values Obtained from the Literature

Item	Costs (AUS $)	Utility Weights
Surveillance (utility decrement)	$89.05	0.010
Disease-free (weekly cost)	$3.20	0.832
Removal of local recurrence	$9,802.00	0.655
Remission (weekly cost)	$3.20	0.752
Distant recurrence (weekly cost)	$147.00	0.443

to adhere to mammography testing, adherence rates of 75% and 90% were tested in separate scenario analyses.

10.5 Implementation

The DES was run as a calibration exercise first. Each replication of the calibration model simulated the life history of 2,000 model entities (women with treated early breast cancer at moderate risk of recurrence), each of which was sampled (with replacement) from the dataset of 407 observed women at moderate risk. For each sampled woman, *age at diagnosis of early breast cancer* was stored as an attribute, and the attribute *number of surveillance mammograms* was set to zero. A mammographic surveillance schedule was defined for each entity, reflecting the observed schedule of surveillance mammograms in the source data. In addition, for each model replication, input values were randomly sampled from the defined ranges for each timing input parameter.

The calibration analysis only replicated the observed surveillance schedules, and so no duplicate sets of entities were required to analyze alternative surveillance strategies at this stage. After the entities enter the simulation, a time to event is sampled for each entity for each possible next event, from which point the model follows the logic as represented in Figure 10.2. As the simulation moves forward in time, individual entities experience new events. After each event, the relevant attributes for the entity experiencing it are updated, and new times to events are sampled for relevant subsequent events.

The outputs of each calibration model replication were compared to the observed results of the cohort of 407 moderate-risk women, with respect to the experience and timing of disease recurrence and mortality at 5 and 10 years (the calibration targets). Each sampled set of input parameter values was defined as convergent if all of the calibration target output values were within the 95% confidence intervals of the observed values.

To boost the efficiency of the calibration process, formal optimization methods could have been applied to guide the sampling process toward local peaks in the parameter space (Vanni et al. 2011). For this study, however, a random sampling procedure was applied. Additional runs of the calibration model were undertaken until 1,000 sets of convergent input parameter values were identified. This number was based on nonsignificant variation in mean output values across consecutive calibration analyses.

The calibration model was then converted to a main analysis model to evaluate the cost-effectiveness of the three alternative mammographic follow-up schedules. This involved the replacement of the observed mammographic surveillance schedules with time points that reflected the alternative mammographic surveillance strategies being evaluated: annual, two-yearly surveillance, and mixed.

The DES was implemented in the software Simul8 (Visual Thinking 1998).

10.6 Analyses

In addition to comparing three alternative mammographic surveillance schedules (annual, two-yearly, and mixed), and the alternative adherence scenarios (75% and 90%), the model also tested the effects of these scenarios in patients aged 50 to 69 years, and patients aged 70 years and older (Table 10.4).

TABLE 10.4

Model Scenarios

Age (Years)	Adherence (%)	Mammographic Schedule
50–69	90	Annual
		Two-yearly
		Mixed
	75	Annual
		Two-yearly
		Mixed
70+	90	Annual
		Two-yearly
		Mixed
	75	Annual
		Two-yearly
		Mixed

Annual (ongoing); two-yearly (ongoing); mixed (annual for 5 years and then two-yearly ongoing).

For each main analysis model run, the age at diagnosis was sampled (with replacement) for 2,000 entities from the corresponding age-based subsets (patients aged 50 to 69 years, and aged 70 years and over) of the observed 407 women with early breast cancer at moderate risk of recurrence. The adherence rate for the scenario being analyzed was set and applied to all the simulated women. Then, the 1,000 sets of convergent input parameters identified through the model calibration process were sampled to obtain a set of input parameter values for the model run. The DES then created three copies of each of the 2,000 entities, one for each of the three surveillance strategies (to reduce the effects of random variation across the runs).

Each model run comprised 2,000 model replications of 1,000 alternative sets of patients and input parameter values across 12 scenarios. This amounted to analyses of the experience of 36 million entities (i.e., $12 \times 2,000 \times 1,000$). Each model run required two to three hours on a 64-bit operating system with 8 GB RAM, depending on the age of the analyzed cohort (older cohorts were quicker). Outputs from each of the 2,000 model replications included total costs and total QALYs across the 1,000 model entities. These were collected in an internal Simul8 spreadsheet and exported to Microsoft Excel® for analysis at the end of each run.

Mean ICERs were estimated between mammographic schedules within each age and adherence scenario analysis (e.g., 50–69-year olds assuming 75% adherence). The results across the 2,000 model replications in each model run were analyzed to estimate mean incremental cost-effectiveness ratios between the three alternative surveillance strategies (arranged in ascending order of effectiveness) as well as to estimate the probability that each surveillance option was the most cost-effective strategy according to the equivalent monetary value placed on the gain of additional QALYs.

In addition, deterministic sensitivity analyses tested the effects of best- and worst-case scenarios with respect to the cost and utility weight input parameter values, requiring another 12 model runs. Total model running time of the main analysis model was around 60 hours, though many more hours were spent, first testing and then calibrating the model.

10.7 Results

The mean cost-effectiveness results (Table 10.5) show that for postmenopausal women with early breast cancer at moderate risk of recurrence, the minimum mean incremental cost per QALY gained is $126,481, which is well above the often quoted threshold of $50,000 per QALY in Australia. At that threshold, a mixed annual to two-yearly surveillance schedule is most cost-effective in three of the four scenarios defined by age and adherence.

The probabilities of cost-effectiveness across the age- and adherence-based scenario analyses are presented in Table 10.6, which shows significant uncertainty regarding the preferred surveillance strategy (from a maximizing QALYs perspective). Current practice, which consists of ongoing annual surveillance,

TABLE 10.5

Cost-effectiveness Results across the Scenarios Analyzed

Age (Years)	Adherence (%)	Schedule	Incremental Costs (AUS $)	QALYs	ICER (AUS $/QALY)
50–69	90	Two-yearly			
50–69	90	Mixed	140	0.005	28,199
50–69	90	Annual	326	0.003	126,481
50–69	75	Two-yearly			
50–69	75	Mixed	138	0.006	21,481
50–69	75	Annual	256	0.002	133,525
70+	90	Two-yearly			
70+	90	Mixed	142	0.002	69,608
70+	90	Annual	184	0.0004	413,230
70+	75	Two-yearly			
70+	75	Mixed	131	0.003	40,706
70+	75	Annual	139	0.0004	377,290

ICER, incremental cost-effectiveness ratio; QALY, quality-adjusted life year; AUS, Australian.

TABLE 10.6

Proportion of Model Replications Yielding ICERs below the Specified Threshold

Age Group (Years)	Adherence (%)	ICER Threshold AUS $/QALY	Annual Screen (%)	Biennial Screen (%)	Mixed (%)
50–69	90	25,000	5.2	58.3	36.5
		50,000	20.8	40.1	39.1
		75,000	29.5	33.4	37.1
	75	25,000	11.4	49.2	39.4
		50,000	26.7	33.1	40.2
		75,000	34.5	27.2	38.3
70–79	90	25,000	5.5	69.7	24.8
		50,000	17.5	48.7	33.8
		75,000	24.4	39.7	35.9
	75	25,000	7.6	62.2	30.2
		50,000	20.8	43.3	35.9
		75,000	26.3	37.6	36.1

Note: ICER threshold, the value attached to the gain of additional QALYs; adherence, % of women adherent with given mammography schedule; annual screen, mammography every year; biennial screen, mammography every 2 years; mixed, annual mammography for 5 years, then biennial thereafter. AUS, Australian; ICER, incremental cost-effectiveness ratio; QALY, quality-adjusted life year.

has the lowest probability of attaining cost-effectiveness in all scenarios other than the analysis of 50–69-year-olds assuming 75% adherence with surveillance.

10.8 Comments

This case study has described a DES that was used to estimate the cost-effectiveness of alternative mammographic surveillance strategies for women remaining disease-free following diagnosis and treatment of early breast cancer. The model structure was not particularly complex, but the use of individual-level data to calibrate the model precluded the use of a cohort-based state transition model. Thus, DES is not solely a modeling technique to be used when model structures become too complex to be feasibly implemented as a state transition model. DES also provides flexibility with respect to the handling of alternative forms of input data.

References

Adewunmi, A., & Aickelin, U. (2012). Investigating the effectiveness of variance reduction techniques in manufacturing, call center and cross-docking discrete event simulation models. In S. Bangsow (Ed.), *Use Cases of Discrete Event Simulation* (pp. 1–26). Berlin, Germany: Springer.

Afzali, H. H., Karnon, J., & Merlin, T. (2013). Improving the accuracy and comparability of model-based economic evaluations of health technologies for reimbursement decisions: A methodological framework for the development of reference models. *Med. Decis. Making, 33*(3), 325–332. doi:10.1177/0272989X12458160.

Akaike, H. (1974). A new look at the statistical model identification. *IEEE Trans. Automat. Contr., 19*(6), 716–723.

Alvarez-Napagao, S., Koch, F., Gómez-Sebastià, I., & Vázquez-Salceda, J. (2011). Making games ALIVE: An organisational approach. In F. Dignum (Ed.), *Agents for Games and Simulations II* (Vol. 6525, pp. 179–191). Berlin, Germany: Springer.

Ambler, S. W. (1998). *Process Patterns: Building Large-Scale Systems using Object Technology*. Cambridge: Cambridge University Press.

Anand, S., & Hanson, K. (1997). Disability-adjusted life years: A critical review. *J. Health Econ., 16*(6), 685–702.

Anderson, K. M., Wilson, P. W., Odell, P. M., & Kannel, W. B. (1991). An updated coronary risk profile. A statement for health professionals. *Circulation, 83*(1), 356–362.

AnyLogic. (2009). *AnyLogic Version 6.4. 0 Manual*. St. Petersburg, Russia: XJ Technologies.

Auchincloss, A. H., & Diez Roux, A. V. (2008). A new tool for epidemiology: The usefulness of dynamic-agent models in understanding place effects on health. *Am. J. Epidemiol., 168*(1), 1–8. doi:10.1093/aje/kwn118.

Australian Bureau of Statistics (ABS). (2007). 3302.0.55.001—life tables, Australia, 2003 to 2005 bureau of statistics. Retrieved April 3, 2015, from http://www.abs.gov.au/AUSSTATS/abs@.nsf/allprimarymainfeatures/6A75A50AE791F95FCA25738D000F462B?opendocument.

Australian Institute of Health and Welfare (AIHW). (2012). *Breast Cancer in Australia: An overview* (Cancer series no. 71, Cat. no. CAN 67). Retrieved April 3, 2015, from http://www.aihw.gov.au/publication-detail/?id=10737423008.

Australian Institute of Health and Welfare (AIHW). (2015). *Australian Cancer Incidence and Mortality (ACIM) Books*. Breast Cancer (ICD-10 code C50), Australia, 1968–2012: Age-specific mortality rates. Retrieved April 3, 2015, from http://www.abs.gov.au/AUSSTATS/abs@.nsf/DetailsPage/3303.02013?OpenDocument.

Avramidis, A. N., & Wilson, J. R. (1996). Integrated variance reduction strategies for simulation. *Oper. Res., 44*(2), 327–346. doi:10.1287/opre.44.2.327.

Baio, G., & Dawid, A. P. (2011). Probabilistic sensitivity analysis in health economics. *Stat. Methods Med. Res.*, 20. doi:10.1177/0962280211419832. Retrieved from http://smm.sagepub.com/content/early/2011/09/17/0962280211419832.full.pdf+html.

Bair, A. E., Song, W. T., Chen, Y. C., & Morris, B. A. (2010). The impact of inpatient boarding on ED efficiency: A discrete-event simulation study. *J. Med. Syst., 34*(5), 919–929. doi:10.1007/s10916-009-9307-4.

Banks, J. (2001). *Discrete-Event System Simulation* (3rd ed.). Upper Saddle River, NJ: Prentice Hall.

Banta, D. (2003). The development of health technology assessment. *Health Policy, 63*(2), 121–132.

Bartels, R. (1982). The rank version of von Neumann's ratio test for randomness. *J. Am. Stat. Assoc., 77*(377), 40–46.

Barton, P., Bryan, S., & Robinson, S. (2004). Modelling in the economic evaluation of health care: Selecting the appropriate approach. *J. Health Serv. Res. Policy, 9*(2), 110–118. doi:10.1258/135581904322987535.

Barton, P. I., & Lee, C. K. (2002). Modeling, simulation, sensitivity analysis, and optimization of hybrid systems. *ACM Trans. Model. Comput. Simul., 12*(4), 256–289. doi:10.1145/643120.643122.

Barua, B., Rovere, M. C., & Skinner, B. J. (2010). Waiting your turn: Wait times for health care in Canada 2010 report. Retrieved March 28, 2014, from http://ssrn.com/abstract=1783079.

Beck, J. R., & Pauker, S. G. (1983). The Markov process in medical prognosis. *Med. Decis. Making, 3*(4), 419–458.

Berg, B., Denton, B., Nelson, H., Balasubramanian, H., Rahman, A., Bailey, A., & Lindor, K. (2010). A discrete event simulation model to evaluate operational performance of a colonoscopy suite. *Med. Decis. Making, 30*(3), 380–387. doi:10.1177/0272989X09345890.

Berger, M. L., Mamdani, M., Atkins, D., & Johnson, M. L. (2009). Good research practices for comparative effectiveness research: Defining, reporting and interpreting nonrandomized studies of treatment effects using secondary data sources: The ISPOR Good Research Practices for Retrospective Database Analysis Task Force Report—Part I. *Value Health, 12*(8), 1044–1052. doi:10.1111/j.1524-4733.2009.00600.x.

Berger, M. L., Martin, B. C., Husereau, D., Worley, K., Allen, J. D., Yang, W., ... Crown, W. (2014). A questionnaire to assess the relevance and credibility of observational studies to inform health care decision making: An ISPOR-AMCP-NPC Good Practice Task Force Report. *Value Health, 17*(2), 143–156. doi:10.1016/j.jval.2013.12.011.

Bessen, T., & Karnon, J. (2014). A patient-level calibration framework for evaluating surveillance strategies: A case study of mammographic follow-up after early breast cancer. *Value Health, 17*(6), 669–678. doi:10.1016/j.jval.2014.07.002.

Beyer, D. (2014). Status report on software verification. In E. Ábrahám & K. Havelund (Eds.), *Tools and Algorithms for the Construction and Analysis of Systems* (Vol. 8413, pp. 373–388). Berlin, Germany: Springer.

Bilcke, J., Beutels, P., Brisson, M., & Jit, M. (2011). Accounting for methodological, structural, and parameter uncertainty in decision-analytic models: A practical guide. *Med. Decis. Making, 31*(4), 675–692. doi:10.1177/0272989X11409240.

Binkley, D., Davis, M., Lawrie, D., & Morrell, C. (2009). To camelcase or underscore. *Paper Presented at the 17th IEEE International Conference on Program Comprehension.* Vancouver, BC, Canada.

Bland, J. M., & Altman, D. G. (1998). Survival probabilities (the Kaplan-Meier method). *BMJ, 317*(7172), 1572.

Blunk, A., & Fischer, J. (2014). A highly efficient simulation core in C++. *Paper Presented at the Proceedings of the Symposium on Theory of Modeling & Simulation—DEVS Integrative.* Tampa, FL.

Borshchev, A., & Filippov, A. (2004, July 25–29). From system dynamics and discrete event to practical agent based modeling: Reasons, techniques, tools. *Paper Presented at the 22nd International Conference of the System Dynamics Society*. Oxford, England.

Brailsford, S., Churilov, L., & Dangerfield, B. (2014). *Discrete-Event Simulation and System Dynamics for Management Decision Making*. Chichester, West Sussex: John Wiley & Sons.

Brailsford, S., & Schmidt, B. (2003). Towards incorporating human behaviour in models of health care systems: An approach using discrete event simulation. *Eur. J. Oper. Res., 150*(1), 19–31. doi:citeulike-article-id:1966440; doi:10.1016/s0377-2217(02)00778-6.

Brailsford, S. C., & Hilton, N. A. (2001). A comparison of discrete event simulation and system dynamics for modelling health care systems. Retrieved from http://eprints.soton.ac.uk/35689/1/glasgow_paper.pdf.

Brandeau, M. L., Sainfort, F., & Pierskalla, W. P. (2004). *Operations Research and Health Care: A Handbook of Methods and Applications*. Boston, MA: Kluwer Academic.

Brennan, A., Chick, S. E., & Davies, R. (2006). A taxonomy of model structures for economic evaluation of health technologies. *Health Econ., 15*(12), 1295–1310. doi:10.1002/hec.1148.

Briggs, A. H., Claxton, K., & Sculpher, M. J. (2006). *Decision Modelling for Health Economic Evaluation*. Oxford: Oxford University Press.

Briggs, A. H., Weinstein, M. C., Fenwick, E. A., Karnon, J., Sculpher, M. J., & Paltiel, A. D. (2012). Model parameter estimation and uncertainty: A report of the ISPOR-SMDM Modeling Good Research Practices Task Force—6. *Value Health, 15*(6), 835–842. doi:10.1016/j.jval.2012.04.014.

Buss, A. H. (1996). Modeling with event graphs. *Proceedings of the 28th conference on Winter simulation* (pp.53–160). Coronado, CA: IEEE Computer Society press.

Canadian Agency for Drugs and Technologies in Health (CADTH). (2006). *Guidelines for Economic Evaluation of Health Technologies: Canada*. Retrieved June 12, 2014, from http://www.cadth.ca/en/products/methods-and-guidelines/overview.

Cardoen, B., Demeulemeester, E., & Beliën, J. (2010). Operating room planning and scheduling: A literature review. *Eur. J. Oper. Res., 201*(3), 921–932. doi:10.1016/j.ejor.2009.04.011.

Caro, J. J. (2005). Pharmacoeconomic analyses using discrete event simulation. *Pharmacoeconomics, 23*(4), 323–332.

Caro, J. J. (2015). Psst, have I got a model for you. *Med. Decis. Making, 35*(2), 139–141. doi:10.1177/0272989X14559729.

Caro, J. J., Briggs, A. H., Siebert, U., & Kuntz, K. M. (2012). Modeling good research practices—Overview: A report of the ISPOR-SMDM Modeling Good Research Practices Task Force–1. *Med. Decis. Making, 32*(5), 667–677. doi:10.1177/0272989x12454577.

Caro, J. J., Caro, G., Getsios, D., Raggio, G., Burrows, M., & Black, L. (2000). The migraine ACE model: Evaluating the impact on time lost and medical resource use. *Headache, 40*(4), 282–291.

Caro, J. J., Eddy, D. M., Kan, H., Kaltz, C., Patel, B., Eldessouki, R., … ISPOR-AMCP-NPC Modeling CER Task Forces. (2014). Questionnaire to assess relevance and credibility of modeling studies for informing health care decision making: An ISPOR-AMCP-NPC Good Practice Task Force Report. *Value Health, 17*(2), 174–182.

Caro, J. J., Flegel, K. M., Orejuela, M. E., Kelley, H. E., Speckman, J. L., & Migliaccio-Walle, K. (1999). Anticoagulant prophylaxis against stroke in atrial fibrillation: Effectiveness in actual practice. *CMAJ, 161*(5), 493–497.

Caro, J. J., Getsios, D., & Möller, J. (2007). Regarding probabilistic analysis and computationally expensive models: Necessary and required? *Value Health, 10*(4), 317–318; author reply 319. doi:10.1111/j.1524-4733.2007.00176.x.

Caro, J. J., & Ishak, K. J. (2010). No head-to-head trial? Simulate the missing arms. *Pharmacoeconomics, 28*(10), 957–967. doi:10.2165/11537420-000000000-00000.

Caro, J. J., Ishak, K. J., & Migliaccio-Walle, K. (2004). Estimating survival for cost-effectiveness analyses: A case study in atherothrombosis. *Value Health, 7*(5), 627–635. doi:10.1111/j.1524-4733.2004.75013.x.

Caro, J. J., & Möller, J. (2014). Decision-analytic models: Current methodological challenges. *Pharmacoeconomics, 32*(10), 943–950. doi:10.1007/s40273-014-0183-5.

Caro, J. J., Möller, J., & Getsios, D. (2010). Discrete event simulation: The preferred technique for health economic evaluations? *Value Health, 13*(8), 1056–1060. doi:10.1111/j.1524-4733.2010.00775.x.

Caro, J.J., Möller, J., Getsios, D., Coudeville, L., El-Hadi, W., Chevat, C., ... Caro, I. (2007). Invasive meningococcal disease epidemiology and control measures: A framework for evaluation. *BMC Public Health, 7,* 130. doi:10.1186/1471-2458-7-130.

Cellier, F. E., Elmqvist, H., Otter, M., & Taylor, J. (1993). Guidelines for modeling and simulation of hybrid systems. *Paper Presented at the IFAC 12th Triennial World Congress,* Sydney, Australia.

Chambless, L. E., Folsom, A. R., Sharrett, A. R., Sorlie, P., Couper, D., Szklo, M., & Nieto, F.J. (2003). Coronary heart disease risk prediction in the atherosclerosis risk in Communities (ARIC) study. *J. Clin. Epidemiol., 56*(9), 880–890.

Chan, W.-H., & Lu, M. (2012, December 9–12). Construction operations simulation under structural adequacy constraints: The stonecutters bridge case study. *Paper Presented at the Proceedings of the 2012 Winter Simulation Conference.* Berlin, Germany.

Chapman, G. B., & Sonnenberg, F. A. (2003). *Decision Making in Health Care: Theory, Psychology, And Applications.* Cambridge: Cambridge University Press.

Chen, E. J. (2013). Some insights of using common random numbers in selection procedures. *Discrete Event Dyn. Syst., 23*(3), 241–259. doi:10.1007/s10626-012-0142-2.

Chhatwal, J., & He, T. (2015). Economic evaluations with agent-based modelling: An introduction. *Pharmacoeconomics, 33,* 1–11. doi:10.1007/s40273-015-0254-2.

Claeskens, G., & N. L. Hjort. (2008). *Model Selection and Model Averaging.* Cambridge: Cambridge University Press.

Clark, D. E., Hahn, D. R., Hall, R. W., & Quaker, R. E. (1994). Optimal location for a helicopter in a rural trauma system: Prediction using discrete-event computer simulation. *Proc. Annu. Symp. Comput. Appl. Med. Care,* 888–892.

Clarke, L. D., Plevritis, S. K., Boer, R., Cronin, K. A., & Feuer, E. J. (2006). A comparative review of CISNET breast models used to analyze U.S. breast cancer incidence and mortality trends. *J. Natl. Cancer Inst. Monogr.* (36), 96–105. doi:10.1093/jncimonographs/lgj013.

Clarke, P. M., Gray, A. M., Briggs, A., Farmer, A.J., Fenn, P., Stevens, R.J., Matthews, D.R., Stratton, I.M., Holman, R.R., & UK Prospective Diabetes Study Group. (2004). A model to estimate the lifetime health outcomes of patients with type 2 diabetes: The United Kingdom Prospective Diabetes Study (UKPDS) Outcomes Model (UKPDS no. 68). *Diabetologia, 47*(10), 1747–1759. doi:10.1007/s00125-004-1527-z.

Claxton, K., Sculpher, M., McCabe, C., Briggs, A., Akehurst, R., Buxton, M., Brazier, J., & O'Hagan, T. (2005). Probabilistic sensitivity analysis for NICE technology assessment: Not an optional extra. *Health Econ., 14*(4), 339–347. doi:10.1002/hec.985.

Claxton, K. P., & Sculpher, M. J. (2006). Using value of information analysis to prioritise health research: Some lessons from recent UK experience. *Pharmacoeconomics, 24*(11), 1055–1068.

Cleary, P. W., & Prakash, M. (2004). Discrete-element modelling and smoothed particle hydrodynamics: Potential in the environmental sciences. *Philos. Trans. A Math. Phys. Eng. Sci., 362*(1822), 2003–2030. doi:10.1098/rsta.2004.1428.

Collett, D. (2003). *Modelling Survival Data in Medical Research* (2nd ed.). Boca Raton, FL: Chapman & Hall/CRC Press.

Collins, G. S., & Altman, D. G. (2012). Predicting the 10 year risk of cardiovascular disease in the United Kingdom: Independent and external validation of an updated version of QRISK2. *BMJ, 344*, e4181. doi:10.1136/bmj.e4181.

Comas, M., Castells, X., Hoffmeister, L., Roman, R., Cots, F., Mar, J., Gutierrez-Moreno, S., & Espallargues, M. (2008). Discrete-event simulation applied to analysis of waiting lists. Evaluation of a prioritization system for cataract surgery. *Value Health, 11*(7), 1203–1213.

Cooper, K., Davies, R., Raftery, J., & Roderick, P. (2008). Use of a coronary heart disease simulation model to evaluate the costs and effectiveness of drugs for the prevention of heart disease. *J. Oper. Res. Soc., 59*(9), 1173–1181.

Coudeville, L., Van Rie, A., Getsios, D., Caro, J.J., Crepey, P., & Nguyen, V.H. (2009). Adult vaccination strategies for the control of pertussis in the United States: An economic evaluation including the dynamic population effects. *PLoS One, 4*(7), e6284. doi:10.1371/journal.pone.0006284.

Cox, D. R. (1972). Regression models and life tables. *JR Stat. Soc. B, 34*(2), 187–220.

Cox, E., Martin, B. C., Van Staa, T., Garbe, E., Siebert, U., & Johnson, M.L. (2009). Good research practices for comparative effectiveness research: approaches to mitigate bias and confounding in the design of nonrandomized studies of treatment effects using secondary data sources: The International Society for Pharmacoeconomics and Outcomes Research Good Research Practices for Retrospective Database Analysis Task Force Report—Part II. *Value Health, 12*(8), 1053–1061. doi:10.1111/j.1524-4733.2009.00601.x.

Crane, G. J., Kymes, S. M., Hiller, J. E., Casson, R., Martin, A., & Karnon, J.D. (2013). Accounting for costs, QALYs, and capacity constraints: Using discrete-event simulation to evaluate alternative service delivery and organizational scenarios for hospital-based glaucoma services. *Med. Decis. Making, 33*(8), 986–997. doi:10.1177/0272989X13478195.

D'Agostino, R. B., Sr., Vasan, R. S., Pencina, M. J., Wolf, P.A., Cobain, M., Massaro, J.M., & Kannel, W.B. (2008). General cardiovascular risk profile for use in primary care: The Framingham Heart Study. *Circulation, 117*(6), 743–753. doi:10.1161/CIRCULATIONAHA.107.699579.

Dahl, O.-J., Nygaard, K., & Norsk R (1966). *SIMULA. A Language For Programming and Description of Discrete Event Systems*. Oslo, Norway: Norwegian Computing Center.

Davis, S., Stevenson, M., Tappenden, P., & Wailoo, A. (2014). NICE DSU technical support document 15: Cost-effectiveness modelling using patient-level simulation. Retrieved from http://www.nicedsu.org.uk/TSD15_Patient-level_simulation.pdf.

De Boer, A. G. E. M., Van Lanschot, J. J. B., Stalmeier, P. F. M., Van Sandick, J.W., Hulscher, J.B.F., De Haes, J.C.J.M., & Sprangers, M.A.G. (2004). Is a single-item visual analogue scale as valid, reliable and responsive as multi-item scales in measuring quality of life? *Qual. Life Res., 13*(2), 311–320.

de Koning, H. J., van Ineveld, B. M., van Oortmarssen, G. J., de Haes, J.C., Collette, H.J., Hendriks, J.H., & van der Maas, P.J. (1991). Breast cancer screening and cost-effectiveness; policy alternatives, quality of life considerations and the possible impact of uncertain factors. *Int. J. Cancer, 49*(4), 531–537.

de Lara, J. (2005). Distributed event graphs: Formalizing component-based modelling and simulation. *Electron. Notes Theor. Comput. Sci., 127*(4), 145–162. doi:http://dx.doi.org/10.1016/j.entcs.2004.08.052.

Diggle, P., & Diggle, P. (2002). *Analysis of Longitudinal Data* (2nd ed.). Oxford, NY: Oxford University Press.

Djanatliev, A., Kolominsky-Rabas, P., Hofmann, B., Aisenbrey, A., & German, R. (2014). System dynamics and agent-based simulation for prospective health technology assessments. In M. S. Obaidat, J. Filipe, J. Kacprzyk, & N. Pina (Eds.), *Simulation and Modeling Methodologies, Technologies and Applications* (Vol. 256, pp. 85–96). New York: Springer.

Dolan, P., Gudex, C., Kind, P., & Williams, A. (1996). The time trade-off method: Results from a general population study. *Health Econ., 5*(2), 141–154. doi:10.1002/(SICI)1099-1050(199603)5:2<141::AID-HEC189>3.0.CO;2-N.

Donovan, J. J., & Jones, M. M. (1968). A graphical facility for an interactive simulation system. In A. J. H. Morrell (Ed.), *Information Processing 1968: World Congress Proceedings*. Amsterdam, the Netherlands: North-Holland.

Doubilet, P., Begg, C. B., Weinstein, M. C., Braun, P., & McNeil, B. J. (1984). Probabilistic sensitivity analysis using Monte Carlo simulation. A practical approach. *Med. Decis. Making, 5*(2), 157–177.

Drummond, M. F., Schwartz, J. S., Jonsson, B., Luce, B.R., Neumann, P.J., Siebert, U., & Sullivan, S.D. (2008). Key principles for the improved conduct of health technology assessments for resource allocation decisions. *Int. J. Technol. Assess. Health Care, 24*(3), 244–258.

Duguay, C., & Chetouane, F. (2007). Modeling and improving emergency department systems using discrete event simulation. *Simulation, 83*(4), 311–320. doi:10.1177/0037549707083111.

Eddy, D. M. (2005). Evidence-based medicine: A unified approach. *Health Aff. (Millwood), 24*(1), 9–17. doi:10.1377/hlthaff.24.1.9.

Eddy, D. M., Hollingworth, W., Caro, J. J., Tsevat, J., McDonald, K.M., & Wong, J.B. (2012). Model transparency and validation: A report of the ISPOR-SMDM Modeling Good Research Practices Task Force—7. *Value Health, 15*(6), 843–850. doi:10.1016/j.jval.2012.04.012.

Eldredge, D. L., McGregor, J. D., & Summers, M. K. (1990). Applying the object-oriented paradigm to discrete event simulations using the C++ language. *Simulation, 54*(2), 83–91. doi:10.1177/003754979005400205.

Emanuel, E. J., Fuchs, V. R., & Garber, A. M. (2007). Essential elements of a technology and outcomes assessment initiative. *JAMA, 298*(11), 1323–1325. doi:10.1001/jama.298.11.1323.

Enders, C. K. (2010). *Applied Missing Data Analysis*. New York: Guilford Press.

Epstein, J. M. (2009). Modelling to contain pandemics. *Nature, 460*(7256), 687. doi:10.1038/460687a.

European Network for Health Technology Assessment (EUnetHTA). (2011). *HTA Core Model Online: Handbook* (version 1.3). Retrieved June 12, 2014, from http://meka.thl.fi/htacore/ViewHandbook.aspx.

Feig, S. A. (2014). Screening mammography benefit controversies: Sorting the evidence. *Radiol. Clin. North Am.*, 52(3), 455–480. doi:10.1016/j.rcl.2014.02.009.

Fenwick, E., Claxton, K., & Sculpher, M. (2001). Representing uncertainty: The role of cost-effectiveness acceptability curves. *Health Econ.*, 10(8), 779–787.

Fishman, G. S., & Huang, B. D. (1983). Antithetic variates revisited. *Commun. ACM*, 26(11), 964–971. doi:10.1145/182.358462.

Fone, D., Hollinghurst, S., Temple, M., Round, A., Lester, N., Weightman, A., Roberts, K., Coyle, E., Bevan, G., & Palmer, S. (2003). Systematic review of the use and value of computer simulation modelling in population health and health care delivery. *J. Public Health Med.*, 25(4), 325–335.

Forsyth, K. D., Wirsing von Konig, C. H., Tan, T., Caro, J., & Plotkin, S. (2007). Prevention of pertussis: Recommendations derived from the second Global Pertussis initiative roundtable meeting. *Vaccine*, 25(14), 2634–2642. doi:10.1016/j.vaccine.2006.12.017.

Fujimoto, R. M., Malik, A. W., & Park, A. (2010). Parallel and distributed simulation in the cloud. *SCS M&S Magazine*, 3, 1–10.

Gagnon, M. P., Lepage-Savary, D., Gagnon, J., St-Pierre, M., Simard, C., Rhainds, M., Lemieux, R., Gauvin, F.P., Desmartis, M., & Legare, F. (2009). Introducing patient perspective in health technology assessment at the local level. *BMC Health Serv. Res.*, 9, 54. doi:10.1186/1472-6963-9-54.

Galea, M. H., Blamey, R. W., Elston, C. E., & Ellis, I. O. (1992). The Nottingham prognostic index in primary breast cancer. *Breast Cancer Res. Treat.*, 22(3), 207–219.

Garrido, J. M. (2013). *Object-Oriented Discrete-Event Simulation with Java: A Practical Introduction.* New York: Springer.

Getsios, D., Blume, S., Ishak, K. J., Maclaine, G., & Hernandez, L. (2012). An economic evaluation of early assessment for Alzheimer's disease in the United Kingdom. *Alzheimers Dement.*, 8(1), 22–30. doi:10.1016/j.jalz.2010.07.001.

Getsios, D., Caro, I., El-Hadi, W., & Caro, J. J. (2004). Assessing the economics of vaccination for Neisseria meningitidis in industrialized nations: A review and recommendations for further research. *Int. J. Technol. Assess. Health Care*, 20(3), 280–288.

Goldhaber-Fiebert, J. D. (2012). Accounting for biases when linking empirical studies and simulation models. *Med. Decis. Making*, 32(3), 397–399. doi:10.1177/0272989X12441398.

Goldsman, D., Nance, R. E., & Wilson, J. R. (December 5–8, 2010). A brief history of simulation revisited. *Paper Presented at the Proceedings of the 2010 Winter Simulation Conference.* Baltimore, MD.

Greasley, A. (1998). An example of a discrete-event simulation on a spreadsheet. *Simulation*, 70(3), 148–162. doi:10.1177/003754979807000302.

Greasley, A. (2008). 3: Acquiring the resources for simulation. *Enabling a Simulation Capability in the Organisation* (pp. 21–29). London: Springer.

Griffin, S., Claxton, K., Hawkins, N., & Sculpher, M. (2006). Probabilistic analysis and computationally expensive models: Necessary and required? *Value Health*, 9(4), 244–252.

Groot Koerkamp, B., Hunink, M. G. M., Stijnen, T., Hammitt, J.K., Kuntz, K.M., & Weinstein, M.C. (2007). Limitations of acceptability curves for presenting uncertainty in cost-effectiveness analysis. *Med. Decis. Making*, 27(2), 101–111.

Guo, S., Pelligra, C., Saint-Laurent Thibault, C., Hernandez, L., & Kansal, A. (2014). Cost-effectiveness analyses in multiple sclerosis: A review of modelling approaches. *Pharmacoeconomics, 32*(6), 559–572. doi:10.1007/s40273-014-0150-1.

Gustavsson, A., Van Der Putt, R., Jonsson, L., & McShane, R. (2009). Economic evaluation of cholinesterase inhibitor therapy for dementia: Comparison of Alzheimer's disease and Dementia with Lewy bodies. *Int. J. Geriatr. Psychiatry, 24*(10), 1072–1078. doi:10.1002/gps.2223.

Habbema, J. D., van Oortmarssen, G. J., Lubbe, J. T., & van der Maas, P. J. (1985). The MISCAN simulation program for the evaluation of screening for disease. *Comput. Methods Programs Biomed., 20*(1), 79–93.

Haji Ali Afzali, H., Gray, J., & Karnon, J. (2013). Model performance evaluation (validation and calibration) in model-based studies of therapeutic interventions for cardiovascular diseases: A review and suggested reporting framework. *Appl. Health Econ. Health Policy, 11*(2), 85–93.

Haji Ali Afzali, H., & Karnon, J. (2015). Exploring structural uncertainty in model-based economic evaluations. *Pharmacoeconomics, 33*(5), 435–443.

Hammersley, J. M., & Morton, K. W. (1956). A new Monte Carlo technique: Antithetic variates. *Math. Proc. Cambridge, 52*(03), 449–475. doi:10.1017/S0305004100031455.

Hawkes, N. (2008). Why is the press so nasty to NICE? *BMJ, 337*(7673), 788.

Health Technology Assessment International (HTAi). (2014). *Resources: Information, Tools and Publications.* HTAi. Retrieved June 13, 2014, from http://www.htai.org/index.php?id=573.

Hiligsmann, M., Ethgen, O., Bruyere, O., Richy, F., Gathon, H.J., & Reginste, J.Y. (2009). Development and validation of a Markov microsimulation model for the economic evaluation of treatments in osteoporosis. *Value Health, 12*(5), 687–696. doi:10.1111/j.1524-4733.2008.00497.x.

Hill, R. R., Miller, J. O., & McIntyre, G. A. (2001, December 9–12). Applications of discrete event simulation modeling to military problems. *Paper Presented at the Proceedings of the 2001 Winter Simulation Conference.* Arlington, VA.

Hippisley-Cox, J., Coupland, C., Vinogradova, Y., Robson, J., Minhas, R., Sheikh, A., & Brindle, P. (2008). Predicting cardiovascular risk in England and Wales: Prospective derivation and validation of QRISK2. *BMJ, 336*(7659), 1475–1482. doi:10.1136/bmj.39609.449676.25.

Hivon, M., Lehoux, P., Denis, J. L., & Tailliez, S. (2005). Use of health technology assessment in decision making: Coresponsibility of users and producers? *Int. J. Technol. Assess. Health Care, 21*(2), 268–275.

Hlupic, V., & Robinson, S. (1998). Business process modelling and analysis using discrete-event simulation. In D. J. Medeiros, E. F. Watson, J. S. Carson, & M. S. Manivannan (Eds.), *Proceedings of the 30th Conference on Winter Simulation* (pp. 1363–1370). Los Alamitos, CA: IEEE Computer Society Press.

Hofmann, B. M. (2008). Why ethics should be part of health technology assessment. *Int. J. Technol. Assess. Health Care, 24*(4), 423–429. doi:10.1017/S0266462308080550.

Homer, J. B., & Hirsch, G. B. (2006). System dynamics modeling for public health: Background and opportunities. *Am. J. Public Health, 96*(3), 452–458. doi:10.2105/AJPH.2005.062059.

Hoogendoorn, M., Feenstra, T. L., Asukai, Y., Borg, S., Hansen, R.N., Jansson, S.A., Samyshkin, Y., Wacker, M., Briggs, A.H., Lloyd, A., Sullivan, S.D., & Rutten-van Molken, M.P. (2014). Cost-effectiveness models for chronic obstructive pulmonary disease: Cross-model comparison of hypothetical treatment scenarios. *Value Health, 17*(5), 525–536. doi:10.1016/j.jval.2014.03.1721.

Hoyle, M. W., & Henley, W. (2011). Improved curve fits to summary survival data: Application to economic evaluation of health technologies. *BMC Med. Res. Methodol., 11*, 139. doi:10.1186/1471-2288-11-139.

Hupert, N., Mushlin, A. I., & Callahan, M. A. (2002). Modeling the public health response to bioterrorism: Using discrete event simulation to design antibiotic distribution centers. *Med. Decis. Making, 22*(Suppl. 5), S17–S25.

IBM Corporation. (2011). *FORTRAN: The Pioneering Programming Language*. Retrieved March 27, 2014, from http://www-03.ibm.com/ibm/history/ibm100/us/en/icons/fortran/.

Iman, R. L., & Conover, W. J. (1982). A distribution-free approach to inducing rank correlation among input variables. *Commun. Statist.—Simul. Computa., 11*(3), 311–334. doi:10.1080/03610918208812265.

Indrayan, A. (2012). *Medical Biostatistics* (3rd ed.). Boca Raton, FL: CRC Press.

International Network of Agencies for Health Technology Assessment (INAHTA). (2014). *Publications*. INAHTA. Retrieved June 13, 2014, from http://www.inahta.org/Publications/.

Ishak, K. J., Kreif, N., Benedict, A., & Muszbek, N. (2013). Overview of parametric survival analysis for health-economic applications. *Pharmacoeconomics, 31*(8), 663–675. doi:10.1007/s40273-013-0064-3.

Jackson, C. H., Bojke, L., Thompson, S. G., Claxton, K., & Sharples, L. D. (2011). A framework for addressing structural uncertainty in decision models. *Med. Decis. Making, 31*(4), 662–674. doi:10.1177/0272989X11406986.

Jackson, C. H., Jit, M., Sharples, L. D., & De Angelis, D. (2015). Calibration of complex models through Bayesian evidence synthesis: A demonstration and tutorial. *Med. Decis. Making, 35*(2), 148–161. doi:10.1177/0272989X13493143.

Jackson, C. H., Sharples, L. D., & Thompson, S. G. (2010). Survival models in health economic evaluations: Balancing fit and parsimony to improve prediction. *Int. J. Biostat., 6*(1), Article 34.

Jacobson, S. H., Hall, S. N., & Swisher, J. R. (2006). Discrete-event simulation of health care systems. In R. W. Hall (Ed.), *Patient Flow: Reducing Delay in Healthcare Delivery* (Vol. 91, pp. 211–252). New York: Springer.

Jacobson, S. H., Hall, S. N., & Swisher, J. R. (2013). Discrete-event simulation of health care systems. In R. Hall (Ed.), *Patient Flow: Reducing Delay in Healthcare Delivery* (2nd ed., Vol. 206, pp. 273–310). New York: Springer.

Jahn, B., Pfeiffer, K. P., Theurl, E., Tarride, J. E., & Goeree, R. (2010). Capacity constraints and cost-effectiveness: A discrete event simulation for drug-eluting stents. *Med. Decis. Making, 30*(1), 16–28. doi:10.1177/0272989X09336075.

Jain, S., Sigurðardóttir, S., Lindskog, E., Andersson, J., Skoogh, A., & Johansson, B. (December 8–11, 2013). Multi-resolution modeling for supply chain sustainability analysis. *Proceedings of the 2013 Winter Simulation Conference*. (pp. 1996–2007). Washington, DC: IEEE.

Janssen, M. F., Birnie, E., Haagsma, J. A., & Bonsel, G. J. (2008). Comparing the standard EQ-5D three-level system with a five-level version. *Value Health, 11*(2), 275–284. doi:10.1111/j.1524-4733.2007.00230.x.

Jennett, B. (1992). Health technology assessment. *BMJ, 305*(6845), 67–68.

Johnson, M. E. (2013). *Multivariate Statistical Simulation: A Guide to Selecting and Generating Continuous Multivariate Distributions*. New York: John Wiley & Sons.

Jun, J. B., Jacobson, S. H., & Swisher, J. R. (1999). Application of discrete-event simulation in health care clinics: A survey. *J. Oper. Res. Soc., 50*, 109–123.

Kagan, A. (1996). *The Honolulu Heart Program: An Epidemiological Study of Coronary Heart Disease and Stroke*. Amsterdam, the Netherlands: Harwood Academic Publishers.

Kanis, J. A., Borgstrom, F., De Laet, C., Johansson, H., Johnell, O., Jonsson, B., Oden, A., Zethraeus, N., Pfleger, B., & Khaltaev, N. (2005). Assessment of fracture risk. *Osteoporos. Int., 16*(6), 581–589.

Kanis, J. A., & Hiligsmann, M. (2014). The application of health technology assessment in osteoporosis. *Best Pract. Res. Clin. Endocrinol. Metab, 28*(6), 895–910. doi:10.1016/j.beem.2014.04.001.

Karnon, J. (2003). Alternative decision modelling techniques for the evaluation of health care technologies: Markov processes versus discrete event simulation. *Health Econ., 12*(10), 837–848. doi:10.1002/hec.770.

Karnon, J. (2007). Cost-effectiveness of letrozole, anastrozole and exemestane for early adjuvant breast cancer. *Expert Rev. Pharmacoecon. Outcomes Res., 7*(2), 143–153. doi:10.1586/14737167.7.2.143.

Karnon, J., & Brown, J. (1998). Selecting a decision model for economic evaluation: A case study and review. *Health Care Manag. Sci., 1*(2), 133–140.

Karnon, J., Goyder, E., Tappenden, P., McPhie, S., Towers, I., Brazier, J., & Madan, J. (2007a). A review and critique of modelling in prioritising and designing screening programmes. *Health Technol. Assess., 11*(52), iii–iv, ix–xi, 1–145.

Karnon, J., & Haji Ali Afzali, H. (2014). When to use Discrete Event Simulation (DES) for the economic evaluation of health technologies? A review and critique of the costs and benefits of DES. *Pharmacoeconomics, 32*(6), 547–558.

Karnon, J., Kerr, G. R., Jack, W., Papo, N. L., & Cameron, D. A. (2007b). Health care costs for the treatment of breast cancer recurrent events: Estimates from a UK-based patient-level analysis. *Br. J. Cancer, 97*(4), 479–485. doi:10.1038/sj.bjc.6603887.

Karnon, J., Stahl, J., Brennan, A., Caro, J.J., Mar, J., & Möller, J. (2012). Modeling using discrete event simulation: A report of the ISPOR-SMDM Modeling Good Research Practices Task Force—4. *Value Health, 15*(6), 821–827. doi:10.1016/j.jval.2012.04.013.

Karnon, J., & Vanni, T. (2011). Calibrating models in economic evaluation: A comparison of alternative measures of goodness of fit, parameter search strategies and convergence criteria. *Pharmacoeconomics, 29*(1), 51–62. doi:10.2165/11584610-000000000-00000.

Kelton, W. D., Sadowski, R. P., & Sturrock, D. T. (2007). *Simulation with Arena* (4th ed.). Boston, MA: McGraw-Hill.

Kelton, W. D., Sadowski, R. P., & Swets, N. B. (2010). *Simulation with Arena* (5th ed.). Boston, MA: McGraw-Hill.

Kenett, R., Zacks, S., & Amberti, D. (2013). *Modern Industrial Statistics: With Applications in R, MINITAB and JMP*. Chichester, UK: John Wiley & Sons.

Khan, K., Kunz, R., Kleijnen, J., & Antes, G. (2011). *Systematic Reviews to Support Evidence-Based Medicine*. Boca Raton, FL: CRC Press.

Kimmel, A. D., Fitzgerald, D. W., Pape, J. W., & Schackman, B. R. (2015). Performance of a mathematical model to forecast lives saved from HIV treatment expansion in resource-limited settings. *Med. Decis. Making*, 35(2), 230–242. doi:10.1177/0272989X14551755.

Kin, W., & Chan, V. (2010). Foundations of Simulation Modeling. In J. J. Cochran (Ed.), *Wiley Encyclopedia of Operations Research and Management Science*. Hoboken, NJ: John Wiley & Sons, Inc.

Kjærulff, U. B., & Madsen, A. L. (2012). *Bayesian Networks and Influence Diagrams: A Guide to Construction and Analysis* (2nd ed., Vol. 22). New York: Springer.

Klein, M. G., & Reinhardt, G. (2012). Emergency department patient flow simulations using spreadsheets. *Simul. Healthc.*, 7(1), 40–47. doi:10.1097/SIH.0b013e3182301005.

Konrad, R., DeSotto, K., Grocela, A., McAuley, P., Wang, J., Lyons, J., & Bruin, M. (2013). Modeling the impact of changing patient flow processes in an emergency department: Insights from a computer simulation study. *Oper. Res. Health Care*, 2(4), 66–74.

Korpanty, G. J., Graham, D. M., Vincent, M. D., & Leighl, N. B. (2014). Biomarkers that currently affect clinical practice in lung cancer: EGFR, ALK, MET, ROS-1, and KRAS. *Front. Oncol.*, 4, 204. doi:10.3389/fonc.2014.00204.

Korsunsky, I., McGovern, K., LaGatta, T., Olde Loohuis, L., Grosso-Applewhite, T., Griffeth, N., & Mishra, B. (2014). Systems biology of cancer: A challenging expedition for clinical and quantitative biologists. *Front. Bioeng. Biotechnol.*, 2, 27. doi:10.3389/fbioe.2014.00027.

Krahn, M., & Gafni, A. (1993). Discounting in the economic evaluation of health care interventions. *Med. Care*, 31(5), 403–418.

Kristensen, F. B., Hørder, M., & Poulsen, P. B. (2001). *Health Technology Assessment Handbook*. Copenhagen, Denmark: Danish Institute for Health Technology Assessment.

Kwon, C., & Tew, J. D. (1994). Strategies for combining antithetic variates and control variates in designed simulation experiments. *Manage. Sci.*, 40(8), 1021–1034. doi:10.1287/mnsc.40.8.1021.

Langston, P., Tüzün, U., & Heyes, D. (1994). Continuous potential discrete particle simulations of stress and velocity fields in hoppers: Transition from fluid to granular flow. *Chem. Eng. Sci.*, 49(8), 1259–1275.

Latimer, N. (2011). NICE DSU technical support document 14: Undertaking survival analysis for economic evaluations alongside clinical trials—extrapolation with patient-level data. Retrieved November 14, 2014, from http://www.nicedsu.org.uk.

Law, A. M. (2011). How to select simulation input probability distributions. In *Proceedings of the Winter Simulation Conference* (pp. 1394–1407), Phoenix, AZ.

Law, A. M. (2013, December 8–11). A tutorial on how to select simulation input probability distributions. *Paper Presented at the Proceedings of the 2013 Winter Simulation Conference*. Washington, DC.

Law, A., & Kelton, W. (2000). *Simulation Modeling and Analysis*. Retrieved from http://www.mhhe.com/engcs/industrial/lawkelton/.

LeBaron, B. (2006). Chapter 24 agent-based computational finance. In L. Tesfatsion & K. L. Judd (Eds.), *Handbook of Computational Economics* (Vol. 2, pp. 1187–1233). North Holland: Elsevier.

L'Ecuyer, P. (1994). Efficiency improvement and variance reduction. *Paper presented at the proceedings of the 26th conference on winter simulation*. Orlando, FL.

L'Ecuyer, P., & Lemieux, C. (2000). Variance reduction via lattice rules. *Manage. Sci.*, 46(9), 1214–1235. doi:10.1287/mnsc.46.9.1214.12231.

Levy, D. (2014). The use of simulation models in public health with applications to substance abuse and obesity problems. In Z. Sloboda & H. Petras (Eds.), *Defining Prevention Science* (pp. 405–430). New York: Springer.

Liberati, A., Sheldon, T. A., & Banta, H. D. (1997). EUR-ASSESS project subgroup report on methodology. Methodological guidance for the conduct of health technology assessment. *Int. J. Technol. Assess. Health Care*, 13(2), 186–219.

Lim, M. E., Worster, A., Goeree, R., & Tarride, J. E. (2013). Simulating an emergency department: The importance of modeling the interactions between physicians and delegates in a discrete event simulation. *BMC Med. Inform. Decis. Making*, 13, 59. doi:10.1186/1472-6947-13-59.

Luce, B. R., & Elixhauser, A. (1990). Estimating costs in the economic evaluation of medical technologies. *Int. J. Technol. Assess. Health Care*, 6(1), 57–75.

Lutzenberger, M., & Albayrak, S. (December 8–11, 2013). Can you simulate traffic psychology? An analysis. *Paper Presented at the Proceedings of the 2013 Winter Simulation Conference*. Washington, DC.

Mandelblatt, J., Schechter, C., Levy, D., Zauber, A., Chang, Y., & Etzioni, R. (2012). Building better models: If we build them, will policy makers use them? Toward integrating modeling into health care decisions. *Med. Decis. Making*, 32(5), 656–659. doi:10.1177/0272989X12458978.

Mandelblatt, J. S., Cronin, K. A., Berry, D. A., Chang, Y., de Koning, H.J., Lee, S.J., Plevritis, S.K., Schechter, C.B., Stout, N.K., van Ravesteyn, N.T., Zelen, M., & Feuer, E.J. (2011). Modeling the impact of population screening on breast cancer mortality in the United States. *Breast*, 20(Suppl. 3), S75–S81. doi:10.1016/S0960-9776(11)70299-5.

Matloff, N. (2011). *The Art of R Programming: A Tour of Statistical Software Design*. San Francisco, CA: No Starch Press.

McEwan, P., Bergenheim, K., Yuan, Y., Tetlow, A. P., & Gordon, J. P. (2010). Assessing the relationship between computational speed and precision: A case study comparing an interpreted versus compiled programming language using a stochastic simulation model in diabetes care. *Pharmacoeconomics*, 28(8), 665–674. doi:10.2165/11535350-000000000-00000.

McLeish, D. L., & Rollans, S. (1992). Conditioning for variance reduction in estimating the sensitivity of simulations. *Ann. Oper. Res.*, 39(1), 157–172. doi:10.1007/BF02060940.

Meng, Y., Davies, R., Hardy, K., & Hawkey, P. (2010). An application of agent-based simulation to the management of hospital-acquired infection. *J. Simul.*, 4(1), 60–67.

Merrifield, B., Richardson, S., & Roberts, J. (1990). Quantitative studies of discrete event simulation modelling of road traffic. In *Proceedings of the SCS Multiconference on Distributed Simulation* (Vol. 22, no. 1, pp. 188–193). San Diego, CA.

Miller, A. B., Wall, C., Baines, C. J., Sun, P., To, T., & Narod, S.A. (2014). Twenty five year follow-up for breast cancer incidence and mortality of the Canadian National Breast Screening Study: Randomised screening trial. *BMJ*, 348, g366. doi:10.1136/bmj.g366.

Möller, J., Nicklasson, L., & Murthy, A. (2011). Cost-effectiveness of novel relapsed-refractory multiple myeloma therapies in Norway: Lenalidomide plus dexamethasone vs bortezomib. *J. Med. Econ.*, *14*(6), 690–697. doi:10.3111/13696998.2011.611841.

Moore, J. H., & Weatherford, L. R. (2001). *Decision Modeling with Microsoft Excel®*. Upper Saddle River, NJ: Prentice Hall.

Murphy, D. R., Klein, R. W., Smolen, L. J., Klein, T. M., & Roberts, S. D. (2013). Using common random numbers in health care cost-effectiveness simulation modeling. *Health Serv. Res.*, *48*(4), 1508–1525. doi:10.1111/1475-6773.12044.

Naimark, D. M. J., Kabboul, N. N., & Krahn, M. D. (2013). The half-cycle correction revisited: Redemption of a Kludge. *Med. Decis. Making*, *33*(7), 961–970. doi:10.1177/0272989x13501558.

National Center for Biotechnology Information, & U.S. National Library of Medicine. (2014). *Health Services/Technology Assessment Texts (HSTAT)*. U.S. National Library of Medicine. Retrieved June 13, 2014, from http://www.ncbi.nlm.nih.gov/books/NBK16710/

National Center for Health Statistics. (2005). *NHANES 2003–2004 Public Data General Release File Documentation*. Retrieved April 1, 2014, from http://www.cdc.gov/nchs/data/nhanes/nhanes_03_04/general_data_release_doc_03-04.pdf.

National Institute for Health and Care Excellence (NICE). (2013). *Guide to the Methods of Technology Appraisal*. NICE. Retrieved June 13, 2014, from http://www.nice.org.uk/article/PMG9/chapter/Foreword.

National Institute for Health and Care Excellence (NICE). (2014). *Advanced Breast Cancer: Diagnosis and Treatment (Clinical Guideline 81)*. Retrieved April 3, 2015, from http://www.nice.org.uk/guidance/CG81.

Nelson, C. L., Sun, J. L., Tsiatis, A. A., & Mark, D. B. (2008). Empirical estimation of life expectancy from large clinical trials: Use of left-truncated, right-censored survival analysis methodology. *Stat. Med.*, *27*(26), 5525–5555. doi:10.1002/sim.3355.

Ness, R. M., Holmes, A. M., Klein, R., & Dittus, R. (2000). Cost-utility of one-time colonoscopic screening for colorectal cancer at various ages. *Am. J. Gastroenterol.*, *95*(7), 1800–1811. doi:10.1111/j.1572-0241.2000.02172.x.

Nutaro, J. J. (2011). *Building Software For Simulation: Theory and Algorithms, With Applications in C++*. Hoboken, NJ: John Wiley & Sons.

O'Hagan, A., Stevenson, M., & Madan, J. (2005). *Monte Carlo probabilistic sensitivity analysis for patient level simulation models*. Sheffield, England: University of Sheffield, Department of Probability & Statistics.

O'Hagan, A., Stevenson, M., & Madan, J. (2007). Monte Carlo probabilistic sensitivity analysis for patient level simulation models: Efficient estimation of mean and variance using ANOVA. *Health Econ.*, *16*(10), 1009–1023.

O'Reilly, D., Campbell, K., Goeree, R., & I. Programs for Assessment of Technology in Health Research. (2009). Basics of health technology assessment. *Methods Mol. Biol.*, *473*, 263–283.

Parnell, G. S., Bresnick, T. A., Tani, S. N., & Johnson, E. R. (2013). Appendix B: Influence diagrams. In *Handbook of Decision Analysis* (pp. 374–380). Hoboken, NJ: John Wiley & Sons.

Peasgood, T., Ward, S. E., & Brazier, J. (2010). Health-state utility values in breast cancer. *Expert Rev. Pharmacoecon. Outcomes Res.*, *10*(5), 553–566. doi:10.1586/erp.10.65.

Pitman, R., Fisman, D., Zaric, G. S., Postma, M., Kretzschmar, M., Edmunds, J., & Brisson, M. (2012). Dynamic transmission modeling: A report of the ISPOR-SMDM Modeling Good Research Practices Task Force—5. *Value Health*, 15(6), 828–834. doi:10.1016/j.jval.2012.06.011.

Pratt, J. W., Raiffa, H., & Schlaifer, R. (1995). *Introduction to Statistical Decision Theory*. Cambridge, MA: MIT Press.

Rachner, T. D., Khosla, S., & Hofbauer, L. C. (2011). Osteoporosis: Now and the future. *Lancet*, 377(9773), 1276–1287. doi:10.1016/S0140-6736(10)62349-5.

Railsback, S. F., & Grimm, V. (2011). *Agent-Based and Individual-Based Modeling: A Practical Introduction*. Princeton, NJ: Princeton University Press.

Ramwadhdoebe, S., Buskens, E., Sakkers, R. J., & Stahl, J. E. (2009). A tutorial on discrete-event simulation for health policy design and decision making: Optimizing pediatric ultrasound screening for hip dysplasia as an illustration. *Health Policy*, 93(2–3), 143–150. doi:10.1016/j.healthpol.2009.07.007.

Ransohoff, D. F., & Feinstein, A. R. (1976). Editorial: Is decision analysis useful in clinical medicine? *Yale J. Biol. Med.*, 49(2), 165–168.

Raunak, M. S., Osterweil, L. J., Wise, A., Clarke, L. A., & Henneman, P. L. (2009). Simulating patient flow through an emergency department using process-driven discrete event simulation. *Paper Presented at the International Conference on Software Engineering Workshop on Software Engineering For Health Care*. Vancouver, BC, Canada. Retrieved from http://www.umass.edu/eei/EEIWebsiteArticles/SimulatingPatientFlowthroughanEmergencyDepartmentUsingProcess-DrivenDiscreteEventSimulation.pdf.

Rauner, M. S., Brailsford, S. C., & Flessa, S. (2004). Use of discrete-event simulation to evaluate strategies for the prevention of mother-to-child transmission of HIV in developing countries. *J. Oper. Res. Soc.*, 56(2), 222–233.

Reitman, J. (1971). *Computer Simulation Applications; Discrete-Event Simulation For Synthesis and Analysis of Complex Systems*. New York: Wiley-Interscience.

Robson, J., Hippisley-Cox, J., & Coupland, C. (2012). QRISK or Framingham? *Br. J. Clin. Pharmacol.*, 74(3), 545–546. doi:10.1111/j.1365-2125.2012.04293.x.

Rockwell Automation. (2013). Arena simulation software by rockwell automation. Retrieved from http://www.arenasimulation.com/.

Roethlisberger, F. J., Dickson, W. J., Wright, H. A., Pforzheimer, C. H., & Western Electric Company. (1939). *Management and The Worker: An Account of A Research Program Conducted By The Western Electric Company, Hawthorne Works, Chicago*. Cambridge, MA: Harvard University Press.

Rubinstein, R. Y., & Kroese, D. P. (2011). *Simulation and the Monte Carlo Method* (Vol. 707). John Wiley & Sons.

Russell, L. B., Gold, M. R., Siegel, J. E., Daniels, N., & Weinstein, M. C. (1996). The role of cost-effectiveness analysis in health and medicine. *JAMA*, 276(14), 1172–1177.

Rutter, C. M., Zaslavsky, A. M., & Feuer, E. J. (2011). Dynamic microsimulation models for health outcomes: A review. *Med. Decis. Making*, 31(1), 10–18. doi:10.1177/0272989X10369005.

Saarni, S. I., Hofmann, B., Lampe, K., Luhmann, D., Makela, M., Velasco-Garrido, M., & Autti-Ramo, I. (2008). Ethical analysis to improve decision-making on health technologies. *Bull. World Health Organ.*, 86(8), 617–623.

Sackett, D. L., Rosenberg, W. M., Gray, J. A., Haynes, R. B., & Richardson, W. S. (1996). Evidence based medicine: What it is and what it isn't. *BMJ*, 312(7023), 71–72.

Sambrook, P., & Cooper, C. (2006). Osteoporosis. *Lancet*, *367*(9527), 2010–2018. doi:10.1016/S0140-6736(06)68891-0.

Saquib, N., Saquib, J., & Ioannidis, J. P. (2015). Does screening for disease save lives in asymptomatic adults? Systematic review of meta-analyses and randomized trials. *Int. J. Epidemiol.*, *44*(1), 264–277. doi:10.1093/ije/dyu140.

Savage, E. L., Schruben, L. W., & Yücesan, E. (2005). On the generality of event-graph models. *INFORMS J. Comput.*, *17*(1), 3–9. doi:10.1287/ijoc.1030.0053.

Schriber, T. J. (1974). *Simulation Using GPSS*. Ann Arbor, MI: Michigan University.

Schriber, T. J. (1977). *Introduction To Simulation. Paper Presented at the Proceedings of the 9th Conference on Winter Simulation* (Vol. 1). Gaitherburg, MD: Winter Simulation Conference.

Schriber, T. J., Brunner, D. T., & Smith, J. S. (2013). Inside discrete-event simulation software: How it works and why it matters. *Paper Presented at the Proceedings of the 2013 Winter Simulation Conference: Simulation: Making Decisions in a Complex World*. Washington, DC.

Schruben, L. (1983). Simulation modeling with event graphs. *Commun. ACM*, *26*(11), 957–963. doi:10.1145/182.358460.

Schwammenthal, Y., Bornstein, N., Schwammenthal, E., Schwartz, R., Goldbourt, U., Tsabari, R., Koton, S., Grossman, E., & Tanne, D. (2010). Relation of effective anticoagulation in patients with atrial fibrillation to stroke severity and survival (from the National Acute Stroke Israeli Survey [NASIS]). *Am. J. Cardiol.*, *105*(3), 411–416. doi:10.1016/j.amjcard.2009.09.050.

Schwarz, G. (1978). Estimating the dimension of a model. *Annals Statistics*, (2), 461–464. doi:10.1214/aos/1176344136.

Schwartz, W. B., Gorry, G. A., Kassirer, J. P., & Essig, A. (1973). Decision analysis and clinical judgment. *Am. J. Med.*, *55*(3), 459–472.

Schwarzer, R., & Siebert, U. (2009). Methods, procedures, and contextual characteristics of health technology assessment and health policy decision making: Comparison of health technology assessment agencies in Germany, United Kingdom, France, and Sweden. *Int. J. Technol. Assess. Health Care*, *25*(3), 305–314. doi:10.1017/S0266462309990092.

Semini, M., Fauske, H., & Strandhagen, J. O. (2006). Applications of discrete-event simulation to support manufacturing logistics decision-making: A survey. *Proceedings of the 38th Winter Simulation* (pp. 1946–1953), Monterey, CA.

Seppanen, M. (1998). Designing simulation models to use Visual Basic® for Applications (VBA). *Paper Presented at the Proceedings of the 12th European Simulation Multiconference on Simulation—Past, Present and Future*. Manchester, England.

Shechter, S. M., Bryce, C. L., Alagoz, O., Kreke, J.E., Stahl, J.E., Schaefer, A.J., Angus, D.C., & Roberts, M.S. (2005). A clinically based discrete-event simulation of end-stage liver disease and the organ allocation process. *Med. Decis. Making*, *25*(2), 199–209. doi:10.1177/0272989X04268956.

Siebert, U., Alagoz, O., Bayoumi, A. M., Jahn, B., Owens, D.K., Cohen, D.J., & Kuntz, K.M. (2012). State-transition modeling: A report of the ISPOR-SMDM Modeling Good Research Practices Task Force—3. *Value Health*, *15*(6), 812–820. doi:10.1016/j.jval.2012.06.014.

Siegel, J. E., Weinstein, M. C., Russell, L. B., & Gold, M. R. (1996). Recommendations for reporting cost-effectiveness analyses. Panel on cost-effectiveness in health and medicine. *JAMA*, *276*(16), 1339–1341.

Signorovitch, J. E., Sikirica, V., Erder, M. H., Xie, J., Lu, M., Hodgkins, P.S., Betts, K.A., & Wu, E.Q. (2012). Matching-adjusted indirect comparisons: A new tool for timely comparative effectiveness research. *Value Health*, 15(6), 940–947. doi:10.1016/j.jval.2012.05.004.

Signorovitch, J. E., Wu, E. Q., Yu, A. P., Gerrits, C.M., Kantor, E., Bao, Y., Gupta, S.R., & Mulani, P.M. (2010). Comparative effectiveness without head-to-head trials: A method for matching-adjusted indirect comparisons applied to psoriasis treatment with adalimumab or etanercept. *Pharmacoeconomics*, 28(10), 935–945. doi:10.2165/11538370-000000000-00000.

Simpson, K. N., Strassburger, A., Jones, W. J., Dietz, B., & Rajagopalan, R. (2009). Comparison of Markov model and discrete-event simulation techniques for HIV. *Pharmacoeconomics*, 27(2), 159–165.

Sonnenberg, F. A., & Beck, J. R. (1993). Markov models in medical decision making: A practical guide. *Med. Decis. Making*, 13(4), 322–338.

Sorenson, C., Drummond, M., & Kanavos, P. (2008). *Ensuring Value For Money in Health Care: The Role of Health Technology Assessment in the European Union*. WHO Regional Office Europe, Copenhagen, Denmark.

Sox, H. C., Higgins, M. C., & Owens, D. K. (2013). *Medical Decision Making* (2nd ed.). Chichester, West Sussex: John Wiley & Sons.

Srinivas, S., & Breese, J. S. (2013). *IDEAL: A Software Package For Analysis of Influence Diagrams* (pp. 212–219). arXiv, http://arxiv.org/abs/1304.1107.

Stahl, J. E. (2008). Modelling methods for pharmacoeconomics and health technology assessment: An overview and guide. *Pharmacoeconomics*, 26(2), 131–148.

Stahl, J. E., Furie, K. L., Gleason, S., & Gazelle, G. S. (2003). Stroke: Effect of implementing an evaluation and treatment protocol compliant with NINDS recommendations. *Radiology*, 228(3), 659–668. doi:10.1148/radiol.2283021557.

Stahl, J. E., Rattner, D., Wiklund, R., Lester, J., Beinfeld, M., & Gazelle, G.S. (2004). Reorganizing the system of care surrounding laparoscopic surgery: A cost-effectiveness analysis using discrete-event simulation. *Med. Decis. Making*, 24(5), 461–471. doi:10.1177/0272989X04268951.

Stahl, J. E., Vacanti, J. P., & Gazelle, S. (2007). Assessing emerging technologies—The case of organ replacement technologies: Volume, durability, cost. *Int. J. Technol. Assess. Health Care*, 23(3), 331–336. doi:10.1017.S0266462307070535.

Stevenson, M., Jones, M. L., De Nigris, E., Brewer, N., Davis, S., & Oakley, J. (2005). A systematic review and economic evaluation of alendronate, etidronate, risedronate, raloxifene and teriparatide for the prevention and treatment of postmenopausal osteoporosis. *Health Technol. Assess.*, 9(22), 1–160.

Stout, N. K., Rosenberg, M. A., Trentham-Dietz, A., Smith, M,A., Robinson, S.M., & Fryback, D.G. (2006). Retrospective cost-effectiveness analysis of screening mammography. *J. Natl. Cancer Inst.*, 98(11), 774–782. doi:10.1093/jnci/djj210.

Sullivan, S. D., Mauskopf, J. A., Augustovski, F., Jaime Caro, J., Lee, K.M., Minchin, M., Orlewska, E., Penna, P., Rodriguez Barrios, J.M., & Shau, W.Y. (2014). Budget impact analysis-principles of good practice: Report of the ISPOR 2012 Budget Impact Analysis Good Practice II Task Force. *Value Health*, 17(1), 5–14. doi:10.1016/j.jval.2013.08.2291.

Swisher, J. R., Jacobson, S. H., Jun, J. B., & Balci, O. (2001). Modeling and analyzing a physician clinic environment using discrete-event (visual) simulation. *Comput. Oper. Res.*, 28(2), 105–125. doi:10.1016/s0305-0548(99)00093-3.

Swisher, J. R., Jacobson, S. H., & Yücesan, E. (2003). Discrete-event simulation optimization using ranking, selection, and multiple comparison procedures: A survey. *ACM Trans. Model. Comput. Simul.*, *13*(2), 134–154. doi:10.1145/858481.858484.

Tako, A. A., & Robinson, S. (2012). The application of discrete event simulation and system dynamics in the logistics and supply chain context. *Decis. Support. Syst.*, *52*(4), 802–815. doi:10.1016/j.dss.2011.11.015.

Tappenden, P., & Chilcott, J. B. (2014). Avoiding and identifying errors and other threats to the credibility of health economic models. *Pharmacoeconomics*, *32*(10), 967–979. doi:10.1007/s40273-014-0186-2.

Taylor, S. J. R., Ghorbani, M., Mustafee, N., Turner, S.J., Kiss, T., Farkas, D., Kite, S., & Strassburger, S. (2011). Distributed computing and modeling & simulation: Speeding up simulations and creating large models. In *Proceedings of the 2011 Winter Simulation Conference* (pp. 161–175). doi:10.1109/WSC.2011.6147748.

Tengs, T. O., & Wallace, A. (2000). One thousand health-related quality-of-life estimates. *Med Care*, *38*(6), 583–637.

Tewoldeberhan, T. W., Verbraeck, A., Valentin, E., & Bardonnet, G. (2002). Software evaluation and selection: An evaluation and selection methodology for discrete-event simulation software. *Paper Presented at the Proceedings of the 2002 Winter Simulation Conference*. San Diego, CA, December 8–11, 2002.

Thulasidasan, S., Kroc, L., & Eidenbenz, S. (2012). Developing parallel discrete event simulations in Python: First results and user experiences with the SimX library (p. 7). Technical Report *LA-UR-12-26739*. Los Alamos, NM: Los Alamos National Laboratory.

Tierney, J. F., Stewart, L. A., Ghersi, D., Burdett, S., & Sydes, M. R. (2007). Practical methods for incorporating summary time-to-event data into meta-analysis. *Trials*, *8*, 16. doi:10.1186/1745-6215-8-16.

Tomin, R., & Donegan, W. L. (1987). Screening for recurrent breast cancer—its effectiveness and prognostic value. *J. Clin. Oncol.*, *5*(1), 62–67.

TreeAge Software. (2014). *TreeAge Pro 2014 User's Manual*. Retrieved from http://installers.treeagesoftware.com/treeagepro/14.1.0/PDF/TP-Manual-2014R1.pdf.

Van Gestel, A., Webers, C. A., Severens, J. L., Beckers, H.J., Jansonius, N.M., Hendrikse, F., & Schouten, J.S. (2012). The long-term outcomes of four alternative treatment strategies for primary open-angle glaucoma. *Acta. Ophthalmol. (Copenh.)*, *90*(1), 20–31.

Vanni, T., Karnon, J., Madan, J., White, R.G., Edmunds, W.J., Foss, A.M., & Legood, R. (2011). Calibrating models in economic evaluation: A seven-step approach. *Pharmacoeconomics*, *29*(1), 35–49. doi:10.2165/11584600-000000000-00000.

Van Volsem, S., Dullaert, W., & Van Landeghem, H. (2007). An evolutionary algorithm and discrete event simulation for optimizing inspection strategies for multi-stage processes. *Eur. J. Oper. Res.*, *179*(3), 621–633.

Viana, J., Brailsford, S. C., Harindra, V., & Harper, P. R. (2014). Combining discrete-event simulation and system dynamics in a healthcare setting: A composite model for Chlamydia infection. *Eur. J. Oper. Res.*, *237*(1), 196–206. doi:http://dx.doi.org/10.1016/j.ejor.2014.02.052.

Visual Thinking. (1998). *SIMUL8 Manual and Simulation Guide*. Glasgow, Scotland: Visual Thinking International Limited.

von Scheele, B., Mauskopf, J., Brodtkorb, T. H., Ainsworth, C., Berardo, C.G., & Patel, A. (2014). Relationship between modeling technique and reported outcomes: Case studies in models for the treatment of schizophrenia. *Expert Rev. Pharmacoecon. Outcomes Res.*, *14*(2), 235–257. doi:10.1586/14737167.2014.891443.

Vose, D. (2014). Review of risk analysis add-ins for microsoft excel®: Which is the best Monte Carlo risk analysis add-in for microsoft excel®? Retrieved March 27, 2015, from https://www.linkedin.com/pulse/20140714140045-483951-review-of-risk-analysis-addins-for-excel.

Wang, W.-N., Sun, H., Sun, Z.-X., & Song, R.-F. (November 11–14, 2010). A cache aware speedup algorithm for discrete event simulations. *Paper Presented at the 2010 12th IEEE International Conference on Communication Technology*. Nanjing, China.

Warner, K. E., Smith, R. J., Smith, D. G., & Fries, B. E. (1996). Health and economic implications of a work-site smoking-cessation program: A simulation analysis. *J. Occup. Environ. Med.*, *38*(10), 981–992.

Watkins, J. B. (2012). Creating models that meet decision makers' needs: A US payer perspective. *Value Health*, *15*(6), 792–793. doi:10.1016/j.jval.2012.03.1386.

Webster, B. (2013). Airbus in fear of full emergency test. Retrieved March 27, 2014, from http://www.iasa.com.au/folders/Safety_Issues/others/airbus_in_fear_of_full_emergency.html.

Wikipedia Contributors. (2014). Law of the instrument. Retrieved August 13, 2014, from http://en.wikipedia.org/wiki/Law_of_the_instrument.

Wilks, S. S. (1938). The large-sample distribution of the likelihood ratio for testing composite hypotheses. *Ann. Math. Stat.*, *9*(1), 60–62.

Wolkewitz, M., Cooper, B. S., Bonten, M. J., Barnett, A. G., & Schumacher, M. (2014). Interpreting and comparing risks in the presence of competing events. *BMJ*, *349*, g5060. doi:10.1136/bmj.g5060.

Wu, I., Borrmann, A., Beißert, U., König, M., & Rank, E. (2010). Bridge construction schedule generation with pattern-based construction methods and constraint-based simulation. *Adv. Eng. Inform.*, *24*(4), 379–388.

Xia, W., Yao, Y., & Mu, X. (2012). An extended event graph-based modelling method for parallel and distributed discrete-event simulation. *Math. Comp. Model. Dyn. Systems*, *18*(3), 287–306. doi:10.1080/13873954.2012.655697.

Index

Note: Locator followed by 'f' and 't' denotes figure and table in the text